ELINT

The Interception and Analysis of Radar Signals

For a listing of recent titles in the *Artech House Radar Library*, turn to the back of this book.

ELINT

The Interception and Analysis of Radar Signals

Richard G. Wiley

ARTECH HOUSE

BOSTON | LONDON
artechhouse.com

Library of Congress Cataloging-in-Publication Data
Wiley, Richard G.
    ELINT: the interception and analysis of radar signals / Richard G. Wiley.
        p. cm.—(Artech House radar library)
    Includes bibliographical references and index.
    ISBN 1-58053-925-4 (alk. paper)
    1. Radar. I. Title. II. Series.

    TK6575.W55   2006
    621.3848—dc22   2005058854

British Library Cataloguing in Publication Data
Wiley, Richard G.
    ELINT: the interception and analysis of radar signals.—(Artech House radar library)
    1. Electronic intelligence   2. Radar   3. Signal detection   I. Title
    621.7'34

ISBN-13: 978-1-58053-925-8

Cover design by Igor Valdman

© 2006 ARTECH HOUSE, INC.
685 Canton Street
Norwood, MA 02062

International Standard Book Number: 1-58053-925-4
Library of Congress Catalog Card Number: 2005058854

10 9 8 7 6 5 4 3 2

Contents

CHAPTER 3
Characteristics of ELINT Interception Systems

CHAPTER 4
Probability of Intercept

CHAPTER 11

Intrapulse Analysis   255

CHAPTER 12

Pulse Repetition Interval Analysis   281

CHAPTER 16

APPENDIX A

APPENDIX B

APPENDIX C

Preface

More than 20 years have passed since the publication of *Electronic Intelligence: The Analysis of Radar Signals* (Artech House, 1983) and *Electronic Intelligence: The Interception of Radar Signals* (Artech House, 1985). Much of what those books contained remains valid today. With this new book, I have again avoided emphasizing specific hardware and software currently available because it changes too rapidly. My hope is that this book remains useful for the next 20 years. This book is partially based on my earlier works, *Electronic Intelligence: The Analysis of Radar Signals* (Artech House, 1983), *Electronic Intelligence: The Analysis of Radar Signals, Second Edition* (Artech House, 1993), and *Electronic Intelligence: The Interception of Radar Signals* (Artech House, 1985).

It is appropriate to list those people who introduced me to the exciting field of ELINT and who continue to provide insights and inspiration. Professor E. M. Williams of Carnegie-Mellon University introduced me to the field. Dr. Thomas F. Curry provided my early training. Both are now gone but often remembered. There is no way to adequately recognize the contributions of Robert B. Shields, my friend and business associate for more than 45 years. He always grasps the "big picture" and asks the right questions at the right time.

For their contributions to this book, I am grateful to Arthur Self for his work on Chapter 4 and to Charles Estrella for his work on Chapter 6.

I am also grateful to the many people who work in ELINT and who have shared their experiences and insights with me. This includes not only my coworkers at Research Associates of Syracuse, Inc., but also the several thousand people who have attended continuing education courses I have presented over the past 20 years—many for the Association of Old Crows. The struggle to answer their questions is one of the motives in writing this book.

The continued support of my wife, Jane, and the good start in life provided by my parents Mildred and Asa (1911–1995) were key to the creation of this book. I am also thankful for the support of my six children, their spouses, and our 13 grandchildren. Many of them are now in their teens and approaching adulthood. It is my hope and prayer that vigilance through ELINT and other means helps keep them and all children everywhere in safety and freedom.

"Lord God Almighty, in whose Name the founders of this country won liberty for themselves and for us, and lit the torch of freedom for nations then unborn: Grant that we and all the people of this land may have grace to maintain our liberties in righteousness and peace. . . ."

—Collect for Independence Day, page 241, *The Book of Common Prayer* (1979)
According to the use of the Episcopal Church

CHAPTER 1

Electronic Intelligence

1.1 Electronic Intelligence Defined

Radar has been called the greatest advance in sensing remote objects since the telescope was invented in 1608 [1]. *Electronic intelligence* (ELINT) is the result of observing the signals transmitted by radar systems to obtain information about their capabilities: it is the remote sensing of remote sensors. Through ELINT, it is possible to obtain valuable information while remaining remote from the radar itself. Clearly, ELINT is most useful in hostile situations; otherwise, the information could be obtained directly from the radar user or designer. The value of ELINT is that it provides timely information about threatening systems, such as radars that guide aircraft or missiles to targets. ELINT also provides information about defensive systems, which is important in maintaining a credible deterrent force to penetrate those defenses.

ELINT refers to the information gained from the interception of signals of interest. In addition to radar, other types of signals referred to as sources of ELINT include beacons and transponders, jammers, missile guidance, some data links, altimeters, navigation emissions, and identification friend or foe (IFF). In this broad sense, ELINT can refer to nearly any noncommunications emission. This book is concerned largely with ELINT derived from radar signals.

Receiving radar signals is often not difficult because the available power density is quite high. The power transmitted by a radar is proportional to the fourth power of the range at which it is to detect a target; whereas the power available at an ELINT receiver is proportional to the reciprocal of the square of the distance from the radar. An analogous situation is a searchlight being used to spot an aircraft. An observer on the aircraft can see the searchlight from far off, even if the light is not pointed directly toward the aircraft. By comparison, an observer on the ground can see the aircraft only when the searchlight beam is aimed at the aircraft and when the plane is closer than it was when the observer on the aircraft first saw the searchlight. This aspect and other radar constraints are explored in detail in Chapter 2, and the topic of low probability of intercept (LPI) radar is explored in Chapter 9.

While the ELINT observer may enjoy a large range advantage over the radar, ELINT signals can still be quite weak, and analyzing them is not a simple matter. The purpose of the chapters that follow is to describe the interception and analysis processes for various radar signals.

1.2   The Importance of Intercepting and Analyzing Radar Signals

The first large-scale use of radar occurred in World War II. No sooner had radar been deployed (by both sides) than the effort to counter it spawned electronic countermeasures (ECM), which included the interception of the radar's signals. In today's terminology, this is called electronic attack (EA). In the early 1940s, "the basic countermeasures principle was simple. A ship or aircraft carried a rapidly tunable receiver which could scan the wavelength bands in which enemy radars might be operating. If one was detected, its strength and characteristics gave clues as to the dangers involved and what the next tactic should be. If the operator was flying over enemy territory, he could switch on a jammer, a powerful transmitter which would overload the radar receiver rendering it ineffective, or he could dump bales of aluminum 'chaff' from the airplane, or he could dump and jam" [2]. A side effect of the large amounts of chaff dropped was that thousands of cows died in Germany from eating the foil as they grazed.

The following illustrates the fundamental role of ELINT in preventing disastrous technological surprise [3]:

> In early 1942, the RAF Coastal Command used L-band radar as an aid for locating German U-boats recharging batteries on the surface. The overall effectiveness of the RAF in this task was quite good until the U-boats began using L-band search receivers. These receivers allowed the submarine to hear transmitted radar signals at a range greater than that over which the radar echo could effectively be returned. The U-boat therefore had time to dive before actually being sighted by the searching aircraft. In turn, general effectiveness of the RAF antisubmarine effort decreased. The Allies, realizing what had happened, installed new S-band search radars aboard their aircraft during early 1943. As a result of the effectiveness of new equipment the intercept rate rose sharply. German submarines sitting on the surface, listening to L-band search receivers, became vulnerable targets for S-band radar directed aircraft.
>
> As the U-boat sinkings increased, the Germans tried frantically to determine the method of detection the Allies were using. Since reports from surviving submarines stated that no radiation had been heard in their L-band search receivers prior to the attack, it was thought that perhaps an infrared detection device of some type was being employed. Considerable effort was spent in an attempt to combat a non-existing infrared threat. U-boat activity was greatly reduced by the time the German High Command realized that new S-band radar was in use.
>
> This is an example of weapon (L-band radar), a countermeasure (L-band search receiver), and an improvement (S-band radar) providing a clear margin of technical supremacy.
>
> There is another point to be considered. To be sure, the use of S-band radar employing magnetrons and extending the useable frequency by a factor often provided a definite advantage. However, had the Germans had information as to what was being used, the time lag until they were able to develop an effective S-band search receiver would have been greatly reduced. It is obvious that the lack of information by the opponent side is the basic requirement of the so-called "secret weapon."
>
> This point is mentioned here because illustrated in this example is one of the important roles of electronic reconnaissance. Had the Germans been conducting

an extensive reconnaissance program at the time, it is probable that they would have intercepted S-band signals from magnetron oscillators in the development and testing stages during flights over England. The development of the magnetron was, of course, the solution to the problem of generating high power for 10 cm radar, and simple crystal receivers for reconnaissance purposes were indeed available, if the Germans had cared to use them in this application. High sensitivity is not necessary for intercepting high-power sources.

Of course, one can find more recent examples of the value of radar interception than World War II. In the Vietnam War, there were heavy losses of aircraft because of the large-scale use of surface-to-air missiles. New tactics and the use of radar warning receivers designed with the aid of ELINT data helped reduce U.S. aircraft losses and allowed the North Vietnamese air defense to be destroyed with much lower losses on the U.S. side [4]. Similarly, the SA-6 missiles supplied by the Soviet Union to the Arabs in the October 1973 conflict with Israel proved a major factor in destroying low flying Israeli aircraft. New tactics and jammers eventually overcame the advantage achieved by the Arab combatants through technological surprise [4].

Radar uses have proliferated, particularly for military applications, but also for civilian air traffic control, harbor surveillance, weather monitoring, and so on. Prudence requires knowledge of the military capabilities of potential adversaries, and this means continued use of receivers to listen for radar transmissions and all of the other aspects of ELINT. For this reason, the accumulation of knowledge of radar signals has a peacetime role in proportion to the development of modern weapons, many of which incorporate radar target detection and tracking.

Consider the following remarks of Soviet authors, Rear Admiral Peronmov and Captain First Rank-Engineer A. Partala from their essay "A Look at Development of Means of Electronic Warfare" [5]:

> The military effectiveness of electronic countermeasures, even with emphasis on real life experimentation, must be regarded as confirmed only under the condition that basic characteristics of the electronic weapons of the opponent have been revealed with sufficient accuracy. But the possibility of dependably uncovering and obtaining this information is an extremely difficult task, according to American specialists; ". . . Experience in battle with application of means of 'electronic warfare' in Southeast Asia," writes the American press, gave evidence that even the most modern reconnaissance [intelligence] is not able to secure timely disclosure of all nomenclatured radio-electronic means, their tactical-technical characteristics, and special military employment. And, therefore, from the beginning of military operations various surprises are probably inevitable. For prevention of large material losses and tactical operational failures, questions about preparation for 'electronic warfare' cannot really be left without attention.

In [5], these remarks are interpreted to mean: "If you don't know the exact details of the electronic threat and fail to prepare effective countermeasures in advance, you will be shot down or sunk."

ELINT has an important role in maintaining defensive capabilities and preventing surprises—in this very real sense, knowledge is power, and where hostile radar is concerned, ELINT provides a great deal of knowledge. The path to that knowledge begins with antennas, receivers, and strategies for search, which are major topics of this book, and continues through the measurement of signal parameters and the recording and reporting of the interceptions for more detailed analysis.

The data gathered is typically about a potential rival's defense network, such as radars, surface-to-air missile (SAM) systems, and aircraft. ELINT can be gathered from ground stations near the hostile country, from ships near the coast, from aircraft near the airspace, or from satellites in space. This may lead to incidents where aircraft or ships stray into hostile waters or airspace. In one particularly dangerous incident, a Chinese fighter collided with a U.S. reconnaissance aircraft and forced it to land in China.

ELINT data is valuable in the event of a conflict. Knowledge of the location and type of SAM and antiaircraft artillery (AAA) systems allows planning to avoid heavily defended areas, making use of flight profiles which will give the penetrator the best chance of avoiding hostile fire. It also enables intelligent jamming of an enemy's defense network. Knowledge of the whereabouts of ships, command and control centers, surface-to-air missile systems, and other assets of the enemy permits them to be attacked if need be. ELINT is important to stealth operations because stealth aircraft are not totally undetectable and also need to know which areas to avoid. Of course, ELINT data is an important part of network-centric warfare. One way it is used is "traffic analysis" of ELINT signals. In ELINT, it is not the message that is of interest as it is in communications intelligence (COMINT); rather, it is the type of signal, the location of its transmitter, and the timing of the transmissions relative to other events which may be taking place.

1.3   Limitations Due to Noise

Although in the usual situation the intercept receiver enjoys a considerable range advantage over the radar, this does not mean that the ELINT signal analyst has the luxury of a high signal-to-noise ratio (SNR). The situation is frequently just the reverse, as the ELINT site attempts to receive signals located at great distances or beyond the radar horizon—possibly due to national boundaries that cannot be crossed. It is therefore typical for an ELINT station to be equipped with very low noise front-end amplifiers. High gain antennas that can be steered may also be used at the price of decreased probability of interception. It is also typical to locate sites at high altitudes and to use airborne platforms. The potentially large ratio of ELINT range–to–radar range for long-range radar systems often means that the limiting factor in receiving signals from a radar is the reduction in signal strength due to the curvature of the Earth.

In many cases, the ingenuity of the ELINT signal analyst is severely taxed by a low SNR. The limitations due to low SNR arise repeatedly in the following chapters. No matter what clever signal analysis technique is devised, the fundamental limits due to noise remain.

1.4   Probability of Intercept Problems

The ELINT receiver has an advantage in the R^2 versus R^4 path loss over the radar. However, the ELINT operator usually does not know exactly how to ensure reception of the signal from the radar. The direction of arrival and carrier frequency (RF) of the signal are not known. It is also not known when the radar's main beam may be pointed in the direction of the ELINT receiving antenna. As a result, the ELINT operator must search for the radar's signal in much the same way the radar operator searches for targets. The ELINT receiver may be tuned in radio frequency (RF) across a band repeatedly as the ELINT antenna is pointed in different directions. Intercepting the radar's main beam then requires that the ELINT receiver be tuned to the proper frequency and have its antenna pointed toward the radar at the same time the radar "looks" in the direction of the ELINT site. This problem makes interception of the radar's sidelobes a great convenience if there is a sufficient signal level to do so. The problems and strategies for searching for radar signals have greatly influenced intercept receiver design, as described in Chapters 3 and 4.

1.5 Direction Finding (DF) and Emitter Location

Of special importance to operational ELINT (OPELINT) is the location of emitters. This is usually approached by means of angle-of-arrival (AOA) measurements for coarse location and also for pulse sorting. More precise location makes use of precise time of arrival (TOA) and frequency of arrival (FOA) measurements at multiple platforms. These aspects are discussed in detail in Chapters 5 and 6.

1.6 Inferring Radar Capabilities from Observed Signal Parameters

The basic function of ELINT is to determine the capabilities of the radar, so that decisions can be made as to what threat it poses. Radar functions are generally to make measurements of the range, velocity, and direction of targets of interest to the radar. Consider radar measurement of target range. A simple pulsed radar has a maximum unambiguous range measurement capability given by the speed of light multiplied by half of the pulse repetition interval (PRI). Therefore, if the PRI is measured, the maximum unambiguous range can be determined. This provides a clue to what detection range the radar achieves. It is to the radar's advantage to use the shortest PRI commensurate with the range measurement capabilities necessary to perform the radar's mission. True assessment of radar maximum range measurement capability requires determining all of the quantities in the radar range equation. Some of these are impossible to know through ELINT (such as the required signal level at the radar receiver). Therefore, intelligent estimates or knowledge from other sources must be used.

Of course, inferring radar capabilities becomes very complicated as the radar becomes more complex. Nevertheless, it is always true that the radar's transmitted waveform determines the ambiguity diagram for the radar. ELINT can probe the character of the ambiguity function by a close examination of the radar's transmitted signal. This brief outline of the procedures useful in inferring radar capabilities from ELINT is intended to point out the motivation for the signal analysis techniques described in Chapters 7, 8, and 10 through 15.

1.7   Receivers for Radar Interception

The subject of intercept receivers is quite broad because there is no single design or type appropriate for all uses. Generally, intercept receivers can be divided into warning receivers and ELINT receivers.

Radar warning receivers are designed to give nearly immediate warning if specific threat signals are received (e.g., illumination of an aircraft's warning receiver by the target tracking radar of a surface-to-air missile system). The warning receiver typically has poor sensitivity and feeds into a near-real-time processor that uses a few parameter measurements to identify a threat. Usually, rough direction (e.g., quadrant or octant) is determined for the threat and the operator has a crude display showing functional radar type, direction, and relative range (strong signals displayed as being nearer than weaker ones). This type of receiver does not provide the kind of output that is analyzed using the methods described here. Rather, the identification algorithms used in the warning receiver might be based on the results of ELINT.

The ELINT receiver is usually more sensitive compared to the warning receiver and often has a variety of filters and demodulation adjuncts that are used to get the best SNR possible. Modern ELINT receivers and radar warning receivers (RWRs) are becoming more nearly equal in sensitivity and capability due to the widespread adoption of digital techniques. The outputs of the receiver may be analyzed using off-line analysis tools based on software. The on-site analysis results may be forwarded to centralized analysis centers for further processing.

ELINT receivers must cope with a wide variety of signal parameters. The signals of interest extend over a wide dynamic range—wider than any receiver design can hope to handle due to the tremendous range differences possible and the different radar effective radiated power (ERP) values that may be encountered. The ELINT receiving equipment must cover a wide frequency range to be able to search for new radars in any part of the spectrum. Coverage to 100 GHz may be needed. The modulation bandwidths of the signals can range from very short pulses to those of frequency modulation (FM) and continuous wave (CW) signals. The receiving equipment may also be operated remotely with automatic digitizing and/ or retransmission to analysis sites. ELINT receivers may be found in a variety of locations, including ships, aircraft, balloons, and ground locations. Often this places them in proximity to local radar sets or jammers. In this case, it is generally necessary to provide for blanking of the ELINT receiver during the transmission cycle of these local emitters. Receivers are discussed in Chapter 3.

1.8   Major ELINT Signal Parameters

The typical radar signal can be analyzed according to its power (Chapter 7), antenna (Chapters 8 and 10), and carrier modulation characteristics (Chapter 14). There are many receiver characteristics of interest as well, but these cannot be obtained directly from ELINT. The power characteristics include any variations in the power. The antenna characteristics include the observed polarization beam patterns and the angular motion (scanning) of the beam. The modulation characteristics can be divided into the more usual pulsed types and CW types. Pulse modulation includes both the envelope and intrapulse frequency or phase modulation (Chapter 11), and pulse interval characteristics (Chapter 12). The radio frequency, or carrier, characteristics are of importance, including the carrier frequency used, as well as any intentional variations such as frequency agility. Also of interest are measures of incidental variations in the frequency (or phase), which are measures of RF coherence.

When measuring any of these parameters, two problems must be addressed: separating the signal of interest from the rest of those signals in the environment, and statistically determining parameter ranges for radar sets of a given type. Brief discussions of these two subjects are included in Chapters 13 and 15. Finally, Chapter 16 briefly describes some uses of ELINT data files including their use in radar warning receivers.

1.9   The Impact of LPI Radar on ELINT

Since ELINT has such value, a natural direction for radar design would be to reduce the ability of a would-be listener to receive the radar signal. Such radar systems are said to have *low probability of intercept* features. These may include low sidelobe antennas, infrequent scanning, reducing the radar power when tracking a closing target (as range is reduced, the radar power is also reduced), making use of waveform coding to provide transmitting duty cycles approaching one (to reduce peak power while maintaining the required average power), and using frequency hopping to force the interceptor to consider more of the spectrum in attempting to characterize the radar.

The problem of LPI radar design is fundamentally quite difficult. Search via radar is analogous to using a searchlight to spot an airplane at night. Energy is sent out, reflected by the target and used by the observer to spot the airplane. An LPI-type searchlight design still would require that the airplane be spotted, but also that the searchlight itself could not be seen by an observer. Because of the fundamental nature of this problem, it requires careful analysis. LPI radar is considered in some detail in Chapter 9.

The problems inherent in LPI radar design are such that electronically intercepted signal information will continue to be available; however, new approaches for interception of LPI-type radars will be needed.

The design and use of intercept receivers is heavily influenced by the design of radar systems: today's intercept receivers reflect the radar signals in use today; tomorrow's intercept receivers will reflect tomorrow's radar signals. From the earliest days of radar until recent times, radar designers have not been directly concerned with the activities of would-be interceptors. Their aim has been to produce well-designed radars capable of performing specified functions without regard to the activities of ELINT operations. Of course, radar designers and operators have always been concerned with chaff and jamming or electronic attack, and have taken steps to reduce their effects through electronic counter-countermeasures (ECCM) or electronic protection (EP) [6].

If the LPI philosophy of radar design influences the signal emissions of tomorrow's radar systems, the designers of intercept receivers must respond (as best they can) to be successful, where possible, in spite of the LPI design approaches which may be adopted. (The LPI radar designer makes a strategic mistake by assuming that intercept receiver designs will not change in response to introducing LPI radar into the signal environment.)

Another influence on radar design is efforts to reduce the radar cross section of targets. This makes LPI design more difficult because the lower the target's radar cross section, the more average power required by the radar to detect the target. Radar systems are evolving in the direction of multiple functions and multiple modes. Future radars may well have both LPI and low target cross section modes among many others in their repertoire.

References

[1]   Barton, D. K., *Radar Systems Analysis*, Dedham, MA: Artech House, 1976, p. 1.

[2]   Kraus, J., *The Big Ear*, Powell, OH: Cygnus-Quasar Books, 1976, Chapter 12, p. 18.

[3]   Schlesinger, R. J., *Principles of Electronic Warfare*, Englewood Cliffs, NJ: Prentice Hall, 1961.

[4]   Johnston, S. L., *Radar Electronic Counter-Countermeasures*, Dedham, MA: Artech House, 1979, Section 1.3.

[5]   Stone, N. L., "Soviet Perceptions: The Soviets on Electronic Countermeasures for Surface Ships," *Defense Science and Electronics*, March 1983, p. 59.

[6]   Maksimov, M. V., et al., *Radar Anti-Jamming Techniques*, Dedham, MA: Artech House, 1979.

CHAPTER 2

ELINT Implications of Range Equations and Radar Constraints

The effects of the one-way range equation of ELINT and the two-way range equation of radar on signal strength must be understood and explored in order to appreciate the typical situations encountered in ELINT and electronic warfare (EW). Similarly, the constraints placed on radar waveforms must be understood in order to correctly interpret the functions and applications of the signals transmitted by radar and also to be aware of the signal characteristics expected to be encountered by ELINT. In many ways, understanding these aspects of ELINT is what separates one who only observes signals from one who both observes and analyzes signals.

2.1 Range Equations

In free space, the signal received from a target by the radar receiver can be expressed as [1]

$$S_R = \frac{P_T G_T G_R \lambda^2 \sigma}{(4\pi)^3 R_R^4 L_T L_R} \tag{2.1}$$

where:

S_R = signal power at radar receiver (W)

P_T = transmitter power (W)

G_T = gain of radar transmit antenna in the direction of the target

G_R = gain of radar receive antenna in the direction of the target

λ = wavelength (m)

σ = target radar cross section (square meters)

R_R = radar range to target (m)

L_T = radar losses from transmitter to antenna

L_R = radar losses from antenna to receiver

The received signal is inversely proportional to the range to the target to the fourth power because the signal undergoes spreading on the way to the target and again on the way back from the target. The radar equation is often referred to as the two-way range equation. In free space, the signal level received by the ELINT receiver from the radar transmitter is given by [2]:

$$S_E = \frac{P_T G_{TE} G_E \lambda^2}{(4\pi)^2 R_E^2 L_T L_E} \tag{2.2}$$

where:

S_E = signal level at ELINT receiver (W)

G_{TE} = gain of radar transmit antenna in the direction of the ELINT receiver

G_E = gain of ELINT antenna in the direction of the radar transmit antenna

R_E = range from radar transmit antenna to ELINT antenna

The other symbols are defined under (2.1).

The signal undergoes spreading from the radar transmitter to the ELINT receiver, and so it is inversely proportional to the second power of the range. This equation is often referred to as the one-way range equation. Properly speaking, these are not range equations but rather received signal level equations. When solved for the range in terms of the signal strength received, they become expressions for the range.

The qualification "in free space" points out that the effects of the atmosphere and the Earth's horizon have not been included. Note that the one-way range equation may also be used to compute the signal level from a jammer at a victim receiver.

A significant aspect of these range equations is that the power level transmitted by pulsed radar transmitters in order to detect targets at long range is very high. This allows ELINT receivers to detect radar signal at very long ranges even when observing the sidelobes of the radar's transmit antenna.

To simplify the discussion, suppose that the ELINT receiver requires a signal level that is a factor δ times the signal level needed by the radar receiver; that is,

$$S_E = \delta(S_R) \tag{2.3}$$

Combining (1.1), (1.2), and (1.3) and solving for the ratio of the ELINT range to the radar range give

$$\frac{R_E}{R_R} = R_R \left[\frac{4\pi}{\delta} \frac{1}{\sigma} \frac{G_{TE}}{G_T} \frac{G_E}{G_R} \frac{L_R}{L_E} \right]^{1/2} \tag{2.4}$$

To get an idea of the typical situation, it is useful to assume some values for the quantities in (2.4). Suppose the radar target cross section is $\sigma = 1$ m^2 and that the losses in the radar and ELINT receive paths are equal. Suppose that the radar uses the same antenna for transmit and receive so that $G_T = G_R$ and that the ELINT receiver is in the sidelobes of the radar's transmit antenna where $G_{TE} = 1$. Also, to eliminate the need to search for ELINT signals at various angles, assume that the ELINT antenna is omni-directional and that $G_E = 1$. Equation (2.4) becomes

$$\frac{R_E}{R_R} = \frac{R_R}{G_R} \left[\frac{4\pi}{\delta} \right]^{1/2} \tag{2.5}$$

In (2.5), R_R is in meters because the radar target cross section of 1 m^2 is included inside the square root.

The factor $1/\delta$ is the sensitivity advantage of the radar receiver over the ELINT receiver. Generally speaking, the ELINT receiver requires a higher level of signal power because it is not matched to the radar signal, whereas the radar receiver is nearly matched to the signal transmitted. There are several reasons for this state of affairs. The ELINT receiver's function is to detect the presence of a radar signal and also to measure its parameters such as amplitude, carrier frequency, angle of arrival, time of arrival, pulse duration, and so forth. The function of the radar receiver is to detect the presence of a target echo and to determine its range, velocity, and angle. The radar receiver often makes use of multiple echoes to detect the presence of a target; today's ELINT receivers, however, often depend on detecting each individual radar pulse and then associate these individual pulses into pulse trains after detection. The result is that the factor δ is generally larger than 1, meaning that the ELINT receiver has poorer sensitivity (requires a stronger signal) than the radar receiver. One major difference is that the ELINT receiver may not know the carrier frequency of the signal and must search over broad bands to find signals of interest (SOI). To minimize search time, the ELINT receiver may choose to use a very wide bandwidth. This degrades its sensitivity relative to the radar receiver—except for small Doppler shifts it can be tuned to the exact carrier frequency. For example, a radar receiver may have a bandwidth of 1 MHz to receive echoes from its 1-μs transmitted pulses. The ELINT receiver may have a bandwidth of 10 MHz to preserve the details of the pulse shape and ease the problem of tuning to the center of the transmitted spectrum. Furthermore, the radar may use the echoes from 10 pulses prior to deciding that a target is present while the ELINT receiver is making a "pulse present" decision on every pulse. In this case, the ELINT receiver sensitivity is approximately 100 times (or 20 dB) poorer than the radar receiver.

Sometimes the ELINT receiver is in the direction of the main beam of the radar antenna. (This is the situation for a radar warning receiver carried by targets of interest to the radar.) In this scenario, the expression for the ratio of the ELINT range to radar range becomes

$$\frac{R_E}{R_R} = R_R \left[\frac{4\pi}{G_R \delta} \right]^{1/2} \tag{2.6}$$

Figure 2.1 shows (2.5) and (2.6) for a typical radar antenna gain of 30 dB and a value of δ of 20 dB. Suppose that the ELINT receiver is in a region where the sidelobes of the radar transmit antenna are $G_{TE} = 1$. As can be seen from point A in Figure 2.1, these sidelobes of a radar able to detect a 1-m^2 target at 100 km have a ratio of ELINT range to radar range of about 35.4. In free space, those sidelobes of the radar are detectable at 3,540 km even when the ELINT receiver is 20 dB less sensitive than the radar receiver. If the ELINT receiver is in the main beam of the radar, the main beam could be detected in free space at a range in excess of 100,000 km.

This example illustrates the situation prevalent in ELINT much of the time: namely that signals from long-range radars are quite large and these signals can be received at long range, often limited only by the line of sight. The range equations provide insight into why ELINT is of interest in a number of aspects of electronic warfare. Low values of the range ratio are the goal of designers of LPI radars. Some aspects of LPI radar are explored in Chapter 9.

2.2 Radar Constraints

ELINT signals of interest include radar signals of all types. Sometimes, people concerned about ELINT attribute properties to radar signals that are contrary to the constraints under which radar systems must function. Avoiding this pitfall is an important aspect of ELINT work. Understanding the fundamental limitations faced by radar designers and the associated ELINT implications is important. Consider this statement: *Radars of the future could transmit noise waveforms over*

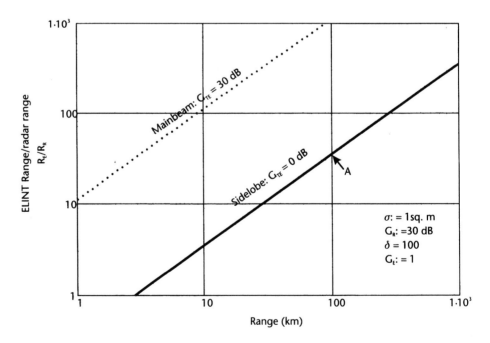

Figure 2.1   ELINT to radar range ratio.

gigahertz bandwidths and be undetectable by ELINT receivers. Should ELINT equipment be developed to intercept and process this kind of signal? Probably not—because signals like this would not be useful for tracking or search radars in military applications. Assume that the technology for transmitting noise waveforms over wide (>1 GHz) bands and the digital signal processing (DSP) hardware to store and correlate the transmitted waveform with the echoes received is available. Will such radars be deployed? The answer depends on the job the radar is designed to do and the characteristics of the target and target motion it must detect or track. To be "undetectable" by ELINT means that the power must be low. After all, a high power noise-like waveform would be detectable like the signal from a jammer. Low peak power means the radar must use a long integration time in order to get sufficient energy back from the target. Long integration time may restrict the radar's ability to keep up with target motion (the target could move out of a range or Doppler cell if the integration time is too long). Transmitting a wide instantaneous bandwidth means that the radar has fine range resolution. If the range cells are smaller than the range extent of the target, the target echo will be spread across many range cells (a bandwidth of 1 GHz provides range cells every 15 cm). This reduces the energy received in a single range cell. For example, if one wanted to recover all the energy from a target 15m long, it would be necessary to coherently add the energy of 100 range cells. Therefore, the ability to transmit and receive a particular waveform does not by itself make it a good choice for a radar designer. ELINT work must be done with these factors in mind.

2.2.1   Range Resolution Related to Bandwidth

Range resolution in radar is inversely proportional to the bandwidth of the signal (assuming that it is processed coherently) [3, 4]. The fundamental relationship is

$$\Delta R = \frac{c}{2B} \qquad (2.7)$$

Here c is the speed of light, and B is the bandwidth of the signal during the coherent processing interval—also called its instantaneous bandwidth.

For example, to distinguish between two fighters in tight formation 30m apart in range, bandwidth must be about 5 MHz. Figure 2.2 shows the relationship of range resolution to bandwidth.

If one postulates a value of $B = 1$ GHz, the radar has a range resolution of 15 cm. This means that the target echoes are resolvable in 15-cm range increments called range cells. The echoes from a 75-m target are spread across 500 range cells.

This spreading of the echoes across a multiplicity of range cells reduces the apparent radar cross section (and thus reduces the SNR available) in a single range cell. For this reason, radar designs generally have range resolution appropriate for their function. This leads to choosing coherent bandwidths of 10 MHz or less. (10 MHz corresponds to range resolution of 15m.) In this sense, there is no such thing as a "spread spectrum" radar—what is transmitted is also received, and the resulting range resolution is determined by the bandwidth. What this means for

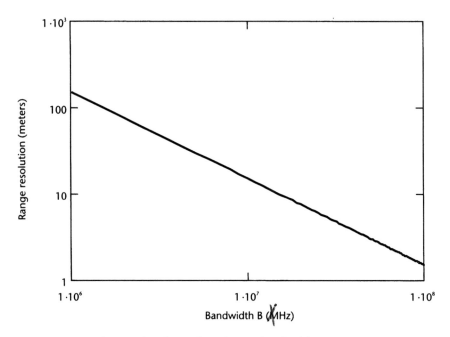

Figure 2.2 Range resolution related to radar coherent bandwidth.

ELINT is that the coherent bandwidth of radar signals is likely to remain the same
as it is now provided the radar performs the same task. (See Table 2.1.)

Tactical radars (early warning, fire control, target acquisition, target track,
airborne intercept) have bandwidth requirements typical of applications 1 and 2
in Table 2.1, that is, under 10 MHz. Some tactical radars and multifunction radars
may have modes for imaging (application 3 in Table 2.1).

2.2.2 Spread Spectrum: Radar Versus Communications

In radar, synchronizing the received signal with the transmitted signal is used to
determine the range. In communication, after synchronization is achieved, the
spreading code can be removed and the narrowband message data may be recovered.
This is the reason for the term "spread spectrum." The transmitted spectrum is
wider than that required by the message. In radar, use of wider coherent bandwidth
than required for its function requires reassembling the target echoes across many
range cells into a single value. It is conceivable that this could be done by future
radar systems; however, such a direction is not evident in current radar signals.
But that is the sort of radar that would qualify as spread spectrum in the sense

Table 2.1 Instantaneous Bandwidth Required for Selected Applications

Range Resolution Required	Resolution (m)	Bandwidth (MHz)
1. Count A/C in attack formation	30	5
	60	2.5
2. Detect missile separation at launch	15	10
3. Imaging of ships, vehicles, and aircraft	0.5–1	150–300
4. High-resolution mapping	0.15	1,000

that a wider bandwidth is transmitted than is used to detect the target. For ELINT today, the radar's coherent bandwidth is determined by the radar's function—bandwidth is not widened or narrowed without altering the radar's range resolution.

2.2.3   Moving Targets and Integration Time Constraints

If a radar is to detect targets moving in a radial direction (toward or away from the radar), the amount of time the target will be present in a given range cell is determined by the target velocity and the range resolution. This limits the coherent integration time of present day radars to

$$T_{CV} = \frac{\delta R}{v} < \frac{\Delta R}{v} \qquad (2.8)$$

Here T_{CV} is the maximum coherent integration time for a constant velocity target with radial velocity v, and δR is the change in range during that time. If the target is accelerating in the radial direction, the maximum integration time is now a quadratic function of both velocity and acceleration

$$T_{ACC} = \frac{v - [v^2 + 2a(\delta R)]^{0.5}}{-a} < \frac{v - [v^2 + 2a(\Delta R)]^{0.5}}{-a} \qquad (2.9)$$

In this expression, T_{ACC} is the coherent integration time available for an accelerating target of velocity v and acceleration a. Figure 2.3 shows range versus time for a constant velocity (dotted curve) and for a target that is accelerating at 2 g's (19.507 m/sec/sec) (solid curve). For integration times on the order of 10 to 100 ms and acceleration of a few g's the effect of acceleration on integration time is small. For example, for 2-g acceleration and 300-m/s velocity, the constant

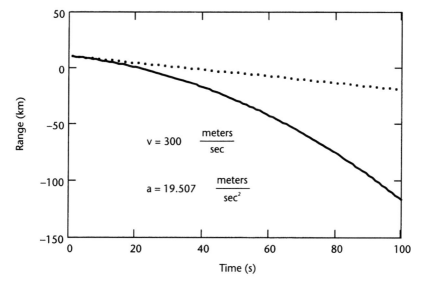

Figure 2.3   Range versus time.

velocity approximation yields 0.5s integration time for 1-MHz radar bandwidth (range resolution of 150m), while the calculation with acceleration included yields an integration time of 0.492s.

What this means for ELINT is that the radar's coherent integration time is limited by the range resolution chosen and the characteristics of the target's motion. During the coherent integration time, the radar waveform bandwidth determines the range resolution of the radar. Furthermore, the center frequency of the carrier must constant over then integration time if Doppler processing is used.

2.2.4 Constraints on Time-Bandwidth Product or Pulse Compression Ratio

Because range resolution is determined by bandwidth and integration time is determined by velocity, there is a natural limit on the product of the instantaneous bandwidth and the duration of the coherent processing interval or pulse width. This is called the *time-bandwidth product*. The radar's pulse compression ratio is limited to no more than its time bandwidth product. By combining (2.7) and (2.8), it is easy to see that this is limited to

$$BT < \frac{B_V}{a}\left(\sqrt{1 + \frac{ac}{B_V^2}} - 1\right) \xrightarrow[a \to 0]{} \frac{c}{2v} \tag{2.10}$$

This limit is usually quite large—if there is no acceleration and v is 300 m/s, then $BT < 500,000$. Most pulse compression systems have a much smaller value of BT—typically between 10 and 1,000. Pulse compression is used to increase average power (or pulse duration) while maintaining the required range resolution. The pulse compression ratio is the ratio of the width of the transmitted pulse to that of the "compressed" pulse at the radar receiver output. For example, $BT = 100$ allows 1-μs pulse duration at the receiver output when the transmitted pulse duration is 100 μs. Some radar systems with pulse compression in excess of 100,000 have been built, but these are long range systems used to observe ICBM flights or for discrimination of warheads from decoys in an ICBM attack scenario. The limit on BT is affected by target acceleration. Equation (2.8) assumes constant velocity. If the target accelerates toward the radar, it will leave the range cell sooner and this affects the maximum BT allowed. Figure 2.4 shows the effect of target acceleration on the BT limit when the acceleration is in the same direction as the velocity. If the target was approaching the radar but slowing down, the limit on BT would increase instead of decrease.

The higher the radial velocity and acceleration of the target, the more constrained the time-bandwidth product. *The time-bandwidth product of deployed radar systems is determined by the nature of the radar's targets, not by the state of the art in signal generation and processing alone.*

2.2.5 Constraints on Doppler Resolution

If the radar coherently integrates the echoes in one range cell for the entire integration time, the minimum Doppler filter bandwidth, B_f, is approximately the recipro-

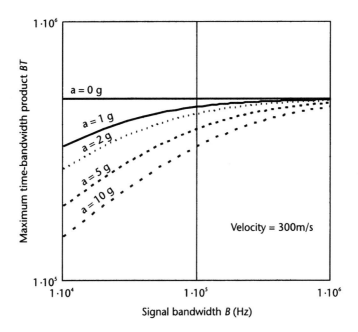

Figure 2.4   Limit on time-bandwidth product.

cal of the integration time T, which is either T_{CV} for constant velocity targets or T_{ACC} for accelerating targets:

$$B_f \approx \frac{1}{T} \tag{2.11}$$

However, if the target is accelerating, the Doppler shift changes. Clearly there is a relationship between acceleration and the time the Doppler shift of the moving target remains within the Doppler filter bandwidth.

$$\Delta f_{acc} = \frac{2aTf_o}{c} = \frac{2aT}{\lambda} < B_f \tag{2.12}$$

Here Δf_{acc} is the spread of Doppler frequencies caused by the target acceleration a during the integration time T at carrier frequency f_o. Because the coherent integration time is approximately equal to $1/B_f$, substituting $B_f = 1/T$ into (2.12) gives the maximum allowable coherent integration time and the minimum Doppler filter bandwidth as

$$T < \sqrt{\frac{\lambda}{2a}}, \qquad B_f > \sqrt{\frac{2a}{\lambda}}$$

To avoid large straddling losses, practical radar designs allow a considerable margin—for example, by moving the factor of 2 from inside the square root to outside of the square root.

Clearly, the dispersion of Doppler frequencies due to acceleration must be no greater than the bandwidth of the Doppler filter during any one coherent integration interval T. Figure 2.5 is a graph of (2.11) and (2.12) to show this relationship for target velocity = 300 m/s, RF = 10 GHz. The spread of Doppler frequencies during the time the target is in one range cell is shown in Figure 2.5 for accelerations of 1, 2, 5, and 10 g's. The Doppler filter bandwidth must be no wider than the spread of Doppler frequencies expected. Figure 2.5 also shows the maximum radar signal bandwidth. For the case where acceleration has a minimal effect on the integration time, the maximum acceleration of the target can be expressed in terms of the radar signal's bandwidth as

$$a_{max} = 2B^2 \frac{v^2}{c(\mathrm{RF})} \tag{2.13}$$

The effect of long integration times is to require small target acceleration values. This means that the radar designer must choose a signal bandwidth that suits the range resolution requirements of the radar function and that the integration time must be limited to values that suit the target motion expected. Long integration time prior to detection implies either slow targets with little acceleration or else very poor range resolution. High acceleration targets require wider signal bandwidths. For example, an aircraft target approaching at 300 m/s and maneuvering at 3 g's needs a radar signal bandwidth of at least 2.5 MHz at 10 GHz. Arbitrarily long integration times cannot be used by practical search radars to detect targets because integration time is limited by expected target radial velocity and acceleration along with the RF used. It means that radar signals exhibit relatively constant characteristics during their coherent integration times—an important consideration for ELINT. Tracking radars can extend the coherent integration time when target velocity and acceleration are approximately known. Correlating with all possible

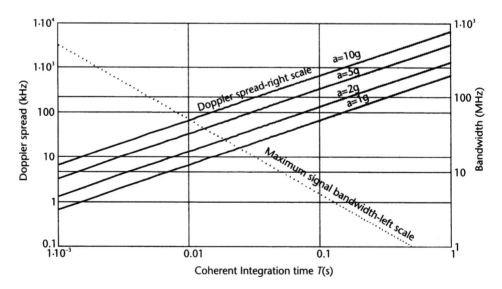

Figure 2.5 Doppler spread and maximum signal bandwidth.

target velocities and accelerations by search radars requires huge processor throughput and is generally not practical today.

2.2.6 Frequency Agility

From one coherent processing interval to the next, the radar can change its carrier frequency without changing its range resolution properties. This type of wideband signal is called frequency agility. The agility band is limited by the radar designer's ability to obtain sufficient transmit power and to maintain the antenna beam width and pointing angle over that band. Typical designs have an agility band of about 10% of the center frequency. (For example, a 1-GHz agility band centered at 10 GHz.) What this means for ELINT is that narrowband receivers have a low probability of intercepting the complete radar transmission. If it is sufficient to intercept only portions of the radar transmission, narrowband receivers can be slowly tuned across the radar band and the entire agility band can still be determined if the signals is present for enough time. If not, then a wideband ELINT receiver is needed. These may be of the channelized type with sufficient channels to cover 500 to 1,000 MHz at once. Frequency agility (FA) is used primarily to make the radar signal more difficult to recognize and to jam, but it also provides a way to reduce fluctuations of target echoes by, in effect, averaging the observed radar target cross section over several different frequencies. Depending on ELINT system architecture, it may be useful to have sufficient instantaneous bandwidth in the ELINT receiver to cover the entire frequency agility band at one setting of the receiver. The FA bandwidth does not affect the radar's range resolution. It does provide enhanced target detection because the target's cross section is examined at several frequencies. Note also that at present the center frequency is not changed during the coherent processing interval or over the time that Doppler processing is performed. That is, if more than one pulse is used for Doppler processing, which is normally the case, the center frequency or RF carrier frequency cannot be changed for the length of the processing interval. And the processing interval determines the Doppler resolution. *When FA is used with Doppler processing, the frequency is changed on a pulse-burst to pulse-burst basis, not a pulse-to-pulse basis.*

2.2.7 PRI Agility

Modern multifunction radar systems make use of multiple pulse repetition intervals values during one look at the target. It is a requirement of today's pulse Doppler radars that the PRI remain constant during each coherent processing interval. For moving target indicating (MTI) radar designs, there is usually a sequence of PRI values that must be completed during one processing interval. This repeated sequence is known as *stagger*, and ELINT analysts call the period of the stagger the *stable sum*. This is because when consecutive PRIs are added, the sum is constant when one adds together the PRIs that make up the stagger period—*regardless of which PRI is selected as the starting point for the sum.*

MTI radars operate by subtracting (in amplitude and phase) the echoes from one PRI from those in the next PRI. Stationary targets have the same phase and amplitude and thus cancel. Echoes from moving targets generally do not have then same amplitude and phase and so do not cancel. However if the target moves an integer multiple of half wavelengths in one PRI, the phase of the second echo is shifted by a multiple of 360° from the first and the echoes cancel. These speeds are called *blind speeds*. Changing the PRI changes the blind speeds and the PRI sequence is selected so that the target can be detected regardless of its speed moving target detection (MTD) radar systems, used at present in air surveillance applications. A Doppler filter bank is used to divides the frequency region between the pulse repetition frequency (PRF) lines into several filter bands (for example, 8). This improves the ability to detect moving targets in clutter. In the original MTD design, the MTI cancellation process was used ahead of the Doppler filter bank due to limited dynamic range of available analog-to-digital (A/D) converters. In modern designs, an MTI canceller need not be included. Of course, there is still a blind speed problem, but now the PRI must remain constant for the number of pulses required to excite the filter bank (and canceller if any). This leads to a series of repeated constant PRIs (say, 10 pulses at one PRI and then 10 pulses at another PRI, and so forth). Multiple PRIs are required to trade between range and velocity ambiguities and also to make visible target ranges and velocities eclipsed by transmitted pulses (in time) or spectral lines (in frequency).

For constant PRI and RF, the maximum unambiguous range (R_u) and the maximum unambiguous velocity (V_u) are given by

$$R_u = \frac{c(\text{PRI})}{2} \tag{2.14}$$

$$V_u = \frac{c}{2(\text{RF})(\text{PRI})} \tag{2.15}$$

Examples at 10 GHz:

PRI 1,000 μs, V_u = 15 m/s and R_u = 150 km
PRI 100 μs, V_u = 150 m/s and R_u = 15 km
PRI 10 μs, V_u = 1,500 m/s and R_u = 1.5 km

As can be seen, the product of unambiguous range and velocity is a constant. This means that the total ambiguity is fixed but changes in PRI can increase the unambiguous range but decrease the unambiguous velocity and vice versa.

$$R_u V_u = \frac{c^2}{4(\text{RF})} \tag{2.16}$$

The product $R_u V_u$ can be made as large as desired by using a low operating frequency; however, low frequencies may make the antenna unwieldy. This relationship is shown in Figure 2.6 for common radar frequencies.

Another reason for PRI agility is that if the range to the target is any integer multiple of unambiguous range, then the echo returns when a pulse is being transmitted and it cannot be received (it is said to be *eclipsed*). Similarly, if the Doppler

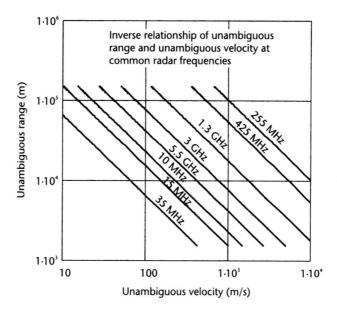

Figure 2.6 Range/velocity related.

shift is any multiple of the PRF (= 1/PRI), the target echo is obscured by the transmitted spectral lines and returns from stationary clutter and it cannot be received (it is said to be at a *blind speed*). Radar designers try to choose PRI sequences that allow the radar to detect targets at all ranges and velocities of interest. Another reason for PRI variation is for protection against electronic attack.

Random variations of PRI (jitter) can be used to prevent jamming through anticipating when a pulse will arrive from the radar. This prevents the generation of false targets at ranges closer than the target. False targets can be generated at ranges beyond the real target range by transmitting a false pulse soon after receiving a pulse from the radar. For this type of jamming, random PRI variations do not matter.

What this means for ELINT is that the PRI sequences observed in pulsed radar systems are highly constrained by the radar functions. Furthermore, the entire sequence of PRIs must be completed during times the antenna beam is illuminating the target. The sequence might be spread over several radar scans. But note that the total time available to detect the target is limited. If the radar performance requirement is to detect the target in one scan, then the entire PRI sequence must be completed during a single beam dwell. Radar designers use a variety of PRIs for different functions. This leads to the multiplicity of modes observed by ELINT.

Radars are said to operate in one of three PRF regimes: low PRF, high PRF, and medium PRF. Low PRF means that target echoes at all ranges of interest return before the next pulse is transmitted. Range measurements are then assumed to be unambiguous. This is the usual case for long-range search radars and all types of simple ranging radars. The actual value of the PRF depends on the range at which the radar is to detect targets. Therefore, there is no particular numerical value for a low PRF. For typical low PRF designs the PRF is in the hundreds of hertz. For this type of radar, velocities of interest usually produce Doppler shifts many times the PRF and are thus highly ambiguous. Of course, if the radar operating frequency is low enough, Doppler shifts of interest may be smaller than the PRF to provide both sufficiently large unambiguous velocity and range.

High PRF means that the target echoes for all velocities of interest have Doppler shifts less than the PRF. Velocity measurements are thus always unambiguous. The values used as PRFs are determined by the RF and by the velocities of interest. For typical designs, the PRF may be in the hundreds of kilohertz. This is the usual case for airborne radars that depend on Doppler filtering to distinguish targets from clutter. For most ranges of interest, the echoes are delayed by many times the PRI and therefore the range measurements are highly ambiguous.

Medium PRF means that several PRIs elapse before the target echoes return and that the Doppler shifts for most velocities are more than several times the PRF. As a result, both the range and velocity measurements are ambiguous.

One radar may have all three of these modes of operation. If the PRF is high, the range measurements are not only ambiguous, but the echoes have some possibility of being eclipsed by the transmitted pulses. The fraction of the ranges eclipsed is given by the ratio of the pulse duration to the PRI. Even if there is no interest in the range itself, the radar designer must make sure that targets are not missed by being eclipsed. For this reason, several PRIs are used so that no matter what the range, for at least some of the PRIs the target will be detectable. An analogous statement can be made concerning blind velocities. There is some band-width (B_e) centered about multiples of the PRF in which the transmitted energy (or ground clutter) obscures the Doppler-shifted target echoes. The fraction of the velocities obscured is the ratio of this bandwidth to the PRF of the radar. Again, the radar makes us of several PRFs. This is necessary to ensure that the Doppler-shifted echoes are visible for at least some of the PRFs. To detect a target it must be visible in both range and velocity. The fraction of the targets visible at an arbitrary range and velocity is given by the product of those not eclipsed in velocity and those not eclipsed in range:

$$F = (1 - \text{PD/PRI})(1 - B_e/\text{PRF})$$

Substituting PRF = 1/PRI, differentiating F with respect to the PRI, setting the result equal to zero, and solving for the optimum PRI gives

$$\text{PRI}_0 = (\text{PD}/B_e)^{1/2}$$

A typical value for B_e at X-band is 3 kHz, and a typical pulse duration is 3 μs. The optimum PRI in this case is 31.6 μs. (In this case, the optimum PRF is 31.6 kHz.) This optimum PRI could be estimated through ELINT analysis. The pulse duration is a standard ELINT parameter and the bandwidth B could be estimated through high-resolution spectral analysis. This is a typical PRF for a medium PRF radar. While the PRI must be switched among several values to detect targets and resolve ambiguities, the PRIs might be near this value. A typical multimode radar might use a set of eight PRIs during one look at a target. The design could be such that the target can be detected on at least three of the eight PRIs regardless of the range and velocity of the target.

2.2.8   Power Constraints

The maximum radar detection range depends on the total energy returned from the target during the integration time. Energy is the product of the average power and the integration time. Average power is the product of the peak power times the duty factor. The radar duty factor is the ratio of the pulse duration to the PRI. However, unless pulse compression is used, radar range resolution is the pulse duration times the speed of light divided by 2. This means that narrow pulses provide fine range resolution *whether you want it or not.*

In addition, narrow pulses reduce the duty factor when the PRI is determined according to the unambiguous range and velocity requirements.

The use of short duration pulses makes it difficult to get sufficient energy back from the target. Pulse compression is used to increase the average power while retaining the maximum range capability of long pulses and the range resolution of short pulses.

As illustrated later, a CW signal can have much lower peak power than a pulsed signal but have the same energy.

2.2.9   Pulse Compression Modulation Constraints

The use of energy detection receivers is necessary if the radar waveform is noise-like to the interceptor. While such waveforms can be used, their time sidelobes and Doppler properties may be far from ideal in any specific time interval; it is only over a long-term average that their ambiguity function approaches the ideal "thumbtack" shape. Up to now, nearly all radar designers prefer to control their ambiguity function by using specially designed, deterministic pulse compression waveforms such as linear FM (LFM) or binary phase shift keyed (BPSK). Polyphase shift keyed and nonlinear FM waveforms are also common in the radar literature but have not been deployed extensively. For example, the "best" BPSK waveforms are the Barker codes. These have time sidelobes at zero Doppler of maximum amplitude of 1. However the longest known binary Barker code is only 13 bits long. It has been shown that even length Barker codes with length greater than 13 do not exist; however, there are longer sequences of polyphase Barker codes.

2.3   Some ELINT Implications of Future Radar Designs

The constraints discussed in the sections above could be somewhat relaxed through the use of sophisticated signal processing (i.e., by combining the returns in many range cells prior to announcing target detection). Faster and cheaper processing is the direction of current developments. Nevertheless there is a cost associated with such changes in radar design and such changes will not occur without an impelling need plus significant development time and sufficient funding. The ELINT community must maintain close contact with radar R&D activities to keep abreast of trends in radar design.

2.3.1 Bistatic and Multistatic Radars

As a general rule, performance of radars systems with receiving antennas located at different places than the transmitting antennas is inferior to monostatic radar. Bistatic and multistatic designs are interesting because of covertness, survivability, flexibility, and economy. The most recent wave of interest was in the mid-1980s, spurred by technological advances. The noncooperative coherent bistatic radar uses a *transmitter of opportunity* (such as a TV transmitter). Through the early 1990s, bistatic radar for surveillance had been reduced to practice but there was no deployment. The use of transmitters of opportunity will always present technical problems because the waveforms are not designed for radar use. In fact, the transition to HDTV will have an adverse effect on using TV transmitters because the HDTV waveform is less suitable for radar use than current TV waveforms.

2.3.2 Radar Trends

Counter stealth and LPI, wideband and ultra wideband are among the trends in radar design. In counter stealth (that is, radars to detect stealthy targets) there is no evidence of a new class of radars currently under development using technological breakthroughs to detect low RCS targets. Most radars today have some counter stealth capability at short range, and greater power-aperture product would improve that capability. The critical system problems are netting of radars deployed more closely in space and the costs associated with it. The many air surveillance radars in existence might be modified to operate with closer spacing and with higher PRFs to provide the Doppler capability to better distinguish small moving targets from birds and other clutter. PRFs are changed on a burst-to-burst basis to uncover blind speeds. (S-band PRFs tend towards 2,500 pps for 60-km unambiguous range and 125-m/s blind speed; L-band PRFs tend towards 1,500 pps for 100-km unambiguous range and 170-m/s blind speed.) The key to modification of existing radar receivers to counter stealth is to maintain full sensitivity at shorter range without saturation. This means that sensitivity time control may need to be removed or modified.

LPI radar trends are to reduce peak power (and increase the duty factor) to avoid detection by current ELINT and ES/ESM receivers designed for single pulse detection (based on peak power). This is discussed further in Chapter 9. Low sidelobes are important to avoid detection of the radar signal except in the main beam; especially as a defense against antiradiation missiles (ARMs). Power management and atmospheric attenuation shielding are often considered as LPI techniques but are of very limited benefit in search radar. Power management is useful in altimeters and some tracking systems. Atmospheric shielding is of limited use due to the two-way path length of the radar. The radar path has more attenuation than the one-way path to the intercept receiver unless the interceptor is located more than twice the range from the radar's target. Of course, if high altitude ELINT collectors are used, the atmospheric effects are largely eliminated as the density of the atmosphere decreases.

The ELINT challenge is frequency agile, noise-like signals of modest bandwidth transmitted with high duty factor, as shown in Figure 2.7. The instantaneous bandwidth is not likely to exceed that of conventional radars having the same

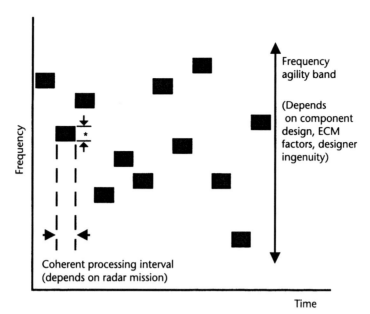

Figure 2.7   Modern frequency agile radar with 100% duty factor.

function and in every application will probably be less than 500 MHz. Randomness in the radar's pulse compression waveform makes it necessary for the ELINT receiver to use energy detection techniques. Today's radars typically transmit the same pulse compression code or waveform in every coherent integration interval during one mode. Tomorrow's radars may choose to transmit different waveforms from one coherent processing interval (CPI) to the next.

Wideband and ultra-wideband (UWB) radar are loosely defined in terms of the fractional bandwidth used by the radar. F_{high} and F_{low} are the upper and lower 3-dB points of the radars bandwidth. The fractional bandwidth, FB, is given by

$$\text{FB} = \frac{2(F_{high} - F_{low})}{F_{high} + F_{low}} \tag{2.17}$$

Wideband radar has a fractional bandwidth greater than 10%, ultra wideband greater than 25% [5]. Applications include underground probing (e.g., mine detection), very short range (<3m) such as proximity sensors and automatic braking devices, and short range (<100m) such as intrusion alarms and through the wall detection. Due to limited energy on target, it is not likely that long range UWB radar will be deployed. The use of ELINT receivers to detect such signals must be weighed against the cost of doing so relative to the threat posed by these signals. Due to their current short range, monitoring these signals is best done at short range also. In this case, the need to sort them from among a myriad of other signals is reduced and their high peak power makes detection relatively easy at short range.

2.3.3 Wideband Active Adaptive Array Radars

The concept of a shared antenna aperture is attractive. Such antennas could provide radar, communications, and ELINT capability in one array. The aperture could be subdivided into sections to allow different beam widths and scan patterns simultaneously in different bands. A complete transmitter and receiver are required at each antenna element along with the processing power to combine them coherently in different ways for different functions. A fully adaptive array of thousands of active elements appears to be too complex at present. There is no way to invert such large matrices in near real time. This area of radar research is called space time adaptive processing (STAP), and is the focus of many research efforts, mainly for airborne radar applications. (One viewpoint is that STAP is a generalized sidelobe canceller). However, partially adaptive arrays are practical. The radar constraints given above apply to active phased array radars. The main purpose of the active array is to provide better control of beam shape and to aim it in a desired direction more quickly (without the effects of mechanical inertia). But the target's range and velocity characteristics still place constraints on the motion of the beam, time on target, and so on. STAP and active arrays will eventually revolutionize radar design. The question is when will it be deployed—that is, when will it become practical and cost effective? If the number of array modules is small enough, applications within the near term are likely and may already be deployed. For large arrays, the answer seems to be probably not in the next decade.

2.4 Summary of Radar Design Constraints and Trends

The one-way range equation provides an advantage to the ELINT system (except for very short range radars). The ability to use nearly matched filter processing in the receiver gives the advantage to the radar. This is explored in more detail in Chapter 9.

Radar signals are designed to achieve certain range, range resolution, velocity, and velocity resolution properties, which are largely determined by the radar function and by the characteristics of its targets. These properties cannot be changed solely for the purpose of denying the opponent's ability to perform the functions of ELINT—radar performance must also be maintained!

The ELINT community must maintain close contact with the radar community in order to anticipate what the radar signals of the future will be like. There is little need to develop ELINT technology (as applied to radar signals) in order to intercept waveforms *unable to perform radar functions*. Note that noise radars and impulse radars have many of the same constraints as conventional radars.

Some of the trends in radar signals are summarized below.

1. Instantaneous bandwidth <100 MHz (more likely <10 MHz for tactical radar).
2. Frequency agility bandwidths ~10% of center frequency typical, 25% upper limit until wideband shared apertures are deployed.
3. PRI agility increases as radars become computer controlled and perform more functions. Dwell/switch (high PRF for pulse Doppler and low PRF

for MTD), stagger for MTI, random jitter for protection from some types of jamming. Programmed or scheduled PRIs when computer control of multifunction radar is used (probably in all new designs for sophisticated radar. Sliding PRIs useful for elevation searching or for adaptive tracking of targets. All of these PRI variations are applicable when computer control of multifunction radar is used.

4. Short duration pulses (<50 ns) are not likely for robust tactical radars, except for short range applications.

5. Bistatic/multistatic deployment will be limited. Even if these radars are deployed, ELINT is only collected from the transmitter. Receive only sites are not targets for ELINT. The principal ELINT concern is if exotic waveforms are chosen for transmission. Note that the constraints described here still apply.

6. Power management is of little use for long-range search radars.

7. Low sidelobe antennas are likely to be increasingly used as protection against ARMs and standoff jamming.

8. A major challenge is high duty factor (or CW) radar with modest coherent bandwidth but wide frequency agility.

2.5   High-Power Microwave Weapons

Leaving the subject of radar for the moment, consider the high power microwave (HPM) weapons being developed to produce field strengths strong enough to damage sensitive receivers. One approach is to transmit very short but high amplitude pulsed signals. Designers of these waveforms need not be concerned with radar constraints; however, the physics of atmospheric effects and breakdown place limitations on the amplitude of such pulses at the source. An explosively pumped flux compression generator (FCG) is the technology used in electromagnetic bombs. A lethal radius of a few hundred meters from a distance of a few hundred meters is likely. The term lethal means that electronic equipment is damaged or at least upset. The terawatts of power in the dc pulse generated is 10 to 1,000 times greater than that in a typical lightning strike.

From the ELINT point of view, there are several questions:

1. Is it necessary to harden ELINT receivers to withstand this type of electronic attack? (This part of EW is called electronic protection or EP.)

2. Is it necessary to intercept and make parameter measurements on microwave bursts from such weapons?

3. Will similar wideband signals made up of very narrow pulses be used to transmit data?

4. Will similar wideband signals made up of very narrow pulses be used as radar waveforms?

The high power density and wide bandwidth of these beam weapons makes it easy for an ELINT receiver to detect them. A typical wideband radiometer could be expected to produce a signal well above the threshold at ranges much greater than the lethal range; the problem to be solved, however, is to distinguish these wideband, low duty factor pulses from lightning, ignition noise, and other types of wideband interference. For this it may be necessary to use multiple-pulse data to detect periodic components. Alternatively, multiple ELINT sites may collect data from a single pulse to locate the source of the emission in the manner currently used by lighting strike detection and location systems.

Wideband communications signals using very short pulses may be used to transmit missile guidance data or otherwise be a part of the ELINT signal environment. These could be very difficult to detect if random pulse spacing known only to the receiver but not to the interceptor is superimposed on the data. Wideband radar signals using short pulses are less of a problem to the ELINT community. In addition to the problem of excessively fine range resolution as the pulse duration becomes short, there is the need to get the same amount of energy back from the target as when a longer duration pulse is used. This means that the transmitted pulses are of higher amplitude and can be more easily detected.

References

[1]   Barton, D. K., *Radar Systems Analysis*, Dedham, MA: Artech House, 1976, p. 111.

[2]   Barton, D. K., *Radar Systems Analysis*, Dedham, MA: Artech House, 1976, p. 199.

[3]   Pace, P. E., *Detecting and Classifying Low Probability of Intercept Radar*, Norwood, MA: Artech House, 2004.

[4]   Wiley, R. G., *Electronic Intelligence: The Interception of Radar Signals*, Dedham, MA: Artech House, 1985, Chapter 2.

[5]   Wiley, R. G., *Electronic Intelligence: The Analysis of Radar Signals*, 2nd ed., Norwood, MA: Artech House, 1993, pp. 2–4.

CHAPTER 3

Characteristics of ELINT Interception Systems

3.1 Intercept System Characteristics and Functions

The characteristics that are important to intercept system designers reflect the functions which such systems are expected to perform, namely, (1) to *intercept* signals, (2) to *recognize* the type of transmitter, and (3) to *measure* signal parameters. The purposes of making detailed parameter measurements include determining:

- Technical features of the radar's design and construction;
- The radar's performance capabilities;
- The radar's weaknesses for jamming (including deception), or other kinds of attack.

The intercept system characteristics of interest include those shown in Table 3.1. In general, interception requires high sensitivity and dynamic range covering large angular regions and frequency bands, but the only information expected is the existence of the signals. The dynamic range need not be excessively wide, since some saturation can be tolerated. For the recognition function, the receiver demands can be relaxed somewhat if the interception process has allowed reducing the angular sector and frequency coverage. Then only a few signals will be received

Table 3.1 Receiver Characteristics for Various Functions

System Characteristic	Function		
	Interception	Recognition	Measurement
Probability of intercept	High	Moderate	Low
Sensitivity	High	Moderate	High (for improved SNR)
Dynamic range	Moderate	Moderate	Large
Coverage band	Wide	Moderate	Moderate (for conventional radars)
Analysis bandwidth	Narrow	Moderate	Wide
Exotic environment performance	High	High	High
Dense environment performance	High	Moderate	Low
Single pulse performance	High	Moderate	Low
Multiple simultaneous signal performance	High	Moderate	Low
Processor/signal sorter requirements	High	Moderate	Low

at once. Usually, recognition requires measuring the basic signal parameters (typically RF, PRI, pulse duration, and scan period) to accuracies on the order of 0.1 to 1%. Naturally, any unusual modulations must be noted to aid in the recognition process.

For the measurement function, the probability of intercept need not be high for signals in general, but must be acceptable for only that particular signal on which measurements are to be made. The sensitivity must be sufficient to provide the high SNR required for precision measurements (generally, the variation due to noise in a parameter measurement is proportional to the reciprocal of the square root of the SNR). The dynamic range should be large in order to permit obtaining a high SNR without saturation in addition to allowing characterization of nulls and lobes in the antenna pattern during scanning. Conversely, the coverage band need only allow for accurately tuning to the carrier of the signal, but the analysis bandwidth should be wide enough to investigate intentional or incidental variations in frequency or phase. Generally, a dense environment can be reduced by using directive antennas and frequency selectivity, so that the measurement receiver need not have the ability to deal with large numbers of signals at once. Likewise, it is not necessary to employ sophisticated signal sorters and processors since the reception process attempts to narrow the field of view, frequency range, and dynamic range to match the signal under scrutiny. Clearly, an intercept system can perform all of these functions, but if it does so simultaneously—that is, attempts to intercept, recognize, and measure all signals from all directions and all frequencies at once—numerous compromises and complications are inevitable.

Describing the requirements for ELINT receivers leads naturally to consideration of various aspects of the signal environment on which the receivers must operate.

This includes some consideration of radar signal modulation as well as the geometry of the intercept situation.

The best approach for an ELINT system is to have very flexible capabilities so that adapting to the functions of interception, recognition, and measurement can be done with ease as the situation demands. This requires a variety of directional and nondirectional antennas with a variety of polarization, several types of wideband and narrowband receivers, a variety of demodulation and analog-to-digital conversion capabilities, and numerous software packages for analysis and display as well as storage devices. In a sense, the ELINT system covers the whole state of the electronic arts, and the system operator should have knowledge in each of these areas. Such diversity is apparent in this chapter.

3.2 Frequency Coverage

Of course, frequency coverage of the antenna system and receiver front-end must include all of the frequencies used by the transmitters of interest. In many cases, the frequencies in use are not known, and thus wide frequency coverage is the rule for intercept systems. Since the total range of frequencies potentially used by radar and radar-related systems is so broad (ranging from below 30 MHz to above 100 GHz), the coverage must be divided into bands compatible with practical components (e.g., antennas, amplifiers, mixers, filters, and detectors). The typical system makes use of octave bands (i.e., the ratio of the upper band edge frequency to the lower band edge frequency is 2:1). This is useful in that any harmonics which may be generated by nonlinearities will be outside the band. Broader bands are possible and practical, particularly in receivers which do not use frequency conversion (heterodyne conversion), such as crystal video or instantaneous frequency measurement (IFM) types (also termed *direct detection* receivers). In the microwave region, coverage of several octaves in one antenna and receiver can be achieved (e.g., 1 to 10 GHz), if desired.

Both radar designers and intercept receiver designers are subject to the constraints of the state of the art in components. However, the designer of a particular radar, who must design a high power transmitter as well as a receiver, is generally subject to more constraints than the intercept receiver designer when it comes to the band covered by the radar's transmitted signals. First, the high power restricts the component performance (e.g., *gain × bandwidth product* of amplifiers tends to be a constant), and second, the need to maintain control over the radiated spectra for electromagnetic compatibility restrains even the most innovative radar designer. Thus, the intercept receiver can usually accommodate the frequencies used by a single radar system or type. The interceptor wishes to maintain enough flexibility to cover a variety of radar types, and this complicates the interceptor's problem considerably. Generally speaking, the interceptor's antenna system bandwidth and receiver coverage bands will be selected based on what can be achieved economically, rather than based on the extremes of a particular radar's tunability or spectrum width.

Common frequency bands and their letter designations are shown in Figure 3.1. The two different letter designations shown are both in use. The older designations have the benefit of many years of common use. The more orderly new band designations are used widely in the electronic countermeasures community and have a logical pattern. Since both designators are in use, such phrases as "I-band-old-X-band" are needed to avoid confusion. The best approach is to simply give the frequency range in hertz (or megahertz or gigahertz) and eliminate any doubt about what is meant.

3.3 Analysis Bandwidth

The analysis bandwidth of a receiver is usually the instantaneous predetection bandwidth available for analysis of the signal spectrum structure. In a superheterodyne receiver, it refers to the bandwidth at the intermediate frequency (IF). In some cases, if the signal envelope is to be analyzed, the term "analysis bandwidth" may refer to the postdetection bandwidth—particularly, in a direct detection receiver. It will be clear from the context of the analysis problem exactly how the analysis bandwidth is defined: *it is the bandwidth which restricts the signal processing or measurement at hand*.

The analysis bandwidth of an intercept receiver is usually not "matched" to the bandwidth of a particular radar signal (which would improve the SNR), but is wider for two reasons:

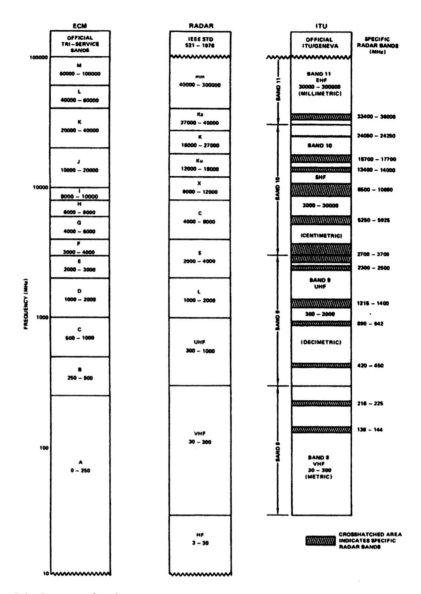

Figure 3.1 Frequency bands.

1. The intercept receiver must be capable of analyzing many different radar signals having a variety of bandwidths. Hence, it usually is as wide as needed for the widest bandwidth signal expected. Modern signal processing approaches may apply multiple bandwidths to the same data.

2. The analysis may be concerned with features of the radar signal incidental to the radar's performance and contained in portions of the signal spectrum beyond the region used by the radar receiver (e.g., pulse top ripples whose period is a fraction of the pulse duration).

Conventional pulsed radars do not normally pose difficult problems for the intercept receiver designer. Once again, since both the radar and interceptor use similar components, the interceptor can maintain an analysis bandwidth at least as wide as the widest radar signal. Radars may make use of both pulse compression and frequency agility, especially wideband radars used for target discrimination.[1] Such radars pose some additional problems for the intercept receiver designer.

3.3.1 Wideband Radar Signal Trends[2]

Ultimately, the bandwidth requirements for intercept receivers are established by the waveform designs used in radar systems. Hence, a discussion of the directions being followed by radar designers is important to the design of intercept receivers.

Wide-bandwidth pulsed signals are used in a number of radar applications. Some of the applications are listed in the approximate order of their popularity in Table 3.2. Wideband signals are used in radar applications to improve radar range resolution, which depends on the reciprocal of the transmitted signal bandwidth. A radar designer can simultaneously increase detection range (average transmitted power) and maintain (or improve) the range resolution through pulse compression techniques without increasing the peak power transmitted by the radar.

As noted in Chapter 2, in a conventional radar, bandwidth (range resolution) is inversely proportional to the pulse width so that the range resolution becomes worse as the pulse becomes wider. For a fixed peak power, then, the designer can improve detection range only by increasing the pulse width, if the PRI is fixed. Pulse compression allows simultaneous detection range and range resolution improvement.

There are several ways to achieve pulse compression, as listed in Table 3.3. One class of techniques is to modulate the frequency within the pulse in one of the ways listed. A second method is to use phase modulation within the pulse. By far the most common frequency modulation technique is to impart a linear frequency ramp on the transmitted pulse.

Frequently used terms in pulse compression systems are *time-bandwidth product* and *pulse compression ratio*. The time-bandwidth product is the product of the transmitted pulse width (T) and the transmitted signal bandwidth (W). For a conventional radar, TW is approximately 1. In a pulse compression radar, it is

Table 3.2 Radar Applications of Wideband Signals

Search-surveillance
Tracking (missile and aircraft)
Missile measurements and instrumentation
Ground mapping
Decoy discrimination, reentry, measurements, ABM
Radar imaging
Ionospheric propagation experiments
Active homing (seekers)
High-resolution battlefield surveillance
Artillery and mortar locating
Missile guidance

---

1. Target discrimination means using the fine structure of the target echo to determine information about the size and shape of the targets.
2. The author is grateful for the help of Mr. Robert B. Shields in preparing this section.

Table 3.3 Wideband-Pulsed Signals

Pulse Compression Techniques
  Frequency modulation on-pulse
    Linear frequency deviation (up or down)
    Stepped frequency
    Nonlinear frequency deviation
    Binary frequency coded
  Phase modulation on-pulse
    Phase reversal (180°)
    Polyphase
  Other Wideband Technique
    Very short pulses (0.5 to 150 ns)

rare to find a time-bandwidth product less than 10, and some radars have products above 100,000. Figure 3.2 is a *scattergram* showing pulse duration versus frequency deviation for a number of radar designs. Lines of constant TW are also shown.

Pulse compression ratio is defined as the ratio of the uncompressed pulse width, τ, to the compressed pulse width, τ_c. The pulse compression ratio is approximately

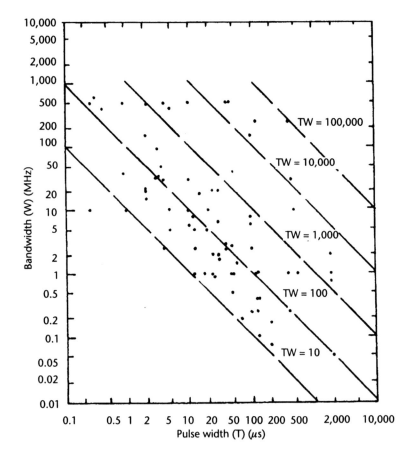

Figure 3.2 Scatter diagram of frequency deviation versus pulse duration for some linear FMOP radars.

equal to the time-bandwidth product. However, it is always somewhat less than TW since, in the processing, amplitude weighting is employed in the receiver to reduce the time sidelobes. The weighting necessarily widens the compressed pulse so that $\tau/\tau_c < TW$.

Some trends in radar design include widespread use of frequency modulation on-pulse (FMOP), with increasing amounts of frequency deviation. Phase modulation on-pulse (PMOP) is also being increasingly employed. Except for cases where the phase modulation is used for missile guidance in a radar-directed weapons system, the purpose of phase coding is generally the same as the frequency modulation: to improve range resolution. Phase reversal systems are the most common, as observed in Table 3.4. Bit lengths as short as 12 ns are being utilized experimentally, although the bit length range of 0.1 to 6 μs is more representative.

Quadrature phase coding (Taylor codes) and Frank polyphase coding represent two other types of coding used to compromise the advantages and disadvantages of frequency modulation and phase modulation. The Frank polyphase code actually uses a quadratic phase progression (modulo 2π) and hence imitates linear frequency modulation [1].

3.4 Dynamic Range

The idea of *dynamic range* is simple to convey: it is the ratio of the largest signal amplitude that the receiver can process to the smallest signal amplitude that the receiver can instantaneously process. However, the words "can process" are troublesome to define. The smallest signal that can be analyzed depends on what the analysis process is like and what its purpose is. Clearly, a precision phase measurement within the pulse requires a larger signal than simply stating that a pulse has

Table 3.4 Parameters for Some Phase Modulation Radars

Phase Coding	RF (GHz)	Pulse Width (μs)	Bit Length (μs)
180° reversal	0.02	2,000	20.0
	0.42	6	0.46
	0.43	390	6.09
	—	780	6.09
	0.60	52	4.0
	5.6	0.2125	0.0125
	9.35	18.1	0.203
	16.25	2.0	0.154
	9.0	CW	125 (511 bits)
	9.0	CW	1.5 and 0.3 (63 and 255 bits)
	10.0	CW	2.19
	10.0	CW	0.0133 (8,000 bits)
	3.0	6.6	0.5
	5.5	6.0	0.1
	8.75	CW	Unknown (range resolution = 25m)
	10.2	CW	Unknown (range resolution = 35m)
Other			
Quadrature	—	25.6	0.2
Frank	—	39.2	0.0625
Polyphase	—	78.4	0.108

crossed a threshold. Likewise, the largest signal that can be tolerated also depends on the measurement being made. Making a precise ERP measurement when the radar's main beam is pointed toward the intercept receiver requires that saturation be minimized; whereas a frequency measurement can be made after a limiter and saturation is of little importance if only one signal is present. Receiver dynamic range must be measured under a variety of conditions, but the conditions must be carefully delineated to avoid confusion and to be certain that a particular receiver can perform a particular function. In a simple test the minimum signal might be determined in the same way as the sensitivity is specified, and the maximum signal level might be that for the "1-dB compression point." This point is illustrated in Figure 3.3. In the linear portion of the input-output transfer characteristic, a 1-dB change in the input level produces a 1-dB change in the output level. As the input level increases, various components in the receiver begin to saturate and the output level is unable to continue to increase in the same proportion to increases in the input level. When the input level reaches a point at which the output level is 1 dB less than it would have been had there been no saturation, the 1-dB compression point has been reached. At this point, for example, the measured power level would be 1-dB less than it would be if the receiver was erroneously considered to follow a linear characteristic. The careful intercept operator frequently inserts calibration signals of known amplitude and stores the receiver output to allow later comparison of the actual signal amplitudes to the known calibration levels.

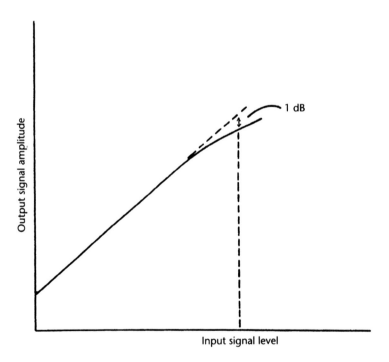

Figure 3.3 1-dB compression point.

3.4.1 Dynamic Range Requirements

Given that there are problems of specifying and measuring dynamic range, the real driving force behind our interest in the subject is the radar signal environment itself and the wide dynamic range that can be encountered. First, consider that the transmitters of interest can have power outputs ranging from a few watts (beacons, altimeters, IFF, and so on) to a few megawatts (long-range search radars). Thus, a 60-dB dynamic range might be expected due to the variety of transmitters alone. Next, the antennas used range from omni-directional types with less than 10-dBi gain to large apertures with gains approaching 40 dBi—another 30 dB of possible dynamic range. Indeed, even for a single scanning radar, the received signal level may be of interest to the interceptor over a 40- to 50-dB dynamic range to allow exploration of the antenna pattern.

Another factor that greatly affects the received signal level is the range from the emitter to the interceptor. The variation in this parameter depends largely on the height of the interceptor (for ground-based emitters). As shown in Figure 3.4, the distance to the horizon for a point at an altitude, h, above the Earth is given by

$$d = \sqrt{(h + R_{\text{Earth}})^2 - R_e^2} \cong \sqrt{2R_{\text{Earth}}h} \tag{3.1}$$

(This is derived in Chapter 7.) The minimum range (h) is when the emitter and interceptor are on a line normal to the Earth's surface. The received power

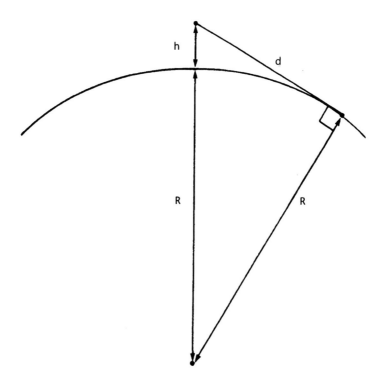

Figure 3.4 Computing the distance to the horizon.

varies as the reciprocal of the square of the range. Therefore, the path loss dynamic range (PLDR) is

$$\text{PLDR}_{dB} = 20 \log \frac{\text{maximum range}}{\text{minimum range}} = 10 \log \frac{2R_{\text{Earth}}}{h} \qquad (3.2)$$

where R_{Earth} is radius of Earth, and h is the height of interceptor above the Earth.

The Earth's radius is about 4,000 miles. Therefore, the PLDR values for a 4/3 Earth-propagation model are as shown in Table 3.5.

The important fact is that the PLDR decreases with altitude so that at high altitudes the PLDR is greatly reduced. This is not surprising since in space, at a large distance from the Earth, all signals, regardless of their terrestrial location, would travel about the same distance. For aircraft operations and typical altitudes of 1 to 10 km, the PLDR is on the order of 32 to 42 dB. For ground stations, the PLDR can be decreased by locating the station at a high elevation point and in an area free from nearby emitters.

Combining all of these dynamic range terms gives a total dynamic range expected at the intercept receiver of

$$60 \text{ dB} + 30 \text{ dB} + \text{PLDR or } 132 \text{ dB to } 122 \text{ dB}$$

for the 1- to 10-km altitude typical of an airborne intercept receiver. Regardless of the altitude, it is clear that a very wide dynamic range will be encountered in the signal environment.

The instantaneous dynamic range of a linear receiver is typically on the order of 50 dB, while that of log-video and limiting receivers (such as IFM) may be 80 dB.

The environment's total dynamic range need not be handled instantaneously in all cases. Several receiver channels can be used to cover different portions of the dynamic range if necessary. Also, a gain control can be used to position the available instantaneous dynamic range at any given portion of the total range to examine a particular signal. Likewise, the gain control may be used to bring different portions of a signal within the linear portion of the receiver's dynamic range at different times. For example, one setting may capture only the main beam and a few sidelobes of radar's scan. Another setting causes saturation when the main beam passes, but the sidelobe structure may be easily observed. Joining separate amplitude versus time plots from both gain settings can provide a total of, let us say, 80 dB of dynamic range for analysis (50 dB from each, setting, with 10 dB of overlap to allow for proper alignment).

Table 3.5 Path Loss Dynamic Range as a Function of Altitude

Altitude h	PLDR (dB)
100m	52.3
1 km	42.3
10 km	32.3
100 km	22.3
1,000 km	12.3

3.5 Sensitivity

The sensitivity of an intercept receiver is defined as the minimum signal power required at the receiver input. "Required" is a tricky word here, for it is necessary to describe the purposes to be fulfilled before the minimum signal level can be determined. At the outset, it should be stated that the reason for the existence of a sensitivity floor (below which the interceptor's purposes cannot be fulfilled) is noise. Generally, it is random, thermal noise. Man-made noise or interference can also be a problem, but it must be considered on a case-by-case basis.

Nearly all noise at radar frequencies is generated in the receiver itself (internal receiver noise). Noise also arrives via the antenna, and is basically a consequence of the thermal radiation reaching the antenna from either terrestrial sources or space. The level of this radiation is a function of the temperature in the direction the antenna is pointing. (The dark night sky has a lower noise level than if the antenna were pointed toward the Earth or the Sun.) This is the principle used in radio astronomy to determine maps of the radiation intensity across the sky at various wavelengths.[3] If the antenna is directed toward a quiet region of the night sky, it might observe an equivalent temperature of only a few degrees Kelvin or, if looking at the Earth's horizon, the observed temperature might be closer to 290 Kelvin (290K is used as the standard or nominal noise temperature expected in an ordinary, noncryogenic environment).

The noise power at the receiver terminals is given by kTB, where k is Boltzmann's constant which is 1.2×10^{-23} watts per hertz per Kelvin, B is the effective noise bandwidth (in hertz) through which the noise is received, and T is the temperature of the noise source (K). For intercept receivers, we may assume an input (antenna terminal) temperature $T = 290K$ and express the noise power as 3.7×10^{-15} W/MHz or −114 dBm per MHz of bandwidth. The ratio of the actual receiver output noise to that which would result from this amount of noise applied to its input is called the noise figure:

$$(NF) = \frac{\text{Noise out of receiver}}{\text{Gain of receiver} \times \text{Thermal noise at input with 290K term}}$$

Clearly, (NF) can be no less than 1. Notice that even if the noise figure were to be reduced to 1, the noise at the input terminals would still depend on what the antenna was observing and in many cases could not be reduced below −114 dBm/MHz. Often, (NF) is expressed in decibels, in which case the equivalent noise level referred to the receiver input terminals is −114 dBm/MHz plus 10 log (NF). It is this noise background against which the intercept receiver attempts to detect signals and measure their characteristic parameters.

Low-noise receivers may have a noise temperature specified rather than a noise figure. For low-noise applications—such as communications satellite ground stations—the receiver may be specified as having a certain noise temperature (e.g., 80°). This means that the amplifier adds 80° to the input noise temperature from

---

3.   The fascinating personal story of John Kraus' pioneering efforts in radio astronomy is contained in *The Big Ear* (Cygnus-Quasar Books, 1976).

the antenna, which may be low since it is directed towards a satellite. Noise figure is defined for an input noise temperature of 290°. The noise figure is, therefore,

$$\frac{290 + 80}{290} \text{ or } 1 \text{ dB}$$

Typical broadband low-noise amplifiers have noise figures in the 2- to 8-dB range at common microwave frequencies.

At the low end of the frequency spectrum in which radars are found, the noise limitation is not receiver noise, but atmospheric noise and man-made interference. This is the limit in the 3- to 30-MHz band, where over-the-horizon (OTH) radars operate. These make use of ionospheric reflections to detect targets along great circle paths around the world. Naturally, the variations in the ionospheric path length and its characteristics make radar range and velocity measurements less precise. Also, highly directive antennas are difficult to build, making the bearing to the target more difficult to determine.

3.5.1 Noise Figure Measurement[4]

Historically, the discussion of noise figure measurement and the many different definitions was important [2–4]. The IEEE definition of noise figure clarified this situation. In this book, the temperature presented by the antenna and feed network to the receiver input is assumed 290K unless otherwise noted. There are essentially two ways to measure the noise figure of a network (amplifier, receiver): (1) using a noise generator, or (2) using a signal generator.

Generally speaking, the noise generator method is more accurate and is used mainly to measure small noise figures on the order of 30 dB or less. (The limitation is the amount of power that the noise source must supply above and beyond that generated internally in the network.) Limitations in the amount of noise power available from noise diodes limit the measurable noise figure using a noise generator.

The two basic methods available to measure the noise figure of a network using a noise generator (source) are: (1) increase the external noise by 3 dB, and (2) measure the Y-factor.

In the 3-dB power-increase method, the device under test and the external amplifier (if one is used) must be capable of supplying a power output 3 dB above the network's internal noise level and remain in linear operation (not saturated). Figure 3.5(a, b) shows typical equipment configurations. Basically, the procedure is to determine a noise level on the power meter or true rms voltmeter [Figure 3.5(a)] or on the detector/voltmeter combination [Figure 3.5(b)] with the 3-dB pad out of the signal path. For Figure 3.5(a), the noise source is turned on and the noise output level is increased until the power indicator reads 3 dB more than with the noise source off. For Figure 3.5(b), switch in the 3-dB pad and adjust the noise source to obtain the same reading on the voltmeter as when the pad was not in the circuit. The noise figure is then determined from the calibration on the noise source for the noise power required to obtain the 3-dB increase.

---

4. The author is grateful for the help of Mr. Grover M. Boose in preparing this section.

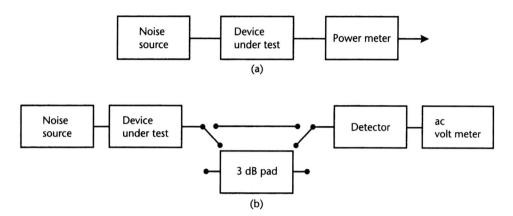

Figure 3.5 Noise figure measurement configuration: (a) 3-dB increase; and (b) 3-dB increase with attenuation substitution.

3.5.2 Y-Factor Measurement

The Y-factor is defined as the ratio of the network's output power with the noise source on to that when the noise source is off. The noise figure in decibels is then

$$\mathrm{NF}_{dB} = 10 \log \frac{\dfrac{T_h}{T_o} - 1}{Y - 1} = 10 \log \mathrm{ENR} - 10 \log(Y - 1)$$

where T_h is the temperature of the noise source when turned on, and T_o is the off noise temperature. The numerator, $T_h/T_o - 1$, is the available excess noise ratio (ENR), which is given as part of the noise source calibration. The equation above assumes that the temperature of the noise source when it is off is the standard temperature, 290K. If more accurate measurements are required, the actual hot and cold temperatures can be measured along with the hot and cold output noise powers. Some precautions to be taken when connecting the equipment for Figure 3.5(a, b) are as follows:

1. The meter in Figure 3.5(a) must either be a power meter or a true rms voltmeter. The meter must also be connected in the circuit prior to any detector (i.e., at the IF output in a receiver).
2. Calibrated and precision attenuators should be used in the attenuator substitution method.
3. The meter in Figure 3.5(b) is only used as an indicator. Accuracy is a function of the calibrated attenuator.
4. Attention must be paid to impedance-matching the noise source to the device under test. In fact, impedance should be matched at all connections (auxiliary amplifier, attenuator) to ensure maximum power transfer and correct noise figure measurements.

Using a noise generator provides a simple way to measure noise figure that does not require a measurement of the noise bandwidth.

The signal-generator method accuracy is determined by the calibration of the generator, and the requirement that the noise bandwidth (not to be confused with the 3-dB bandwidth) must be known to complete the determination of the noise figure. Additionally, it is generally assumed that the input and output noise temperatures are equal to 290K and that the gain of the network is the same for noise as for the signal (i.e., that the system is linear). This method is most applicable to the measurement of the "radar" noise figure, which is defined as the noise figure of a receiver when the signal spectrum is contained in one narrow band. If there are two bands in which a signal can enter a receiver, such as in a superheterodyne receiver with no preselector, the actual noise figure is 3 dB less than that measured by the small-signal method. The small-signal method has won wide acceptance as a means of checking network performance, even if the accuracy is ±2.0 dB or poorer, since the equipment needed to perform the measurement is readily available in most laboratories, namely, a signal generator and some indicating device.

The measurement of a network noise figure using a signal generator follows this procedure:

1. Connect equipment as in Figure 3.5 except that a signal generator is substituted for the noise source.
2. With signal generator turned off, obtain a reading of the noise output power.
3. Turn on the signal generator and tune it to the center of the band of the amplifier or receiver under tests.
4. For Figure 3.5(a), adjust the signal generator power output to obtain a reading exactly twice what it was in step 2 (3 dB higher on a decibel scale). For Figure 3.5(b), insert the attenuator and adjust the signal generator to obtain the same output indication as before.

Under the above conditions and assuming that the input termination temperature is 290K, the noise figure can be determined from

$$NF = \frac{\text{Input signal level}}{k\,(290)B_N}$$

where B_N is the noise bandwidth. The noise bandwidth of a receiver or amplifier can be determined by:

1. Measuring the 3-dB bandwidth and knowing the type and number of coupling circuits comprising the amplifier stages. Table 3.6 is then used to determine the noise bandwidth. If there are more than five stages, using the 3-dB bandwidth as the noise bandwidth results in an error of approximately 10% (or approximately 0.5 dB).
2. The most straightforward way of measuring the noise bandwidth of a circuit (or receiver) is to measure its frequency response using a variable-frequency signal generator, and integrate the area under the response normalized to unity at the point of maximum response.

In most instances, the 3-dB bandwidth is known or easily measured; then the noise bandwidth is determined from Table 3.6. Unless extreme accuracy is required

Table 3.6 Comparison of Noise Bandwidth and 3-dB Bandwidth

Type of Receiver Coupling, Ckt	Number of Stages	Ratio of Noise Bandwidth to 3-dB Bandwidth
Single-tuned	1	1.57
	2	1.22
	3	1.16
	4	1.14
	5	1.12
Double-tuned*	1	1.1
	2	1.04
Stagger-tuned triple	1	1.048
Stagger-tuned quadrature	1	1.019
Stagger-tuned quintuple	1	1.01
Gaussian	1	1.065

*Applies to a transitionally coupled double-tuned circuit or to a stagger-tuned circuit with two tuned circuits.
Source: [5].

for the noise figure, this procedure, with a calibrated signal generator, will be accurate to ±2 dB. Care must be taken in selecting the meter to ensure that it has sufficient bandwidth. A broadband power meter is adequate for applications where the center frequency is above 10 MHz. A true rms meter is useful for frequencies below 10 MHz. If the noise is impulsive in nature, the meter crest factor becomes important; the higher, the better [5]. Impulsive noise is most prevalent at HF (2 to 32 MHz) where thunderstorm activity around the world contributed to atmospheric noise levels.

3.5.3 Some Sensitivity Measures

Receiver sensitivity is measured using several criteria. Figure 3.5(a) shows one measurement scheme for determining the receiver noise. The power meter is like a square-law detector in that the output (dial reading) is proportional to the input power. The integration or averaging of the metering circuit and the tendency of an observer to read the average value of the meter pointer along the dial corresponds to postdetection filtering. If there were little or no filtering, the ideal power meter reading would rapidly fluctuate over a wide range of power values as the noise level changed. Suppose we wanted to detect the presence of one signal against the thermal noise background by observing the power meter. By using very long averaging times, the power meter reading would become quite steady (very small fluctuations). Then a small amount of additional power (which might be several times larger than the amount of fluctuation) could be reliably detected. Under these conditions, it is clear that while the receiver noise figure and bandwidth set the noise level reaching the power meter, our ability to *detect* a signal in the presence of that noise is greatly affected by the averaging done after the RF signal has been converted to a voltage proportional to the power. This postdetection averaging is characterized by a video bandwidth, B_v, which is the 3-dB cutoff frequency of the low pass filter placed after the detector. The important point is that our ability to detect the signal in the presence of noise depends not on the average value to which the power meter needle points, but on how much it is fluctuating. In concept, we may simply offset the meter scale to place the average value at zero and then detect the signal in the presence of the fluctuations only.

Tangential signal sensitivity is a commonly used receiver sensitivity term. In the light of this model, it is defined as follows:

> When the signal is present, the lower limit of the power meter reading fluctuation is just equal to the upper limit of its fluctuation when the noise alone is present. In practice, an oscilloscope is used to display the receiver output and a pulsed signal is used. The tangential condition is when the "top" of the noise alone is at the same level as the "bottom" of the noise present on the top of the pulse.

Making laboratory measurements of the signal level required to meet this condition is somewhat subjective and depends on the judgment of the observer. Usually, the variation from observer to observer is on the order of 2 dB. Tangential signal sensitivity is not a good measure of receiver sensitivity because it is subjective and no receiver function or performance is specified. However, it is useful as a tool for rapidly determining approximate receiver sensitivity.

While the concept of tangential signal sensitivity is traditionally used and easy to measure in the laboratory, receiver sensitivity must be measured in terms of the signal power needed in order to perform some function. The detection function has been analyzed in great detail by radar designers, and is formulated in terms of the probability of detection, P_D (i.e., the probability that the signal plus noise crosses a threshold), and the probability of false alarm, P_{FA} (i.e., the probability that noise alone crosses the same threshold). As the SNR increases, a threshold can be selected such that $P_D \rightarrow -1$ and $P_{FA} \rightarrow 0$.

For the tangential signal condition, one can imagine that a threshold is selected so that the probability of noise alone crossing the threshold is the same as the probability that the signal plus noise fails to cross the threshold; that is,

$$1 - P_D = P_{FA}$$

This condition alone does not define a SNR, since for any SNR a threshold can be selected for which the above condition is satisfied.

3.5.4 Output SNR and Receiver Applications

The signal level at the receiver input determines its output SNR. The output SNR is related to the input SNR by the processing gain:

$$\text{SNR}_{out} = \text{SNR}_{in} \cdot (\text{Processing Gain}) \tag{3.3}$$

The SNR at the input is the ratio of the signal power at the input to the noise level referred to the input. The ELINT processing gain is basically related to the postdetection filtering done prior to any measurement or detection decision; for example, the averaging done by the power meter in Figure 3.5(a). The processing gain in this case is [6]

$$G_P = \left[\frac{B_{RF}}{2B_V}\right]^\gamma$$

where:

B_{RF} = predetection or RF bandwidth

B_V = postdetection or (one-sided) video bandwidth

γ = integration efficiency (typically 0.5 to 0.8)

In this situation, we have

$$R_{in} = \frac{\text{Signal power}}{\text{Noise power referred to input}}$$

Thus,

$$\text{SNR}_{out} = \text{SNR}_{in}\left[\frac{B_{RF}}{2B_V}\right]^\gamma = \frac{\text{Signal input}}{kTB_{RF}}\left[\frac{B_{RF}}{2B_V}\right]^\gamma$$

For $\gamma = 0.5$, we have

$$\text{SNR}_{out} = \frac{\text{Signal input}}{kT(2B_V B_{RF})^{1/2}}$$

The above result explains the use of the term "effective bandwidth" for

$$\sqrt{2B_V B_{RF}}$$

Consider estimating the pulse power. The fluctuation in the measurement depends on the output noise level; in fact, the fractional standard deviation in the measurement will be the reciprocal of the output SNR:

$$\sigma_P = \sqrt{\frac{kT(2B_V B_{RF})^{1/2}}{\text{Signal power}}} = \frac{1}{(\text{SNR}_{out})^{1/2}}$$

This form is characteristic of signal parameter measurements of all types; that is, *the error in the measurement is proportional to the reciprocal of the square root of output SNR*. Examples of parameters so affected include amplitude, angle of arrival, time of arrival, and carrier frequency. (This assumes an output SNR larger than 10 to 15 dB.)

3.5.5 Threshold Detection

Threshold detection plays a central role in automated ELINT receivers, since it is necessary to first make a decision that a signal is present before any parameter measurements can be made. This problem has been considered extensively in the radar literature (e.g., Barton [7]). The only difference is that for radar receivers detecting target echoes, the receiver postdetection bandwidth and predetection bandwidths are matched to the transmitted signal and to each other (i.e., $B_V - 0.5$ B_{RF}). In an ELINT receiver, particularly the wideband types that most often use automatic detection, the ratio of the RF bandwidth to the video bandwidth can be quite large, which affects the statistics of the output noise. This aspect is analyzed by Tsui [8]. Figures 3.6 and 3.7 show the changes in the statistics at the square law detector output for various values of $B_{RF}/2B_V$ for no signal and for an input SNR of 2 (or 3 dB). As the postdetection bandwidth narrows, the output distribution peaks near the expected (or dc) value of unity (no signal) or 3 (signal = 2 units, noise = 1 unit). The resulting detection probability is shown in Figure 3.8 for several values of $B_{RF}/2B_V$ for a false alarm probability of 10^{-8}.

As the ratio $B_{RF}/2B_V$ becomes large, the output noise and signal plus noise distributions become normally distributed [9, 10]. Suppose the noise at the input to the detector consists of in-phase and quadrature components each having a mean of zero and variance σ^2; that is,

$$P(v_{in}) = \frac{1}{\sqrt{2\pi}\sigma} e^{-v_{in}^2/2\sigma^2} \tag{3.4}$$

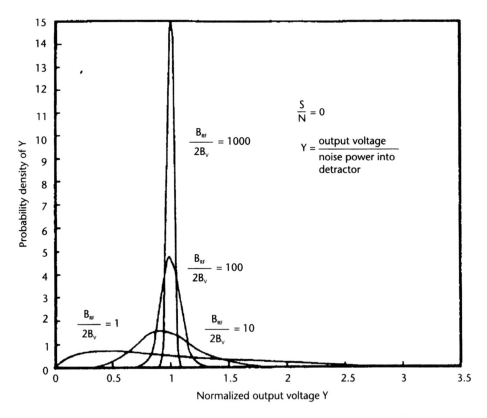

Figure 3.6 Output distribution for the noise voltage for several values of $B_{RF}/2B_V$. (*From:* [9], p. 22. © 1967 IEEE. Reprinted with permission.)

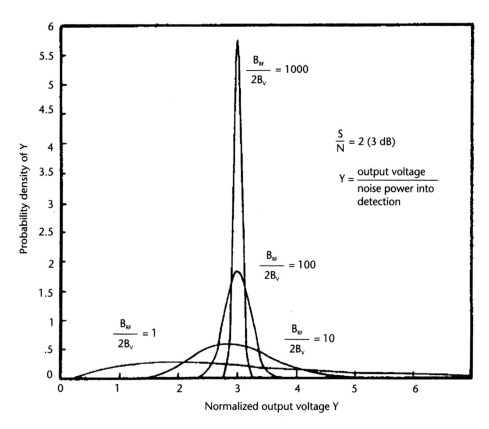

Figure 3.7 Output distribution for SNR = dB for several values of $B_{RF}/2B_V$. (*From:* [9], p. 23. © 1967 IEEE. Reprinted with permission.)

Then the noise voltage after the linear envelope detector has the Rayleigh distribution, namely,

$$P(v_d) = \frac{v_d}{\sigma} e^{-v_d^2/2\sigma^2}, \; v_d > 0 \tag{3.5}$$

Note that $\sigma^2 = N_0 B$ = noise power in bandwidth, B, at the detector input.

The distribution of the *square* of the detector output voltage is

$$P(v_d^2) = \frac{1}{2\sigma^2} e^{-v_d^2/2\sigma^2}, \; v_d^2 > 0 \tag{3.6}$$

$$= 0 \qquad\qquad , \; v_d^2 < 0$$

Then the means and variances for each of these distributions can be computed as:

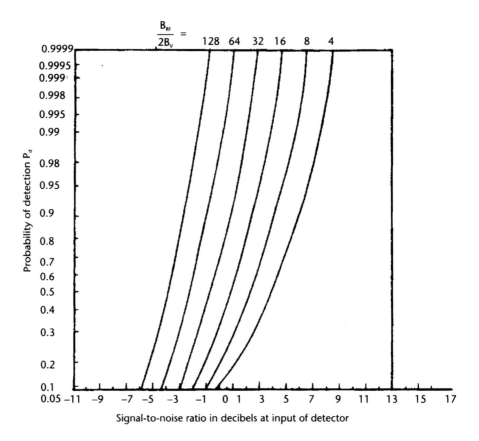

Figure 3.8 Probability of detection for several values of $B_{RF}/2B_V$.

Mean of envelope, $\bar{v}_d = \sigma\sqrt{\pi/2}$

Variance of envelope $= (2 - \pi/2)\sigma^2$

Mean of square envelope, $\bar{v}_d^2 = 2\sigma^2$

Variance of squared envelope $= 4\sigma^4 = (\bar{v}_d^2)^2$

The action of the video low pass filter is to compute the sum of a number of independent samples from the detector output, the number being $B_{RF}/2B_V$. The distribution of the sum of a large number of identically distributed random variables approaches a Gaussian distribution whose mean is the sum of the mean values of the individual samples and whose variance is the sum of the individual variances. For $\sigma = 1$, and $B_{RF}/2B_V$ large, the noise distribution at the output of the low pass filter is, therefore,

$$P(v_{out}) = \frac{1}{\sqrt{2\pi}\,\sqrt{2B_{RF}/B_V}}\, e^{\frac{[v_{out} - (B_{RF}/2B_V)]^2}{4(B_{RF}/B_V)}}, \quad v_{out} > 0 \qquad (3.7)$$

That is, the mean value is $B_{RF}/2B_V$ times the mean value of the squared envelope and the variance is $B_{RF}/2B_V$ times the variance of the squared envelope.

When the signal is present, the means and variance of the squared signal pulse noise are given by [9–11]:

Mean of squared signal plus noise: $(1 + \text{SNR})\sigma^2$

Variance of squared signal plus noise: $4(1 + \text{SNR})\sigma^4$

Using the same reasoning, the output distribution when the signal is present is (again taking $\sigma = 1$)

$$p(v_{out}) = \frac{1}{\sqrt{2\pi}\,\sqrt{2B_{RF}/B_V}\,\sqrt{1 + 2\text{SNR}}}\, e^{\frac{[v_{out} - (B_{RF}/B_V)(1 + \text{SNR})]^2}{4(1 + 2\text{SNR})(B_{RF}/2B_V)}} \qquad (3.8)$$

Note that for $\sigma = 1$, v_{out} is expressed in units of σ^2.

For the case where B_{RF}/B_V is large (say, more than 100), (3.7) and (3.8) can be used to compute the probability of detection and false alarm using the widely tabulated normal distribution and its integral. For the case $B_{RF}/2B_V = 1$, the curves of the radar detection probability can be used [7]. For intermediate values, the published results of Emerson [12], Urkowitz [9], and Tsui [8] are valuable.

An example will serve to illustrate the use of (3.7) and (3.9). Suppose $B_{RF} = 1$ GHz and $B_V = 5$ MHz. Then $B_{RF}/2B_V = 100$. Suppose the predetection SNR is 0 dB. For a false alarm probability of 10^{-5}, the threshold must be set at the mean plus about four times the standard deviation. From (3.7), the mean is 200 and the standard deviation is 14.1. Hence, the threshold should be set at 256.4 and the probability of detection can now be found. Using (3.8), the mean of the signal plus noise distribution is at about 400 and the standard deviation is about 34.6. The threshold at 256.4 is located about 4.1 standard deviations below the mean. Therefore, the probability of detection will be about 0.99995. Using the curves from Barton [7] provides the same results. If the number of samples integrated is 100, the integration efficiency (γ) at $P_{FA} = 10^{-5}$ is approximately 0.75. The processing gain is then $(100)^{0.75} = 31.62$, or 15 dB. Since the input SNR was 0 dB, the output SNR is 15 dB. Referring to the probability of detection curve for a 15 dB SNR shows that the probability of detection should be greater than 0.9999 at $P_{FA} = 5 \times 10^{-5}$. What is not clear from this approach is the close approximation of the output distributions to the normal distribution and the rather easy calculations which result.

Next, consider what would be seen on an oscilloscope if a pulsed signal at 0 dB input SNR were applied to such a receiver, as shown in Figure 3.9(a). In the normal case, the mean value of the noise would be blocked, possibly by using an ac-coupled video amplifier. In that case, with no signal present, normally distributed noise would be observed, with a standard deviation of 14.1 (measured in units of σ^2). Then, when the signal pulse appears, the mean value shifts upward from the nominal baseline of noise by 200 (measured in units of σ^2). The observer might adjust the "peak-to-peak" noise to occupy one division on the oscilloscope screen. If ± 2 standard deviations is chosen as the peak-to-peak noise swing, the noise occupying one screen division represents 56.7 units of σ^2. The pulse height on the screen will then be 200/56.7, or 3.5 oscilloscope divisions above the center of the noise. The noise on the pulse has a standard deviation of 34.6 (units of σ^2). Its

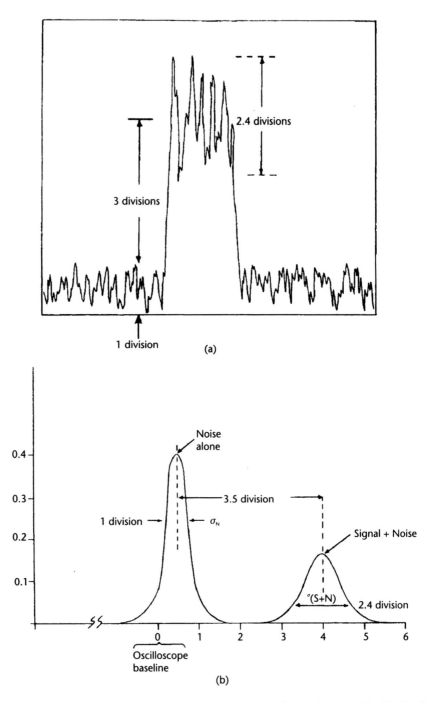

Figure 3.9 (a) Video amplitude versus time (0 dB input SNR for $B_{RF}/2B_V = 100$). (b) Distribution of noise alone and signal plus noise for 0 dB input SNR and $B_{RF}/2B_V = 100$.

peak-to-peak (±2 standard deviations) range will be 138.4/56.7, or 2.4 oscilloscope divisions. Figure 3.9(b) is a sketch of the output signal plus noise and noise distributions with the horizontal scale marked in the same units as the oscilloscope calibration.

3.5.6 Sensitivity and the Received Pulse Density

Once a threshold power level is established for a receiving system, signal pulses that exceed this level are processed by the receiver while those below that level are ignored. For a given emitter and receiver, the maximum range at which that emitter can be received can be determined. For an airborne receiver, suppose that the maximum range is less than the altitude at which the aircraft is flying; then none of that emitter's pulses will activate the receiver. As the aircraft altitude is decreased, a region is defined on the Earth's surface in which all emitters of that type can be received, and the number of received pulses increases as the area of the region increases. As the aircraft altitude continues to decrease, at some point, the distance to the horizon will be less than the maximum range at which the emitter can be received. As the altitude decreases further, the number of pulses which can be received again decreases. Thus, the maximum number of received pulses occurs at that altitude for which the distance to the horizon is equal to the maximum distance at which the emitter can be received. The maximum range at which a signal can be received is [13]

$$R_E = \left[\frac{P_T G_{TE} G_E \lambda^2}{(4\pi)^2 S_E L_T L_E} \right]^{1/2}$$

where:

R_E = range to emitter from ELINT receiver

G_{TE} = gain of transmitter in the direction of the ELINT receiver

G_E = gain of the ELINT receiver in the direction of the emitter

λ = wavelength

S_E = signal power required at ELINT receiver

$P_T T$ = emitter transmit power

L_T = transmitter losses

L_E = ELINT receiver losses

Recall that the distance to the horizon is given by

$$d = \sqrt{2 R_{\text{Earth}} h}$$

Thus, the altitude (h) at which the maximum coverage is achieved is when

$$2 R_{\text{Earth}} h_o = \frac{P_T G_{TE} G_E \lambda^2}{(4\pi)^2 S_E L_T L_E}$$

or

$$h_o = \frac{P_T G_{TE} G_E \lambda^2}{2R_{\text{Earth}}(4\pi)^2 S_E L_T L_E}$$

For sidelobe intercepts, $G_{TE} \sim 1$. (For new radar designs, the average sidelobe level is more typically 0.1 relative to isotropic.) For an omni-directional ELINT antenna $E_E \sim 1$. Assuming $L_T = L_E = 1$, and $\lambda = 0.1$m, the altitude at which the maximum number of pulses may be received as a function of receiver sensitivity is shown in Figure 3.10. To fly at 10,000m and receive sidelobe signals from a 1-kW transmitter at the horizon requires a sensitivity of −83 dBm. (Sensitivity in this case means that power level which permits the receiver to perform those functions required of it.)

3.6 The Ultimate Limits to ELINT Parameter Measurements

In a way, threshold detection only determines the presence of a signal. To provide useful intelligence, signal parameters must be measured. Often, just as in radar, threshold detection provides more information than mere existence of a signal. With varying degrees of precision, the threshold crossing gives some information about the following:

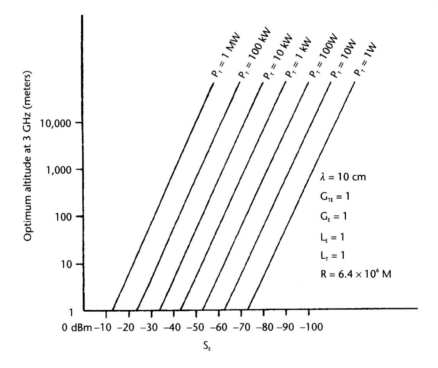

Figure 3.10 Optimum altitude as a function of receiver sensitivity.

1. When the signal begins and ends;
2. The minimum signal amplitude;
3. The bearing angle, if a directional antenna is used;
4. The carrier frequency—at least that it is in the band covered by the receiver at the time the signal occurred.

As Woodward [14] points out, radar systems cannot divorce simple target *existence* information from target *location* information. The radar receiver works best when separate range and velocity resolution cells are provided and when a directional antenna is used. The existence threshold is then the same as the resolution threshold.

An ELINT receiver has somewhat similar properties. A receiver that sought to intercept only one signal in an otherwise empty environment might find excess resolution of time, frequency, angle, and amplitude to be a burden. In practice, in an environment containing many signals, the interceptor needs parameter resolution to make sense of the energy being received.

In signal reception, the word "information" has been given a precise meaning by information theory. In particular, if the initial uncertainty about some signal is U_1, and after some energy has been received the uncertainty is $U_2 < U_1$, we say that an amount of information, I, has been gained, where

$$I \equiv \log_2(U_1/U_2) \text{ bits} \tag{3.9}$$

A consequence of information theory is that to transfer I bits of information from one place to another requires *at the least* an energy exchange of [14]

$$E_{\min} = \frac{N_0 I}{\log_2 e} \tag{3.10}$$

where:

$E_{\min}$ = minimum energy required at receiver

$N_0 = kT$ = noise power density at receiver input

I = information gain (bits)

Of course, this minimum energy can be achieved only in when transmission of the information is intentional and planned and the reception of the information makes use of that knowledge (as in communication systems). Consider the idealized situation where the purpose of the radar's transmission is to transmit information about its signal to the ELINT station. For example, instead of measuring the carrier frequency of an intercepted radar pulse, suppose the radar transmitted a message that said, in effect, "My carrier frequency is now 3167.4285 MHz." According to information theory, information could be received at the ELINT site with transmission of the minimum energy by the radar. It is interesting to think about this admittedly unrealistic situation and to explore the amount of additional energy that is needed to make similar determinations of signal parameters by the traditional methods of radar and ELINT.

The fact that the energy needed at the receiver is determined by the final measurement uncertainty, U_2, is not surprising. However, the required energy also depends on the initial parameter uncertainty, U_1. Consider carrier frequency measurement. If the initial uncertainty is large, a wideband receiving system is needed. Even if a channelized receiver is used, the large number of channels makes the system more susceptible to false alarms and the threshold must be raised in all channels to avoid being distracted from the channel with the real signal by a noise threshold crossing in any one of the other channels. Similarly, if the signal can occur at any time over a long period, the threshold must be raised to provide the very low false alarm probability needed to avoid a false threshold crossing during the time we are searching for the real signal. Thus, the initial uncertainty also determines the signal energy needed at the receiver.

Notice that the energy is the important parameter. For one radar pulse, the energy is the product of the signal power and the pulse duration. Should this energy be too small to achieve the needed resolution, then more than one pulse may be needed and the receiver must provide a means for adding the received energy in a sensible manner.

A most interesting aspect of the information theoretic view of reception is that true resolution can be achieved only by *destroying* part of the received information. Consider an ideal recording of the signal as applied at the receiver input (or its equivalent after frequency translation). While the recording preserves the signal, we remain ignorant of its parameters. The recording contains all the information which was received about both the signal and the noise—in fact, the received signal could be exactly reproduced upon playback (and possibly frequency translation) and the situation would be just as it was at the time the signal was originally recorded. However, nothing would have been learned. This characteristic applies to all *reversible* signal processing operations. Information can be gained only from *irreversible* operations. Since the original signal cannot be recreated after an irreversible operation, some of the original information has been destroyed or lost. The important thing is to destroy the unwanted information (i.e., that which describes the noise) and retain the signal descriptors or interest. For example, an envelope detector destroys the phase information. A threshold detector destroys most of the amplitude information (except for 1 bit), but the resulting signal provides a good time of arrival estimate.

A radar measurement can provide a fractional error of [15, p. 401]

$$\frac{\delta M}{M} = \frac{L}{\sqrt{2E/N_0}} \tag{3.11}$$

where:

δM = error in measurement M

M = resolution in measured coordinate

E = received signal energy

N_0 = thermal noise power density at the receiver input

L = a constant on the order of unity which is dependent on the measurement parameter and signal processing used

Consider a carrier frequency measurement. In radar, the transmitted frequency is known, so the initial uncertainty is the Doppler shift, which can easily be determined to within the reciprocal of the pulse duration, τ. The final frequency uncertainty is [15, p. 407]:

$$\delta F = \frac{\sqrt{3}}{\pi\tau\sqrt{2E/N_0}} \text{ (Rectangular pulse)} \qquad (3.12)$$

The performance projected by (3.11) should certainly be poorer than that predicted by the information theoretic approach. The error in the measurement (δM) is associated with the final uncertainty U_2. The initial uncertainty, U_1, is certainly no smaller than δM (and probably much larger for an ELINT receiver). Solving (3.11) for E and dividing by $E_{\min}$ from (3.10) yields

$$\frac{E}{E_{\min}} = \frac{U_2^2}{\tau^2 M^2} \log_2(U_1/U_2) \leq \frac{U_2^2}{U_1^2} \log_2(U_1/U_2) \qquad (3.13)$$

The term on the right has the form $x^{-2}\log_2 x$. The minimum of this function occurs at $x = e^{-1/2}$, and the value at this point is 0.265, which indeed is less than 1. Thus, the typical radar and ELINT measurements are at least 5.7 dB [5.7 dB = $10\log(0.265)$] worse than the theoretical limits established by information theory—and frequently much worse, as shown in Figure 3.11.

For practical threshold circuits, discriminators, and direction finders, the measurement error is proportional to the reciprocal of the square root of the signal-to-noise energy ratio. On a single-pulse measurement, $E = S\tau$ and $N_0 = N/B$, where τ is the pulse duration and B is the noise bandwidth. Thus, the measurement error on a single rectangular pulse is

$$\frac{\delta M}{M} = \frac{1}{\sqrt{B\tau}\,\sqrt{2\,\text{SNR}}} \qquad (3.14)$$

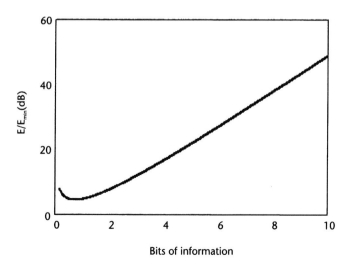

Figure 3.11   Radar compared to information theory limits.

where M is the initial measurement resolution (bandwidth) which may be approximately the reciprocal of the pulse width.

3.7 ECM and ELINT Receivers

In this discussion of ELINT receiver performance there has been an unstated assumption that these receivers are not subject to either noise jamming or deception. Since ELINT, as discussed here, is a peacetime activity,[5] jamming is rarely a problem. Even during a conflict, it is unlikely that noise jamming would be used to hide a radar from an ELINT receiver since that would generally also interfere with the radar's function. Jamming initiated by the forces that were also using ELINT would presumably be done consciously and with knowledge of the consequences.

Spoofing (or deception) of ELINT receivers is another matter. Certainly, during peacetime, spoofing could take place at any of several levels. Radars could simply be operated with restricted sets of parameters that are changed during wartime. Radar-like transmitters could emit large numbers of pulses across a wide band of frequencies, making the ELINT process much more difficult and costly. Such *counters* to ELINT have simply not been judged cost-effective in times past. The radar designers have concentrated on making the radar operate to meet certain performance specifications. Spending resources on spoofing ELINT receivers has not been worthwhile. In tactical situations, such may not be the case. Clearly, there are benefits to deceiving a radar warning receiver, and such activities may become a problem.

The peacetime, strategic, manpower-intensive character of ELINT has meant that interference and deception against it are rare. While one can speculate about the existence of these activities in the future, it is well to remember that until now ELINT has always reacted to the directions established by the radar community. To reverse that and have radar designs be a reaction to ELINT collection activities seems unlikely; *however, the efforts to design and deploy LPI radar* indicate that some activity along these lines is taking place.

3.8 Crystal Video Receivers

The simplest ELINT system is the crystal video type consisting of an antenna, a detector, and a video amplifier, as shown in Figure 3.12(a). Adding a low-noise RF preamplifier, as in Figure 3.12(b), provides additional sensitivity. In practical receivers, there is always some bandpass filtering action ahead of the detector to restrict the receiver output to some band of interest. The band-limiting function may, in some simple receivers, be that provided by the antenna alone. In the low sensitivity receiver, the sensitivity is determined by the detector characteristics and the noise generated in the video amplifier. In the high sensitivity receiver, it is determined by the noise figure of the RF preamplifier. In both cases, the RF and video bandwidths place fundamental limits on the receiver's sensitivity. In some

---

5.   It has been said that ELINT presumes peace on Earth but not goodwill towards men.

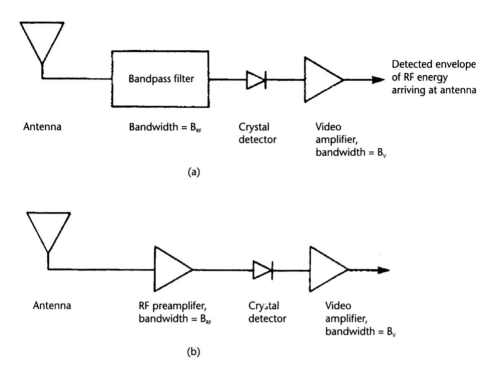

Figure 3.12 Crystal video receivers: (a) low sensitivity receiver; and (b) high sensitivity receiver.

applications, high sensitivity may be undesirable—as for a radar warning application in which an aircraft is to be warned of being tracked at relatively short range by radar associated with a surface-to-air missile system.

The sensitivity of a crystal video receiver has been studied extensively [16]. The sensitivity can be characterized as either gain-limited (when it can be improved by increasing the gain of the RF preamplifier) or noise-limited (when improvement is possible only by reducing the noise figure of the RF preamplifier or the receiver bandwidths). The discussion of sensitivity above was based on a noise-limited receiver in which the detector characteristics are inconsequential. In this case, the sensitivity is determined solely by the RF and video bandwidths and the receiver noise figure. For ELINT applications this is generally the case. In some applications, however, small size and low cost are important, and the sensitivity of the configuration in Figure 3.12(a) [or Figure 3.12(b) with only a small amount of gain ahead of the detector] may be used. In this case, the sensitivity varies in proportion to the square root of the video bandwidth, the proportionality constant being a characteristic of the detector. The tangential sensitivity of the crystal video receiver is [16]

$$\text{TSS} = kTF\left\lfloor 6.31B_V + 2.5\sqrt{2B_{RF}B_V - B_V^2 + AB_V/(GF)^2}\right\rfloor \qquad (3.15)$$

where:

TSS = tangential sensitivity (W)

k = Boltzman's constant 1.38×10^{-23} Ws/K

T = room temperature (290K)

F = noise figure of amplifiers ahead of detector

B_V = video bandwidth (Hz)

B_{RF} = RF bandwidth (Hz) $\geq 2B_V$

A = diode parameter [see (3.16)]

G = gain of amplifiers ahead of the detector

The diode parameter, A, is given by

$$A + \frac{4F_V}{kTM^2} = \frac{4F_V R}{kTC^2} \tag{3.16}$$

where:

F_V = video amplifier noise figure

M = detector figure of merit = $C/\sqrt{R}$

R = dynamic impedance of detector (Ω)

C = detector sensitivity (V/W)

For the gain-limited crystal video receiver, the term containing A is very large, and (3.15) reduces to

$$\text{TSS}_{GL} \cong \frac{kT}{G} 2.5 \sqrt{AB_V} \text{ (gain-limited)} \tag{3.17}$$

If there is no preamplifier, $G = 1$, and we have

$$\text{TSS}_{GL} \cong 2.5kT\sqrt{AB_V} \text{ (no preamplifier)} \tag{3.18}$$

For TSS in dBm and B_V in megahertz, (3.18) becomes

$$\text{TSS}_{\text{No Preamp}} \text{ (dBm)} \cong -110 + 5 \log A + 5 \log B_V$$

If $A = 10^{14}$ and $B_V = 1$ MHz, the tangential sensitivity of the receiver is -40 dBm. In an actual receiver, there will be an RF filter ahead of the detector, and the insertion loss of the filter is equivalent to a gain (G) of less than one (and a noise figure, F, equal to the reciprocal of the insertion loss). This must be taken into account using (3.15). The example used in [15] has a filter loss of 2.1 dB, providing a sensitivity of -38 dBm. These values are typical of a crystal video receiver with no preamplifier and a 1-MHz postdetection bandwidth.

Suppose a radar with 100-kW peak transmit power at 3 GHz and 30-dB antenna gain is of interest to the crystal video receiver. The one-way range equation shows that a −32-dBm signal from the radar main beam will be available at the output of an omni-directional ELINT antenna at a distance 100 km away from the radar. Thus, a simple crystal video receiver is adequate to detect such a radar and could perform in the role of radar warning receiver.

As RF gain (G) is added ahead of the detector, the noise-limited condition is reached and the term containing A can be neglected in (3.15). The tangential sensitivity then becomes

$$\text{TSS}_{\text{NL}} \cong kTF\left(6.31B_V + 2.5\sqrt{2B_{\text{RF}}B_V - B_V^2}\right) \text{ (noise-limited)} \quad (3.19)$$

If $B_{\text{RF}} \gg 2B_V$ in megahertz, (3.19) becomes

$$\text{TSS}_{\text{NL}} \cong kTF\left(2.5\sqrt{2B_{\text{RF}}B_V}\right) \quad (3.20)$$

This is essentially the expression used above for effective noise bandwidth. In (3.20) the factor 2.5 (4 dB) is the predetection SNR empirically determined to produce the TSS condition. If the receiver bandwidth is taken as B_{RF}, then the receiver can be said to exhibit a *processing gain* of $[B_{\text{RF}}/2B_V]$, which reduces the signal level needed to produce the TSS condition compared to a receiver in which $2B_V = B_{\text{RF}}$. Based on the examples from [16], a receiver with $B_{\text{RF}} = 2,000$ MHz and $B_V = 1$ MHz has the TSS shown in Table 3.7 under several conditions for $A = 9.55 \times 10^{13}$.

It should be noted that the SNR at the tangential condition is not sufficient for most ELINT measurements; thus, practical considerations raise the required signal level about 10 dB above the TSS value.

It should also be noted that the sensitivity varies as $\sqrt{B_V}$ in both the gain-limited case when $\sqrt{AB_V/G^2F^2}$ is the predominant term in (3.15), and in the noise-limited case with $B_{\text{RF}} \gg B_V$ when $\sqrt{B_{\text{RF}}B_V}$ is the predominant term in (3.15). It is clear that the use of pulse-to-pulse frequency hopping to force reduced sensitivity on intercept receivers has no effect on a gain-limited crystal video receiver with an RF bandwidth larger than the likely agility band of the radar (only the postdetection or video bandwidth affects its sensitivity).

3.8.1   Crystal Video Applications

The simple crystal video receiver has very poor frequency selectivity—it responds to the total power entering the RF filter. Therefore, applications center on those

Table 3.7   Effect of Front-End Gain on Tangential Sensitivity of a Crystal Video Receiver

	Gain-Limited	Intermediate				Noise-Limited
Gain ahead of detector (dB)	−2.1	10.0	20.0	30.0	40.0	>50.0
Noise figure ahead of detector (dB)	2.1	8.6	8.6	8.6	8.6	8.6
TSS (dB)	−38.0	−50.2	−60.2	70.2	−79.2	−83.4

in which the envelope of the RF energy provides sufficient information. For pulsed signals, if the number of pulses present is not too large, the envelope can be quite informative since it allows examination of the radar PRI. If the sensitivity is low, the number of received pulses is reduced. Also, directional antennas can be used to restrict the angular field of view and hence reduce the number of pulse trains present in any one detection channel. Since a study of the pulse time and amplitude history gives the PRI and pulse duration as well as the scanning characteristics of the radar, it is possible to identify radar types and functions with surprising effectiveness, provided that the interleaving of the pulse trains is not so great that a large fraction of the pulses from any radar are overlapped by other pulses. Excessive overlap can occur either by having a large number (more than 5 to 10) of similar pulse trains present at once, or by having a very short PRI (high PRF) signal present along with a very long PRI (low RPF) signal.

While most ELINT applications would not be well served by a crystal video receiver alone, such receivers may be used in conjunction with other receivers to provide general monitoring or "tip-off" capabilities to bring other receivers into action.

Clearly, there are limits to the number of signals that can be successfully distinguished using PRI alone. Once these limits are reached, additional processing, no matter how sophisticated, can provide only a minor improvement in capability.

For radar warning receiver applications, the crystal video receiver's problem of too many signals is often approached through the addition of multiple channels (e.g., four to eight) to provide angle of arrival information. In other cases, the system design may make use of either an IFM unit (which is basically a discriminator covering the full band) or a narrowband tunable receiver whose selectivity in frequency can resolve complex situations.

3.8.2 Postdetection Signal Recording and Sorting

The crystal video ELINT receiver output can be preserved by digitizing the video signal. In this case video bandwidths on the order of 5 to 10 MHz are required. Should there be more than one signal present (which is quite likely in view of the wide band covered by most crystal video receivers), the amplitude and time of arrival information allows separating the signals (if there are not too many) using the simple techniques of time-gating based on predetermined PRI values coupled with amplitude tests designed to prevent neighboring pulses with widely differing amplitude from being assigned to the same pulse train. The PRI-gating requires wider and wider gates into which the pulse must fall if the signal is intentionally PRI-modulated in some way (e.g., random jitter, periodic stagger, dwell and switch, pulsed Doppler types). This is discussed in Chapters 12 and 13. Even if large PRI variations are present, the amplitude variations due to the radar's antenna scanning may allow the skillful signal analyst to visually separate the interleaved signals.

3.8.3 CV System Design Considerations

The system designer's choice of RF amplifier obviously determines the sensitivity of the crystal video receiver, and the sensitivity required is a function of the system application. The low end of the dynamic range is determined by the sensitivity. The high end of the dynamic range depends on the level at which the RF amplifier, the detector, and the video amplifier saturate. For a single detector, the total receiver dynamic range can be no greater than the dynamic range of the detector. This is obvious in a gain-limited (no RF amplifier) system. If noise-limited operation is chosen, then the noise at the detector output is set by the RF amplifier. The dynamic range will be somewhat reduced from that of the detector alone, since the maximum detector output remains the same, but the noise level with no signal present has increased. Increased total dynamic range can be obtained by using several channels with differing amounts of RF gain and summing the outputs (with some loss of sensitivity).

The video amplifier is often ac-coupled to reduce the effects of low frequency noise (which increases as the reciprocal of the frequency). However, this means that CW signals will not be detectable and may desensitize the receiver (e.g., by saturating the receiver, a single strong CW signal would effectively mask all weaker pulsed signals across the entire band).

Crystal video ELINT receivers often include an RF switch [Figure 3.13(a)] to chop the input as selected by the operator. This allows any CW signals to be detected at the receiver output. A sweeping bandpass filter can achieve the same result (that is, chop a CW signal) with the added advantage of allowing the frequency

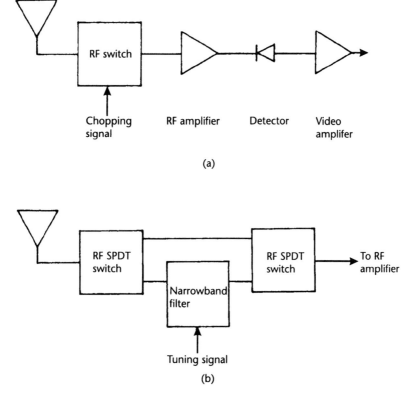

Figure 3.13   CW detection schemes: (a) RF switch to chop the input; and (b) swept bandpass filter as a chopper.

of the CW signal to be determined. A notch filter adds the capability of reducing the CW signal at the detector to possibly allow pulsed signals to pass to the output. The addition of a tunable narrowband filter, as in Figure 3.13(a), creates a receiver of the tuned radio frequency (TRF) type which was extensively used in the early days of radio to provide selectivity. Clearly, a pulsed signal's RF frequency also can be measured with the aid of the filter.

Another system problem associated with crystal video receivers is that of simultaneous signals. Since only the total RF energy present is measured, difficult problems arise if two (or more) pulses are present at the same time. For example, in pulse-gating signal sorting, pulses may be dropped from one or more pulse trains. Likewise, improper pulse amplitudes may be measured. One way to detect simultaneous signals is to make use of the difference frequency generated by the detector. Consider the system shown in Figure 3.14. The RF band covered is from RF_{LOW} to RF_{HIGH}. Any two signals simultaneously present will produce signals at the detector output whose frequency is the difference between the two input frequencies and which must be between zero and ($RF_{HIGH} - RF_{LOW}$). The normal video amplifier looks for the signal envelope and passes frequencies from almost zero to B_v. The upper amplifier covers the frequency band from somewhat above B_V to ($RF_{HIGH} - RF_{LOW}$). The detector then measures the envelope of the energy at the difference frequency at its output and can thereby warn of the presence of multiple signals.

The largest difference frequency considered must be lower than the lower band edge, RF_{LOW}, otherwise signals near RF_{LOW} will pass directly through the multiple-signal channel. In addition, one of the multiple signals must be strong enough to produce a difference frequency component above the noise. Thus, the multiple-signal channel will generally be unable to indicate the presence of two weak signals, if both are near the threshold.

Crystal video systems sometimes use amplitude comparison of the signals from multiple antennas for angle of arrival measurement. In these systems, the antenna patterns must be carefully matched over the band covered. Since there is no fre-

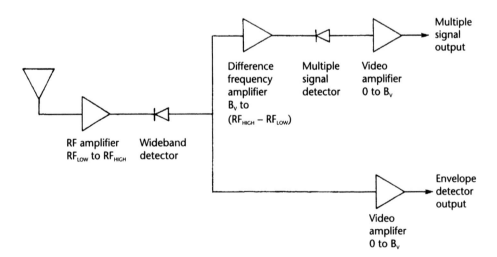

Figure 3.14 Multiple signal detector.

quency information, it is not possible to correct for antenna-to-antenna gain variations as a function of RF. Even if the antennas match perfectly, it is clear that the receiver channels must also match well to obtain good angle-of-arrival performance. This means matching the gain versus frequency characteristics of the receiver channels. One way around errors of this type is to rapidly switch a single receiver channel from one antenna to another. This reduces receiver sensitivity since the switching must be done within the time of the shortest pulse of interest. This increases the video bandwidth (B_v) required by an order of magnitude. Another approach is to switch antennas between pulse arrivals. However, this means that the angle-of-arrival determination cannot be done on a single pulse basis.

In summary, crystal video receivers are useful in low duty cycle pulsed-signal environments. Intercept receiving systems often use these receivers in conjunction with other receivers (e.g., IFMs or superheterodynes). The advantages, disadvantages, and typical characteristics are summarized in Table 3.8.

3.9 Superheterodyne Receivers

The superheterodyne receiver (referred to hereafter as a *superhet*) is the most widely used receiver design for nearly all uses; ELINT applications included. A basic block diagram is shown in Figure 3.15. The idea is to use a local oscillator to convert the incoming signal to a fixed intermediate frequency by the mixing process (heterodyning). Then the IF amplifier need operate at only one frequency and its characteristics can be precisely controlled (gain, center frequency, bandwidth, band edge roll-off, group delay). The bandwidths for typical narrowband ELINT receivers are $B_{RF} \sim B_{IF} \sim 20$ MHz and $B_V \sim 10$ MHz. This allows for good pulse fidelity for normal radar pulses (pulse width ~1 μs). Such receivers were the mainstays of ELINT operations during World War II. More recent designs have additional capabilities. Rapidly sweeping the local oscillator in frequency produces an amplitude versus frequency display, which is created by displaying the amplitude detector output versus the tuning voltage, thus creating an RF spectrum analyzer. Very rapid sweeping creates a rapid sweep superhet, described in Chapter 9, which is useful for detecting high duty cycle, low power signals in the presence of low duty cycle interfering signals. Still more rapid sweeping is the basis of the microscan and metascan compressive receivers described later. Using very wide receiver bandwidths (e.g., 500 MHz to 2 GHz), and using a microwave IF (e.g., 2 GHz) creates

Table 3.8 Crystal Video Advantages and Disadvantages

Advantages	Disadvantages
Low cost	No frequency information
Mature technology broadband	Multiple signals cause errors
Simple	

Typical Characteristics
Sensitivity: −70 dBm (noise-limited) to −30 dBm (gain-limited)
Dynamic range: 50 dB single channel; >60 dB multiple channel
RF bandwidth: Multiple octaves
Video bandwidth: Dependent on application, 10 MHz typical

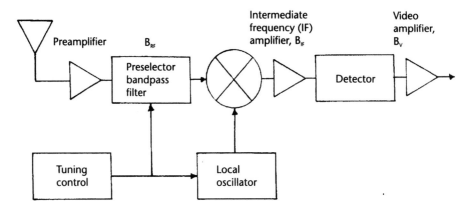

Note: 1. Detector may be an envelope detector, limiter/discriminator, or phase detector.
       2. B_V is typically $B_{IF/2}$ for narrowband receivers. B_{IF} is typically <3 MHz.
       3. Intermediate frequency is typically 30 to 160 MHz.
       4. B_{RF} is typically >50 MHz.
       5. For wideband receivers, B_{RF} may be 500 MHz and B_{IF} may also be 200 to 500 MHz
          with an IF center frequency >1 GHz.

Figure 3.15 Superhet block diagram.

a frequency conversion system which can serve as a front-end to an IFM receiver (IFMs are discussed later). Combining microsecond tuning capabilities with computer control creates a receiver that can tune to a given frequency just in time to collect a given pulse with a known PRI. This kind of "time-shared" operation allows a single receiver to intercept many pulses from a number of different emitters in spite of its single narrowband channel design.

The wide variety of superhet designs shows that there are still many ways to make use of this old but powerful concept. As shown in Appendix C, a slow sweep, narrowband superhet even has some capability against a random frequency hopped radar signal if it is properly used.

Historical Note.   The superheterodyne was invented by Edwin H. Armstrong. RCA owned the patent and produced superhet radios for sale to the public in 1924 [17]. These were very popular because they allowed tuning via a single knob. To get similar performance, other manufacturers had multiple-tuned stages that required separate adjustment (this was the tuned radio frequency receiver, or TRF). The superhet required only one RF filter ahead of the mixer and a tuning control for the local oscillator. The remainder of the stages could be fixed-tuned at the IF to provide very good selectivity and sensitivity compared to the competing TRF designs, which had up to five tuned stages at RF. The TRF manufacturers tried a variety of mechanical linkages using complex cams to attempt to tune all of the stages with a single knob. The advantages of the superhet design nearly forced RCA's competitors out of the receiver field until, threatened with antitrust action, RCA granted licenses to the others in 1930. The ELINT receivers of the 1940s were superhets with gang-tuning of the RF preselector and the local oscillator. Complex mechanical adjustments were necessary to maintain tracking between the two circuits, much like the earlier generation of broadcast receivers.

3.9.1 Superhet Performance

The sensitivity of a narrowband superhet is the power required at the input to produce a given SNR at the IF amplifier output. It is given by

$$RSL = (SNR)(NF)(kTB_{IF}) \qquad (3.21)$$

where:

RSL = required signal level

SNR = signal-to-noise power ratio required at IF to perform specified functions

NF = noise figure

k = Boltzmann's constant

T = equivalent noise temperature at the input (K)

B_{IF} = IF bandwidth (Hz)

It is common to express this in decibels referred to a milliwatt and to use a temperature of 290K, in which case (3.21) becomes

RSL (dBm) = SNR (dB) + NF (dB) − 114 dBm/MHz + 10 log B_{IF} (MHz)

A wideband superhet may exhibit processing gain. It would be rare to make the video bandwidth much wider than 10 to 20 MHz, whereas the RF and IF bandwidths may be several hundred megahertz. In this case the SNR to be used in (3.21) is the *output* SNR, which is the IF SNR increased by the processing gain of

$$PG \cong \sqrt{B_{IF}/2B_V}$$

The required signal level at the input to the wideband receiver, therefore, is increased in proportion to the square root of the factor by which the IF bandwidth is increased beyond $2B_V$. For a 14-dB IF SNR, a receiver having B_{IF} = 20 MHz and B_V = 10 MHz has a required signal level of

$$RSL \ (dBm) = -114 \ dBm/MHz + 10 \log 20 \ MHz + 14 \ dB$$
$$= -87 \ dBm$$

Increasing the RF bandwidth to 200 MHz increases the required signal level to about −82 dBm since

$$\sqrt{200/20} = \sqrt{10}, \text{ or } 5 \ dB$$

The dynamic range of the superhet is set by the saturation of the RF preamplifier at the high power end and by its output noise level at the low end. For maximum dynamic range and nearly the best sensitivity, the noise level at the detector output must increase slightly when the RF amplifier is placed in the circuit.

The probability of intercept depends on the tuning speed and the RF and IF bandwidth. Generally, for CW or high duty cycle signals, the probability of intercept is quite high. For pulsed signals, the dwell time at a given frequency should be longer than several PRIs in the conventional design approach using slow-speed tuning or sweeping. Wide bandwidth superhets with programmed scan strategies can have fairly high probability of intercept even for low duty cycle signals. Frequency hopped signals are a special case. Generally low probability of intercept is expected; however, proper use of the superhet still provides useful data in some situations, as shown in Appendix C. For low duty cycle scanning emitters, the high sensitivity of a narrowband superhet often allows detection of the radar's sidelobes, thus enhancing the probability of intercept. The superhet has good performance in many applications, and it provides this performance with low cost and modest size.

The ability to measure frequency depends on the IF filter bandwidth and whether a discriminator or IF spectrum analyzer is available to aid in tuning. The resolution of signals in frequency depends on the local oscillator tuning steps (if any) and on the IF filter skirts.

For applications in which the coherence or RF stability of the incoming signal is to be measured, the stability of all of the local oscillators in the receiver must be better than the stability of the signal being received.

The conversion of the incoming signal to an intermediate frequency can be accomplished by heterodyning (mixing) the signal with a local oscillator whose frequency is either above or below that of the signal; that is,

$$IF = \text{Signal Frequency} - \text{Local Oscillator Frequency}$$

or

$$IF = \text{Local Oscillator Frequency} - \text{Signal Frequency}$$

If the local oscillator frequency is below that of the signal, the signal spectrum is shifted directly to be centered at the IF. If the local oscillator frequency is above that of the signal, the signal spectrum is inverted and shifted to be centered at the IF. Spectrum inversion may be undesired. If a second frequency conversion takes place, a second spectrum inversion may be used to recreate the original signal spectrum.

Dual (or triple, or multiple) conversion superhets are useful to achieve translation to a low IF, particularly if a narrow IF bandwidth is necessary. The ratio of the 3-dB IF bandwidth to the IF center frequency is proportional to the "Q" of the IF filter, hence very narrow band filters are more easily constructed at a low IF. On the other hand, tuning a local oscillator at a microwave frequency with enough precision to center the incoming signal in a narrow IF may be quite difficult. For this reason, a dual conversion receiver with the first IF relatively high (e.g., 160 MHz) and with a wide bandwidth (e.g., 20 MHz) may be used to initially locate a signal. Conversion to a second IF (e.g., 21.4 MHz) allows for narrowband filtering and demodulation for those signals for which it is appropriate. For step-tuned local oscillators, this arrangement allows coarse frequency steps by the first LO, and fine frequency steps by the second LO.

In some ELINT applications, the signal is digitized at IF before detection (predetection digitizing). In this case, it may be necessary to translate a relatively wide spectrum down to a low center frequency. In translating to a low IF, it is important to make sure that filtering at each step prevents a part of the spectrum from being translated to negative frequencies. This portion folds back into the positive frequency region and may cause distortion, as shown in Figure 3.16. The ideal filter indicated in Figure 3.16(b) is impossible to realize. The actual filter skirts must be considered in determining how to place the IF to maintain any folded components at sufficiently low amplitude to be less than the quantizing noise of the A/D converter.

3.9.2 Sweeping Superhet Receivers

The concept of a receiver which sequentially tunes through the band to search for signals has been important for ELINT applications since the beginning. The early receivers used motorized tuning in which electric motors were used to repeatedly

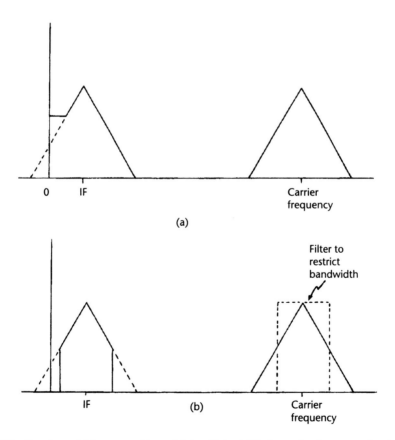

Figure 3.16 Spectrum fold-over problem: (a) insufficient filtering prior to conversion to low IF; and (b) filtering prior to final conversion to eliminate fold-over.

tune through the band when searching. Later, electronically tuned filters and synthesized local oscillators allowed for much more rapid tuning. The natural question to ask is how fast should the tuning occur for various applications and what limits are imposed by the receiver bandwidth.

The analysis of the behavior of a sweeping superhet as the sweep speed increases is accomplished most easily for the assumption of a linear frequency sweep and an IF filter having a Gaussian amplitude and linear phase response [18, 19]. For a CW signal, it is easy to see that for a slow sweep, the envelope of the IF signal will trace out the IF bandpass forming a pulse-like signal whose duration is determined by the filter bandwidth divided by the sweep speed of the local oscillator; the sweep speed being the extent of the band swept divided by the time required to sweep through it. As the sweep speed increases, the duration of the IF pulse decreases until at some point the amplitude of the pulse is reduced and its duration remains fixed. At very fast sweep speeds, the envelope of the IF pulse becomes essentially the impulse response of the filter and it is basically unchanged by the sweep speed. For the Gaussian IF bandpass, the amplitude of the envelope and the frequency resolution as a function of the sweep speed is shown in Figure 3.17. The amplitude response is relatively unaffected by the sweeping as long as

$$\frac{\Delta F}{T} < B^2 \qquad (3.22)$$

where:

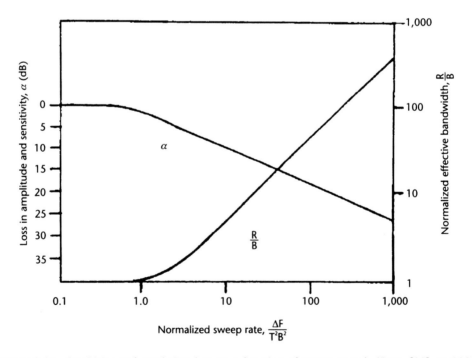

Figure 3.17 Sensitivity and resolution loss as a function of sweep speed. (*From:* [19]. © 1981 Horizon House, Inc. Reprinted with permission.)

ΔF = the extent of the band being swept (Hz)

T = the time to sweep the band (sec)

B = the IF 3-dB bandwidth (Hz)

For a step-tuned system, the dwell time at a frequency should be longer than the reciprocal of the IF bandwidth. If steep filter skirts are needed and a Gaussian IF filter is not suitable, then reducing the sweep speed to a fraction of B^2 may be necessary.

For an ELINT application in which a 1-GHz band is to be searched using a 10-MHz IF filter bandwidth, the time to sweep the band should be longer than

$$\frac{10^9}{(10^7)^2} = 10 \ \mu s$$

If step tuning were used, the dwell time at any frequency should be longer than $1/10^7 = 0.1 \ \mu s$. Thus, a step-tuned LO which increased its frequency in 10-MHz steps could also tune across a 1-GHz band in 10 μs.

ELINT receivers generally take much longer times to sweep across the band since it is important to dwell at least several radar pulse repetition intervals at each frequency. This improves the probability of intercept and also allows PRI measurement to occur. (The PRI is an important parameter used to help decide whether the signal is of interest.) At 10 ms per frequency, it would require about 1 second to search a 1-GHz band in 100 steps of 10 MHz each. As a result, the "speed limit" imposed by the IF bandwidth has generally not been an important factor in the design of ELINT receivers. Of course, the receiver described in Chapter 9 for detection of high duty cycle signals while ignoring pulsed signals is an exception to this rule. For that receiver, the sweeping should take place at close to the speed limit. [In order to sweep even faster, it is necessary to use the microscan (compressive) approach described later.]

It must be noted that as the sweep speed approaches the limit of (3.22), the IF filter characteristics become an important part of the receiver design. Clearly, if the impulse response of the IF filter does not die away in a time comparable to the reciprocal of its bandwidth, a strong signal can obscure a weak signal nearby. While traditional slow-scan receiver designs can choose IF bandpass characteristics with steep skirts for good selectivity, as the sweep speed increases, it is necessary to reduce the slope of the filter skirts to approximate a Gaussian shape. The Gaussian IF bandpass that provides the best resolution is [18]

$$B_o = 0.66\sqrt{\Delta F/T} \tag{3.23}$$

For this bandwidth, the resolution is $0.94 \sqrt{\Delta F/T}$ and the amplitude of the envelope at IF is reduced by a factor of 0.84 from the nonswept receiver. This amounts to a sensitivity loss due to sweeping of about 1.5 dB. Returning to the previous example, to sweep a 1-GHz band in 10 μs, the optimum IF bandwidth is 6.6 MHz. This can resolve signals of equal power separated by about 9.33 MHz.

For the optimum resolution case, if a wider IF bandwidth is used, the frequency resolution is degraded by virtue of the larger bandwidth. If a narrower IF bandwidth is used, the frequency is also degraded because at this sweep rate the impulse response of the filter is important. Making the bandwidth narrower increases the duration of the ringing time of the filter. As it rapidly sweeps through a CW signal, the output, in the time domain, approaches the filter impulse response as the bandwidth is reduced. The ringing reduces the ability to resolve signals in frequency. The optimum IF bandwidth balances these two opposing forces in the way that gives the best frequency resolution.

3.9.3 Tuning Considerations

Successful use of a narrowband, slowly tuned superhet as an intercept receiver requires attention to several important details. One is the tuning process itself. As the operator tunes across the band and encounters a signal, proper tuning to center the signal spectrum within the receiver's passband is essential for proper signal analysis and recording.

Tuning aids in the form of a discriminator and spectrum analyzer (pan display) at IF are very helpful in tuning the signal to the center of the IF filter. Of course, the task is much easier if the signal spectrum has a narrower bandwidth than the receiver's IF bandpass. In a well-designed receiver, there may be a choice of pre- and postdetection filters to help tailor the receiver to the signal characteristics. In most ELINT applications, the IF bandwidth should be a multiple on the order of 1 to 10 times the reciprocal of the pulse duration. If there is intrapulse modulation, then the IF (predetection) bandwidth should be somewhat larger than the peak-to-peak frequency deviation on pulse or the reciprocal of the phase modulation "chip" duration.

It is also important to consider the problems of image responses in tuning the receiver. A very strong signal may enter the receiver's image response. The apparent RF will then be in error by twice the IF frequency, even when carefully tuned to the center frequency. This condition can be detected using either the discriminator or pan display, since tuning the receiver LO causes the frequency within the IF passband to change in the opposite direction from that expected. (Image rejection is provided by a tuned preselector, which "tracks" the frequency of the LO or by an image rejection mixer.) In some systems, there is no image rejection scheme. Both responses are processed as valid signals and the proper RF is determined by slightly changing, (dithering) the LO, and noting the resulting phase relationship of the frequency changes on the IF signal.

Inaccurate tuning of a conventional constant RF pulsed signal can severely distort the pulse envelope as observed at the AM detector output. To understand why, recall that differentiation in the time domain corresponds to multiplication of the frequency spectrum by $2\pi f$; that is,

$$\frac{d^n[s(t)]}{dt^n} = (2\pi if)^n S(f) \qquad (3.24)$$

Therefore, if the signal is tuned so that the center of its spectrum is located on the skirts of the IF filter, differentiation occurs and the output of the envelope detector will develop "rabbit ears" as sketched in Figure 3.18.

Such rabbit ears may also occur in some systems if the received signal is so strong that severe saturation occurs. For example, receivers with low noise traveling wave tubes as preamplifiers have the characteristic that the output of the tube actually decreases when it is strongly driven. This causes the rise and fall of the pulse to appear at the output at higher amplitudes than the center part of the pulse.

Such effects are more serious in automated receiving systems since their threshold circuits may erroneously report a group of two short duration pulses in place of one conventional pulse.

Accurate tuning is especially difficult when fine resolution spectrum analysis of the portion of the spectrum near the centerline is necessary, as in some coherence measurements and measurement of very small RF modulations. (The later can be found in some pulsed Doppler radar systems as part of a range ambiguity resolution scheme. RF variations on the order of 0.01% may be used in some cases.) Discerning such frequency variations may require conversion to a low frequency (with its attendant aliasing or spectrum fold over problems) followed by high-resolution spectral analysis.

For a scanning signal, one may be limited to using the center portion of the main beam, since the sidelobes may have an insufficient SNR for such fine-grain analysis (recall that the error due to noise is proportional to the square root of the reciprocal of the SNR). As a result, only a few pulses may be available for measurement of the carrier frequency. The error in the frequency measurement is also proportional to the reciprocal of the square root of the number of pulses available. In such situations, it is best to test the receiver and analysis system using a signal with known properties. In this way, the absolute limits to the ability to measure

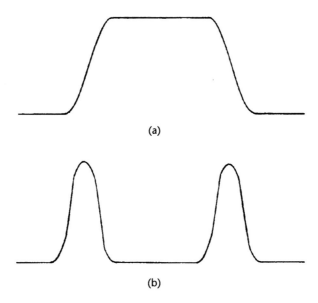

(a)

(b)

Figure 3.18 Effect of inaccurate tuning on pulse envelope: (a) center-tuned; and (b) inaccurately tuned.

the RF characteristics of short bursts of pulses can be clearly established before it is necessary to deal with the fleeting signals of unknown character so often encountered in the world of ELINT.

Extremely accurate tuning of an oscillator to the carrier frequency of a pulsed signal may arise if the frequency stability (coherence) or precise average value of the RF is to be determined. The basic dependence of frequency measurement accuracy was given above as

$$\sigma_f = \frac{\sqrt{3}}{\pi\tau\sqrt{2E/N_0}} = \frac{\sqrt{3}}{\pi\tau\sqrt{2\,\mathrm{SNR}}\,\sqrt{B\tau}} \tag{3.25}$$

where:

σ_f = rms frequency error due to noise

τ = duration of signal or measurement time

E = signal energy = $S\tau$

N_0 = noise spectral density = kT

B = receiver noise bandwidth

SNR = signal-to-noise ratio

Note that in many situations, $B\tau$ may be approximately 1. For a single pulse measurement, τ is the pulse duration, and for a 20-dB SNR, and a 1-μs pulse duration and a 1-MHz bandwidth, this relationship gives an rms frequency error due to noise of 100 kHz. If the carrier frequency error was required to be 100 Hz, simple postmeasurement averaging of the single pulse values would require 10^6 pulses! In most ELINT applications, the signal can be stored for later analysis. Then the frequency estimation can take place off-line, using multiple pulses and a variety of spectrum analysis techniques. If coherent processing can be performed on multiple pulses, the rms frequency error can be reduced in proportion to the number of pulses instead of the square root of the number of pulses. If coherent processing can be used, only 1,000 pulses would need to be processed in the previous example—but note that the signal must be coherent from pulse to pulse for 1,000 received pulses.

3.9.4 Other Heterodyne Receivers

A zero-frequency IF is sometimes used. In this type of receiver, the local oscillator is tuned exactly (or nearly so) to the center frequency of the signal being received. Unlike the envelope detector, the phase information is retained in the baseband signal when the zero-IF technique is used.

Of course, precisely *zero-beating* a pulsed signal is quite difficult. Manually tuned zero-IF receivers often produce a nonzero IF of hundreds of kilohertz. One receiver configuration is shown in Figure 3.19. The in-phase and quadrature outputs

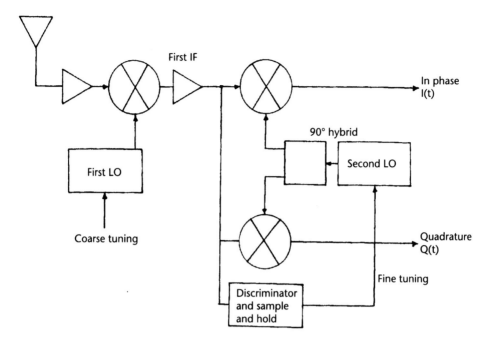

Figure 3.19 Zero IF receiver.

are particularly convenient for predetection digitizing. The envelope can be reconstructed from

$$A(t) = \sqrt{I^2(t) + Q^2(t)} \tag{3.26}$$

and the instantaneous phase can be reconstructed from

$$\phi(t) = \tan^{-1}\frac{Q(t)}{I(t)} = \sin^{-1}\frac{Q(t)}{A(t)} \tag{3.27}$$

A kind of "wide open" receiver that coverts any input signal to a fixed IF signal (sometimes called a homodyne) is shown in Figure 3.20 [20]. Any frequency modulation on the input signal will be removed in the mixing processes. The IF signal, therefore, contains the frequency instabilities of the local oscillator. Note, however, that any differential delay through the two channels means that the input signal at one time is being mixed with the input signal (offset by the IF) from a slightly different time. This effect can be used to ascertain the carrier frequency by measuring the phase of the output signal with respect to the internal LO phase much like an IFM (see below). This receiver has few applications since the broadband mixing processes are difficult to implement and the functions of the receiver can be better performed by crystal video or IFM systems. With low noise preamplifiers, there is no sensitivity advantage to the homodyne over a crystal video receiver.

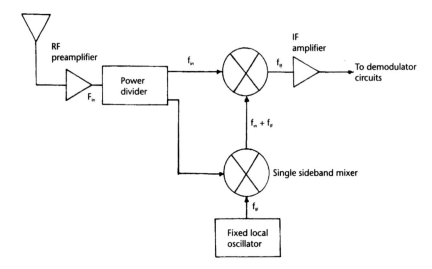

Figure 3.20   Wide-open homodyne receiver.

3.10   Instantaneous Frequency Measurement Receivers

The need to measure the frequency on each pulse motivated the development of the IFM receiver. The IFM is basically a crystal video receiver with the addition of a frequency sensing method. The usual frequency sensing method is to divide the signal into two paths with a short delay inserted in one path. The phase of the signals is compared and the phase shift is proportional to the carrier frequency

$$\theta = 2\pi f \tau \tag{3.28}$$

where:

θ = phase shift (radians)

f = carrier frequency (Hz)

τ = differential delay (sec)

The phase difference is proportional to the frequency, which means that for a given delay, the unambiguous band covered is that for which the phase change is $1/\tau$. (For an unambiguous bandwidth covered equal to 1 GHz, the required delay is 1 ns.)

There are several problems in applying the IFM. Some are the same as for CV receivers. Perhaps the most pressing problem is caused by simultaneous signals. When more than one signal is present at the same instant, the phase shift above is not a meaningful measure of the frequency of any of them.

3.10.1   Limiters Applied to IFMs

The IFM discriminator output is proportional to the signal amplitude. In view of its use to measure frequency, a natural addition (ahead of the discriminator) is a limiting amplifier. Adding a true limiter makes the discriminator output independent of the input signal level (which means that, in the absence of any signal, noise alone creates a full scale output). In such a system, noise limited sensitivity is obtained. However, a separate crystal video amplitude channel is required to operate a threshold detector, which determines the times at which there is sufficient signal level for the discriminator output to be valid. With a limiter, accurate frequency measurement over a dynamic range of 70 dB or more is practical.

A limiter impacts the operation of the IFM in ways other than increasing its dynamic range. Consider frequency accuracy. Because the microwave components in the IFM are not perfectly matched, a constant signal amplitude at the input makes it possible to calibrate the signal frequency only as a function of the angle of the output vector. The same calibration is valid over a broad range of input signal amplitudes. Without the limiter, separate frequency versus angle calibration tables would be needed at each input amplitude level.

Another effect of the limiter concerns operation of the IFM in the presence of simultaneous input signals. The familiar capture effect of the FM broadcast receiver applies in a similar way to the IFM: if more than one signal is present, the strongest one tends to suppress the others.

It should be noted that by making multiple frequency measurements on a single pulse, intrapulse frequency modulation can be detected and characterized using an IFM. Phase reversal signals may also be detected by this procedure since large positive and negative frequency excursions will appear at the IFM output. The phase detection circuitry will indicate the proper frequency during the time the phase is constant. When a 180° phase reversal occurs, the phase detector will momentarily indicate a frequency that is 180° away from the actual one. This disturbance will last as long as the differential delay within the IFM.

3.10.2 The Simultaneous Signal Problem

An IFM system can have many advantages, such as:

- 100% probability of intercept;
- Reasonably good sensitivity;
- Frequency accuracy to 0.1%, or better;
- Digital output;
- Rapid response time;
- Wide dynamic range.

In an environment of pulsed signals in which there is usually one pulse present at a time, the IFM is nearly an ideal solution to many intercept problems. As the environmental pulse density increases, and if high duty cycle or CW signals are present, the problem of how to process simultaneous signals becomes very important. Because the simultaneous signal problem is usually cited as a major reason for this lack of IFM deployment, this problem requires careful examination.

There are two aspects to the problem. Consider the static case with two signals simultaneously present, each at a constant amplitude. (In the dynamic case, one signal pulse begins or ends during the critical time when the frequency of the other is being measured.) In both cases, the error depends on the relative amplitudes of the two signals. Of course, the relative amplitudes at the discriminator input are affected by the action of the limiter (if used).

Static Case Without Limiter

In the case where there are two constant amplitude signals present during the frequency measurement and no limiter is used, the situation is easily analyzed graphically, as shown in Figure 3.21. The signals are represented as phasors. Without loss of generality, the strongest signal, S_1, is drawn along the vertical axis. The weaker signal S_2, is shown adding vectorially to S_1 to produce a resultant signal S displaced by an angle $\Delta\phi$ from the strongest signal S_1. Clearly, $\Delta\phi$ is zero if S_2 is at the same angle as S_1 and has a maximum value if the angle of S_2 is such that the resultant S is just tangent to the circle formed by rotating S_2. This worst-case angle occurs when

$$S^2 = S_1^2 - S_2^2 \tag{3.29}$$

in which case

$$\Delta\phi = \arcsin(S_2/S_1) \tag{3.30}$$

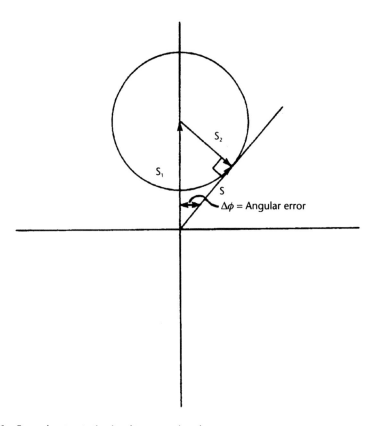

Figure 3.21 Error due to static simultaneous signals.

If S_2/S_1 is small, $\Delta\phi \sim S_2/S_1$. In general, the worst-case error as a function of S_1/S_2 is shown in Figure 3.22. To put these results in perspective, suppose that the finest resolution used in a digital IFM is 1/64 of 360°, or 5.6°. Thus, if the amplitude ratio S_1/S_2 is 10 dB, the presence of the weaker signal has little effect on the measured frequency value.

Static Case with Limiter
When a limiter is placed ahead of the discriminator, the capture effect of the limiter acts to decrease the amplitude of the smaller signal relative to the larger one. This helps to make a nearly correct measurement. The highly nonlinear limiter also generates various intermodulation products at frequencies which are, for example, the differences between twice the frequency of the stronger signal and the frequency of the weaker signal.

When such intermodulation products are within the passband of the filter following the RF limiter, they act like additional simultaneous signals. However, because the total power out of the limiter is constant, these intermodulation products only *reduce* the power of the weaker signal and hence *reduce* any frequency measurement error. Thus, a limiter always acts to *decrease* the frequency measurement error for the static simultaneous signal case. How much the error is reduced depends on the SNR of the stronger signal at the input to the limiter. According to [21], the output signal ratio $(S_2/S_1)_{out}$ relative to the input signal ratio $(S_2/S_1)_{in}$ for a hard limiter and a high SNR (30 dB) varies, as shown in Figure 3.23. This

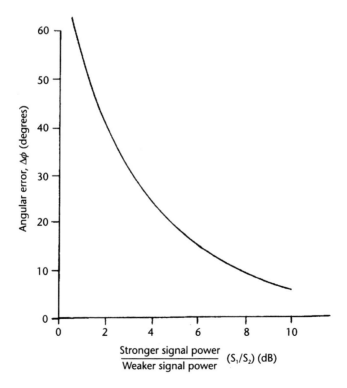

Figure 3.22   Maximum angular error due to a simultaneous signal (static case, no limiting).

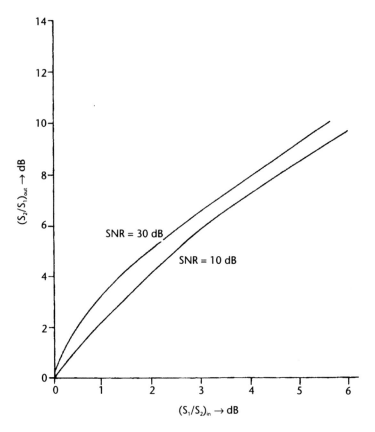

Figure 3.23   Limiter capture effect.

means that the worst case angular error can be obtained from Figure 3.22 using the value of $(S_1/S_2)_{out}$ obtained from Figure 3.23.

Dynamic Case
The problem caused by exactly coincident pulses is not too serious, as such a case is rare in a low density environment. The real problem is the case when a second signal pulse arrives shortly after the leading edge of the first, as shown in Figure 3.24. This can cause large frequency errors due to transients in the IFM. Digital IFMs consist of several discriminators with differing delay line lengths. To see why there may be a large error, notice that at any instant the delayed arm of the discriminator may have only the first signal, while the undelayed arm has the combined signal at its output. With different delay lines in use, one module may have a different condition than another. In tests of an IFM with no circuitry to protect against this condition, significant probability of error (e.g., 20%) occurred even if S_1/S_2 was in excess of 20 dB for a 60-ns delay between the two pulses. These errors can be greatly reduced by circuitry that detects the presence of such a transient and inhibits the frequency measurement during those times.

Several methods for detecting simultaneous signals are available. One of the earliest was to detect the difference frequency between the two signals produced in the detection process. This operates by filtering the detector output from above the video band of interest (e.g., 20 MHz) to some maximum difference frequency

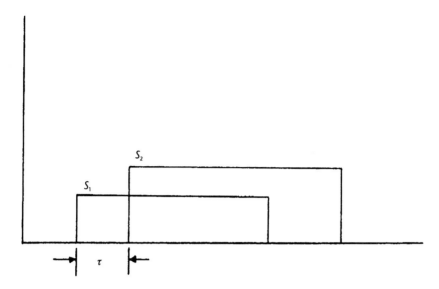

Figure 3.24 Overlapping pulsed signals.

of interest (e.g., 500 MHz), then detecting and low pass filtering the output of this bandpass filter, and comparing the result to a threshold. If two or more strong signals are present, their simultaneous presence is detected. (These circuits were incorporated in the IFMs of the early 1960s.) Unfortunately, such circuits are not useful for simultaneous signal detection when one signal is weak relative to the other, which can still result in transient frequency errors.

A more effective method is to require the video voltages out of the detectors to remain constant during the measurement process. If it does not remain constant, then a "no-read" signal can be generated to scrap that particular measurement. This method can reduce the probability of error to less than a few percent if $|S_1/S_2|$ exceeds 2 dB and can maintain probability of error of less than 20%, even for $S_2 = S_1$ for $\tau = 30$ ns. There are three cases of interest:

1. For $S_1 > S_2$ there are no transient problems and the frequency of S_1 will be correctly measured if S_1/S_2 exceeds a few decibels.
2. If $S_2 > S_1$, the frequency of S_2 will be correctly determined if the delay is less than that required by the IFM video bandwidth to give a stable video voltage (e.g., <100 ns typical). The circuitry to process the video can detect the transients and simply delays the reading until the video voltage stabilizes [provided $(S_2/S_1) >$ a few decibels].
3. If the delay between the pulses exceeds that for producing a stabilized video voltage after the second pulse arrives, the frequency of S_1 is correctly determined, and the frequency of S_2 is correctly determined if S_2/S_1, exceeds a few decibels.

In many other cases, the frequency reading is inhibited. Most of the effort to surmount the simultaneous signal problem has been devoted to analyzing the detected video signals at the IFM output ports, after the limiter. If the signal envelope ahead of the limiter were used, it seems possible to do additional processing (e.g., it would be possible to actually measure the ratio of S_1/S_2 in cases where the delay allows such a measurement). This information would contribute to correctly interpreting the digitized RF values. Of course, adding wide dynamic range AM detection, as well as the necessary high speed amplitude measuring and processing circuitry, would increase the cost, size, weight, and power consumption of the IFM.

The probability of occurrence of overlapped pulses is a key question. The methods of Chapter 4 can be used to compute the probability that pulses from any pair of radar are coincident. Also, one can use a probabilistic model to estimate that probability. Given that a pulse has begun from some radar, it is of interest to compute the probability that a pulse from none of the other radars begins within some time τ after the start of the first pulse. Let the PRIs be denoted T_1, T_2, T_3, ..., T_N for the N radars in the environment. The probability that a pulse from anyone of them begins within time τ is approximately τ/T_k. The desired case is for none of the other radars to produce a pulse in time τ. Assuming statistical independence and random pulsing at average PRFs of $1/T_k$, this is

$$\text{Probability of no other pulses occurring} = 1 - \frac{\prod_{i=1}^{N} (1 - \tau/T_i)}{1 - \tau/T_k} \qquad (3.31)$$

If all the pulse trains have the same PRI, this reduces to $1 - (\tau/T)^{N-1}$ and some simple examples can be explored. Figure 3.25 is a plot of $1 - (1 - \tau/T)^{N-1}$ versus the number of emitters N for τ/T equal to 10^{-4}. For instance, if $\tau = 100$ ns, which is typical of the time required to make a valid frequency measurement by a digital IFM, and if the average PRI (T) is 1 ms, $\tau/T = 10^{-4}$. Note that each emitter is assumed to produce 1,000 pps, so that the pulse density at the receiver is simply the number of emitters multiplied by 1,000 pps. When the probability reaches a level of about 0.01 to 0.1, the problem of overlapped pulses becomes significant. For this model of the environment, the overlapped pulse problem becomes significant when between 100 and 1,000 emitters are being received by the IFM (or 100,000 to 1,000,000 pps). Clearly, there is always a limit to how dense the environment can be and have the IFM remain effective. Roughly speaking, τ/T times the number of pulses per second to which the IFM responds should be less than 0.1. If the environment is too dense, then the sensitivity of the IFM can be decreased; otherwise, its bandwidth can be restricted by using bandpass filters at the input to the IFM. This probabilistic model clearly is inexact for radars with nearly constant PRIs. The synchronization problems discussed in Chapter 4 may cause additional pulse overlaps.

Even for very dense environments, IFMs should remain useful to cover RF bandwidths of 0.5 to 1 GHz. However, methods of detecting simultaneous signals will be even more important as the pulse density increases. Reducing the time (τ) required to make a frequency measurement would also be helpful. However, this can be reduced only to a certain point established by the longest delay line used in the IFM and by the video bandwidths it uses. Another limit is imposed by the

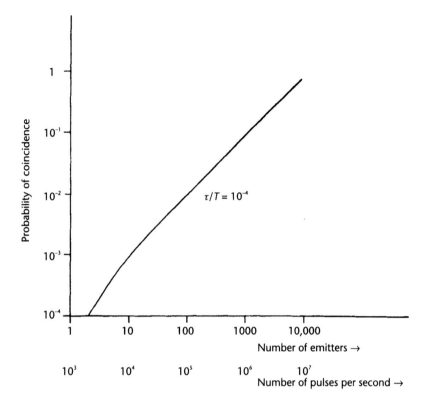

Figure 3.25   Probability of pulse coincidence versus number of emitters.

rise time of typical radar pulses. Therefore, it is unlikely that reductions in τ can be achieved while retaining good performance for all types of pulses.

3.10.3   CW Signals and IFMs

In a signal environment consisting entirely of pulsed signals, the preceding discussion of IFM performance in the case of simultaneous signals would describe the situation, and the pulse density in the band of interest would determine the expected faction of coincident pulses. However, if there is even one CW signal present, then *all* pulses are coincident with that signal. In the usual case, the CW signal is significantly weaker than the pulsed signals, and therefore, the frequency of the pulsed signals can be accurately determined. Indeed, ac-coupled video circuitry can remove any residual bias created by a weak CW signal. Of course, a strong CW signal would change the situation drastically insofar as any pulsed signals weaker than the CW signal will be masked by it.

There are several methods of coping with the CW signal problem. Knowing about the presence of the CW signal is the first step. One way is to insert an RF chopper ahead of the limiter. Then the frequency of the CW signal can be measured in the same way as the RF of pulsed signals. Once the CW frequency has been measured, a tunable notch filter can be used (ahead of the limiter) to remove it from the receiver. Of course, multiple CW signals are still a problem. Generally, the frequency of the strongest one can be measured; then, when it is removed by a notch filter, the frequency of the next strongest can be measured, and another notch filter can be used to remove it. This process can continue until either all of the CW signals are removed or all of the system's notch filters have been brought into action.

Another method of dealing with multiple CW signals is to create a narrowband channel with a swept bandpass filter. As this filter sweeps across the band, all of the CW signals can be detected and their frequencies measured. The sweeping bandpass filter output can be compared with the signals in the full bandwidth to determine when the filter output is actually a CW signal, or if it is just a pulse that happened to occur at the frequency to which the filter is tuned.

Other methods of detecting strong CW signals include measuring the average dc level from a video detector (which can be caused by either CW signals or a dense environment.). Another way is to combine the IFM with a superhet receiver, which then performs the CW detection and measurement function while also controlling a notch filter.

Clearly, an IFM cannot perform well in a system application without taking into account the CW problem. As noted, in the usual situation the power of the CW signals is weak compared to that of the peak pulse power, and therefore, the IFM continues to operate even though there are some weak CW signals present.

3.10.4 Digitizing the IFM Output

Reading the frequency using a multiple delay line IFM is similar to reading a watt-hour meter with several dials rotating at different speeds as power is consumed. The slower-rotating, most significant dial is used to resolve the ambiguity of the next most significant and so on. The IFM is different because, instead of precise gear ratios, the phase indicated by one channel varies as a function of signal amplitude and temperature as well as frequency. Furthermore, the phase versus frequency curve is not precisely linear due to variations in the response of the discriminator across the band. A typical phase (or frequency) versus phase plot for two IFM modules with a 4:1 delay line ratio is shown in Figure 3.26. Also shown are the "handoff corridors" used to unambiguously separate the four sets of curves. While the delay line ratio is 4:1, it is necessary to resolve the phase from the coarse discriminator to better than one part in four so as to avoid ambiguities near the frequencies at which the fine-discriminator phase crosses the 360°/0° boundary.

In general, the resolution required for the coarse phase measurement must be about as fine as the variations due to power and temperature changes, which in this case is about 1/16 of 360°—that is, the coarse discriminator must be digitized to 4 bits, even though, in the final frequency word, it furnishes only about 2 bits of real information. This *excess resolution* is needed to provide digitizing free of ambiguity. Consider a three-channel IFM with 4:1 delay line ratios and a 32-cell resolution on the fine resolution channel. The fine channel resolution will be approximately (1/32)(1/4)(1/4), or one part in 512 (9 bits). However, the two coarse channels each require digitizing to one part in 16, for a total of 5 + 4 + 4 = 13 bits in the raw digital word. Read-only memory (ROM) is generally used to translate from the raw digital word to a frequency value. Many of the $2^{13} - 1$ possible combinations of the raw data bits never occur.

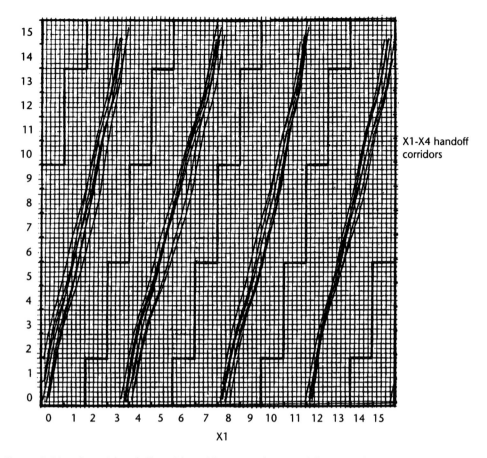

Figure 3.26 ×1 – ×4 handoff corridors. (Courtesy of Anaren Microwave.)

The net result can be thought of as *handoff corridors*. The coarse (×1) IFM along the horizontal axis is digitized to one part in 16. For ×1 segments 0, the fine (×4) discriminator must be between 0 and 9, and so on, as shown in Table 3.9.

This information is sufficient to resolve the RF band (i.e., the band over which the ×1 phase detector has a phase change of 360°) to one part in 64. There is one part in 16 resolution from ×4 digitizer and it is used four times across the band. To see this, assume that the input frequency is such that the ×4 discriminator is in frequency cell #5. According to Table 3.9, if the ×1 discriminator is in cells 0, 1, 2, or 15, the ×4 discriminator is on the first of the four 360° rotations across the band. If the ×1 discriminator output is 3, 4, 5, or 6, ×4 discriminator is on the third rotation. Finally, if ×1 is 11, 12, 13, or 14, then ×4 is on the fourth rotation. Thus, while ×1 is digitized to one part in 16, it serves only to identify unambiguously which of the four fine discriminator tracks applies, thereby providing 2 bits of information.

The linearity of the phase measuring circuitry determines the maximum delay line ratio that can be used. As the slope and the number of tracks of the fine discriminator increases, Figure 3.26 shows that it would be more and more difficult to find corridors to separate them. In the limit, curves from one track would cross those of another and no amount of increased resolution in the ×1 measurement

Table 3.9 Handoff Corridors for Coarse (×1) and Fine (×4) Discriminators

Coarse (×1)	First Rotation	Second Rotation	Third Rotation	Fourth Rotation
		Fine (×4)		
0	0–9			
1	0–13			
2	2–15	0–1		
3	6–15	0–5		
4	14–15	0–13		
5		0–15		
6		2–16	0–1	
7		6–15	0–5	
8		10–15	0–9	
9		14–15	0–13	
10			2–15	0–1
11			6–15	0–5
12			10–15	0–9
13			14–15	0–13
14	0–1			2–13
15	0–5			6–15

would resolve the ambiguity. IFM manufacturers choose different ratios, typically in the range of 2:1 to 8:1. A small ratio (2:1) requires more discriminator modules, but coarser resolution can be used in digitizing each one; whereas a large ratio reduces the number of discriminators, but each requires a higher resolution digitizer. The optimum configuration depends somewhat upon the application.

Analysis of the problem of ambiguity resolution in such a multiple vector system results in the following relationship for discriminators with a delay line ratio of $p:n$ [21]:

$$|n\Delta\theta_p + \Delta\theta_n| < \pi$$

where:

$\Delta\theta_p$ = phase error at output of discriminator with delay line p

$\Delta\theta_n$ = phase error at output of discriminator with delay line n

As an example, the formula can be applied to the hands of a clock. In this case $p = 1$, $n = 12$. If we assume no error in the minute hand ($\Delta\theta_n = 0$), then maximum hour hand error is

$$12\Delta\theta_p \, \pi, \text{ or } \Delta\theta_p < 15°$$

which amounts to half an hour, or half of one of the usual 5-minute major divisions on the minute scale.

For typical IFM delay line ratios, the maximum allowable tolerance is given in Table 3.10, assuming that each discriminator contributes an equal phase error (i.e., $\Delta\theta_p = \Delta\theta_n = \Delta\theta$).

The single pulse frequency measuring capability of the IFM must be combined with other features to make a useful ELINT receiving system. Clearly, a system

Table 3.10 The Maximum Allowable Tolerances for Delay Line Ratios

Delay line ratio	2	3	4	5	8
Phase tolerance (A8)	60°	45°	36°	30°	20°

that measures RF, time of arrival, angle of arrival, pulse duration, pulse amplitude, and that gives an indication of intrapulse FM has indeed rather completely characterized a single radar pulse. Such a system, with recording capability to preserve these single pulse values, could record (for later analysis) a complete description of the signal environment in its band. This would be especially useful in characterizing radar systems with pulse-to-pulse RF agility and varying pulse intervals. Such a system, which captures 50 to 60 bits of information per pulse, creates a difficult recording problem in an environment of perhaps 100,000 to 1,000,000 pulses per second (5 to 50 megabits per second). For this reason, it may be very important to reduce the recorded data to bursts of pulses for those radars having constant RF and pulse intervals. This reserves the recording and analysis capabilities for the unusual (agile) signals.

3.11 Other Receivers

The narrowband superhet receiver has high sensitivity and wide dynamic range. It is not troubled by simultaneous signals. However, it does not cover a broad frequency band, and thus has low probability of intercept in some situations and cannot adequately handle wideband signals (e.g., frequency hop) in others. The IFM provides 100% probability of intercept over broad bands, but it is troubled by time overlapped pulsed signals and CW signals. The receivers described here attempt to overcome these deficiencies.

3.11.1 Channelized Receivers

A straightforward approach to utilizing the properties of a narrowband superhet, while also providing wide frequency coverage, is to use a number of parallel channels, each one tuned to a slightly different RF center frequency. The frequency separation between the channels can be roughly equal to the 3-dB IF bandwidth used. Then, no matter what the carrier frequency, it will be detectable with only a small loss in sensitivity compared to a properly tuned narrowband receiver. Actually, the sensitivity of each channel must be degraded slightly by raising the detection threshold to maintain the same total false alarm rate as a single-channel receiver.

Consider an example in which there are 100 parallel receiver channels. If the tolerable false alarm probability is $P_{FA,1}$ in a single channel, the probability of false alarm probability for N independent channels is

$$P_{FA,N} = 1 - [1 - P_{FA,1}]^N \tag{3.32}$$

If $P_{FA,N}$ is small, such that $NP_{FA,N} \ll 1$, this is approximately

$$P_{FA,1} \cong N P_{FA,1}$$

For a single channel, for 90% probability of detection and 10^{-6} probability of false alarm, the required SNR is 13 dB in the IF bandwidth. If 100 channels are to provide the same overall performance, a 10^{-8} false alarm probability is required on each channel. This raises the threshold about 1.3 dB, making the SNR for 90% probability of detection about 14.3 dB at IF [22]. This is true at the center of each IF band. If the signal RF is exactly halfway between two of the channels, the sensitivity may be reduced by the *straddling loss*, typically an additional 3 or 4.3 dB total. As can be seen, in practice there is a sensitivity loss connected with channelization. The sensitivity loss is less than the penalty incurred by a wideband IFM or crystal video receiver, but there is nonetheless a significant cost and complexity penalty to be paid for channelization.

There is also a severe signal processing problem associated with channelization. If the signal bandwidth is large compared to the channel bandwidth, a number of adjacent channels will be excited simultaneously. Even if the signal bandwidth is small compared to the channel bandwidth, a strong signal will excite a number of channels whose outputs will have distorted pulse shape envelopes because of inaccurate tuning. This makes correct interpretation of the channelized receiver outputs difficult, especially in a dense environment.

These problems are addressed in the design of a channelized receiver by various thresholding schemes [23, Ch. 5] and careful filter skirt design.

A difficult problem is to detect simultaneous signals of widely differing amplitudes but with small carrier frequency separation. The detection of such signals depends, in part, on the character of the signal spectra. (Some pulses may have very sharp rise times, and thus the signal spectrum falls off as the reciprocal of the frequency separation from the carrier. If the pulse has a long rise time, the signal spectrum falls off as the reciprocal of the square of the frequency separation from the carrier.) In this discussion, rectangular pulses are assumed. If a 1-μs pulse is received, its spectrum will be down only 33.7 dB, 15.5 MHz away from the carrier. Therefore, it is not possible to detect another simultaneous signal more than about 35 dB below the first, if its carrier frequency is less than 15 MHz away.

The main lobe spectrum width of a rectangular pulse (which contains about 90% of the energy) is twice the reciprocal of the pulse duration. Thus, good pulse fidelity requires that the channel processing bandwidth be no smaller than twice the reciprocal of the minimum pulse duration. In practice, it is advisable to make the channelizer filter bandwidth about 3.2 times the reciprocal of the pulse duration [24].

A filter bank (such as used by a channelized receiver) provides, in effect, an instantaneous spectrum analysis capability in which the channel bandwidth is a measure of the frequency resolution and the impulse response of the processing filter is a measure of the time resolving power. The product of the channel bandwidth and the impulse response is no less than 2. This gives rise to a view of the channelized receiver as an attempt to measure the frequency-time-amplitude-characteristics of the signal environment, as shown in Figure 3.27. Other receivers may be characterized by their coverage of this frequency-time plane (as indicated, a narrowband receiver covers a single frequency channel for a long period of time). A swept

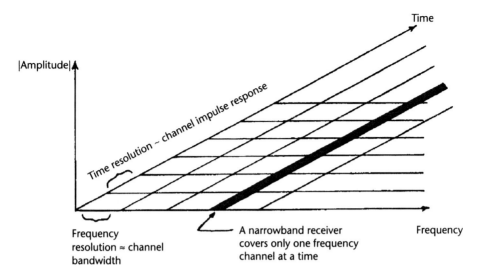

Figure 3.27   Time-frequency-amplitude signal space.

receiver covers the entire frequency band, but spends only a short time at each frequency. The crystal video covers the entire frequency band, but it provides no frequency resolution. The IFM also covers the entire frequency band, and provides good frequency resolution, but the measured value is a function of the amplitude as well as the frequency of signals which overlap along the time axis.

Channelization Schemes
The use of contiguous parallel receiver channels is clearly a very expensive approach if wideband coverage is needed using narrowband channels. In practice, the contiguous filters are constructed at an intermediate frequency, for example, 10 channels, each 20 MHz wide, centered at about 300 MHz. Then, 200-MHz bands are selected in the microwave region for mixing down to the broadband IF filter bank. If a wider instantaneous band is to be covered, several portions of the microwave spectrum can be simultaneously mixed into the *same* IF band. The true frequency can be determined by detecting the energy at RF prior to conversion, as shown in Figure 3.28. Here, only 15 filters are needed to obtain 20-MHz channel spacing across a 1-GHz band. This would take 50 filters using the brute force approach. There are two problems. First, the indication of which RF band contains the signal is accomplished via broadband detection (200-MHz bandwidth in this example), which reduces the sensitivity. Second, this approach introduces a simultaneous signal problem. If two signals occur at the same time in different RF bands, the output of the narrowband filters cannot properly indicate the frequency. Even if different filters in the contiguous filter bank are excited, there can still be confusion as to which response was the result of a particular input pulse. Also, a CW signal can render one of the narrowband channels useless, which, in effect, causes five different RF channels to be lost.

Many of the problems associated with channelizing equipment are greatly reduced if the channel bandwidth is significantly larger than the bandwidth of the signal of interest. For example, a channel bandwidth of 500 MHz greatly reduces

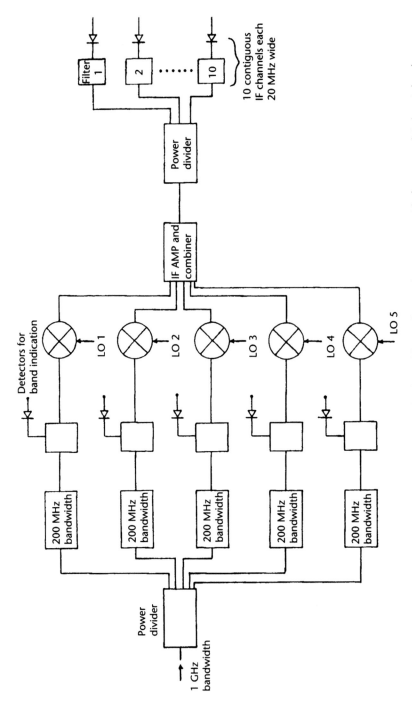

Notes: 1. Local oscillators (LO) 1–5 may be derived from a harmonic generator with the appropriate input signal.
2. 20-MHz resolution over a 1-GHz band is achieved with only 15 filters.

Figure 3.28 Channelized receiver with 1-GHz band coverage.

the pulse density compared to that of a wideband receiver covering, say, 1 to 10 GHz. The 500-MHz channel can be fed to a variety of additional receiving and processing devices such as IFMs, narrowband sweeping receivers, and even additional (narrowband) channelizing equipment. Hence, a wideband channelizer in conjunction with other types of receivers is a very useful system design approach. The use of high-speed A/D converters followed by a fast Fourier transform (FFT) or other digital filter bank is a modern approach to channelizer design.

3.11.2 Acousto-Optic (Bragg Cell) Receivers

The cost and complexity of wideband channelizers led to the development of acousto-optic receivers. These use the angle of deflection of a laser beam as it passes through a Bragg cell to provide a measure of frequency. The amplitude of the signal is determined by the light level striking an array of photocells, as shown in Figure 3.29. Because multiple signals produce multiple output deflections, conceptually, this receiver is a replacement for the channelized receiver. The frequency resolution is a function of the angular spacing of the photocell array that measures the output light intensity as a function of deflection angle. The time resolution is a function of the bandwidth of the photocell detectors. Narrowband photocells (in the range of a few kilohertz) have been used as wide-open high-frequency resolution spectrum analyzers. These have poor ability to reproduce the envelope of pulsed signals for ELINT applications. Wideband photocells (–10 MHz) with individual outputs are needed if this technique is to replace the channelized approach. In terms of the frequency-time plane of Figure 3.27, narrowband photocells provide good frequency resolution, but they sum the spectra of the events along the time axis during their millisecond impulse response time.

The principles of Bragg cell operation are well understood; however, there are many design problems associated with component shortcomings as well as the inherent problems of channelization. One promising application [23, Ch. 7] is to use a two-dimensional Bragg cell (along with a photocell planar array) to resolve both the angle of arrival and the frequency of pulse. In this approach, the Bragg cell is modulated by the signals from separate antennas, and the interferometry effects provide an (x, y) deflection of the beam, which is a function of both frequency

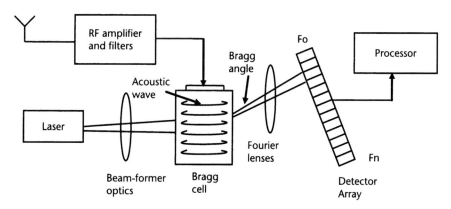

Figure 3.29 Acousto-optic receiver.

and angle of arrival. To achieve the same function with a conventional channelized receiver requires multiple phase-matched channelizer banks with phase comparison circuitry at the outputs of the channel filters.

All of the problems associated with processing the multiple outputs of channelized receivers are present in the acousto-optic implementation as well, with the added problems associated with the less mature technology of the components. The significant advantage of the single Bragg cell replacing the filter bank of the channelized receiver provided an incentive for development. This incentive decreases as high-speed A/D converters and real-time DSP developments become better and cheaper.

3.11.3 Microscan Receivers

The microscan receiver is a fast sweeping superhet that makes use of a pulse compression or *chirp* filter at the intermediate frequency. This makes it possible to sweep the band at sweep speeds greater than the square of the frequency resolving power of the receiver. As noted in Section 3.10.2, in a conventional swept superhet with IF bandwidth (and, hence, frequency resolving power) B, the sweep speed limit is repeated from (3.22)

$$\frac{\Delta F}{T} < B^2$$

where:

ΔF = band swept (Hz)

T = time to sweep the band (sec)

B = IF bandwidth (Hz)

As one sweeps the conventional receiver local oscillator at a faster and faster rate, the duration of the IF filter output due to a CW signal in the band becomes shorter and shorter until it reaches the minimum value, which is that due to the impulse response of the IF filter. To permit more rapid sweeping, it is necessary to make the impulse response shorter (i.e., to widen the IF bandwidth). Unfortunately, this degrades the frequency resolving ability of the receiver. The microscan solves this problem by using the equivalent of pulse compression in radar.

The sweeping local oscillator creates a *chirped* signal at IF as a result of mixing with an incoming signal of constant frequency, as shown in Figure 3.30. (In a conventional swept receiver, this chirping also occurs, but it is small enough to ignore.) These chirped IF signals can be fed into a dispersive delay line, which has the opposite slope as the sweep rate of the local oscillator. The total time of the delay line should be at least equal to the time the signal is present at the IF amplifier output, while the bandwidth must be equal to the selected IF bandwidth. Under these conditions, the duration of the output pulse can be compressed by an amount approximately equal to the product of the IF bandwidth and the duration of the dispersive delay.

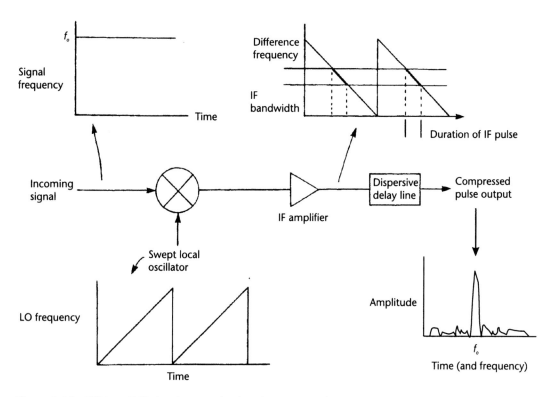

Figure 3.30   "Chirped" IF signal generation in microscan receiver.

Within certain constraints, the microscan receiver can be thought of as a swept superhet receiver that is able to sweep the band as fast as desired and yet achieve good frequency resolution. The design of such receivers is no simple matter [23, Ch. 6; and 25, pp. 161–166]. However, the main topic of interest here is the novel ways such a receiver can be used to assist in the interception of radar signals. Conceptually, its output is like that of a channelized receiver with the output of each channel being sampled sequentially at the local oscillator's sweep repetition rate. In this situation, examining the envelope of a pulsed signal can be done, but the fidelity achieved is limited by the sampling rate (i.e., a single pulse duration and PRI can be determined only to the resolution provided by the sweep rate sampling interval). A design using a 1-GHz band swept in 1 μs achieves 100% probability of intercept for all pulses 1 μs or greater in duration with PRI and PD resolution of 1 μs.

Although the microscan can process simultaneous signals (at the output of the delay line, they are separated in time), precise determination of their PRIs and PRI variations is degraded when compared to the capabilities of the superhet or IFM.

A microscan receiver can achieve frequency resolution of approximately the reciprocal of the duration of the signal stored in the delay line (i.e., a 1-μs delay line provides approximately 1 MHz of resolution). Suppose the full band is to be swept in the same time as the duration of the delay line. It is clear that the output of the delay line will not be completed until one sweep time later. Under these conditions, the microscan receiver can only examine signals 50% of the time. Coverage for 100% of the time requires two channels with interlaced scans. Such a two-channel receiver with interlaced scans can provide samples of each signal spaced by the duration of the dispersive delay line.

A major problem with microscan receivers is processing their high-speed output signals. A 1-GHz sweep using a 1-μs dispersive delay for 1-MHz resolution provides 1,000 frequency channels in 1 μs of time, or 1 ns per frequency resolution cell.

Mathematically, the principle of the microscan receiver is to compute the Fourier transform of the input signal by means of chirp (linear FM) waveforms. The usual Fourier transform notation is

$$F(\omega) = \frac{1}{2\pi} \int\limits_{\infty}^{\infty} f(t) e^{j\omega t} \, dt \tag{3.33}$$

where $F(\omega)$ is the Fourier transform of the input signal $f(t)$. In a chirped system the frequency ω is a linear function of time τ (i.e., $\omega = 2\,\mu\tau$, where μ is the slope of the chirp and μ is a constant). Thus, (3.33) can be rewritten

$$F(\tau) = \int\limits_{\infty}^{\infty} f(t) e^{-j(2\mu\tau)t} \, dt \tag{3.34}$$

The exponent can be formally written as

$$-2(\mu\tau)t = \mu(\tau - t)^2 - \mu\tau^2 - \mu t^2 \tag{3.35}$$

so that (3.34) becomes

$$F(\tau) = \left[\int\limits_{\infty}^{\infty} f(t) e^{j\mu t^2} \, e^{j\mu(t-\tau)^2} \, dt \, e^{j\mu\tau^2} \right] \tag{3.36}$$

The terms $\exp(j\mu t^2)$ and $\exp(j\mu\tau^2)$ are recognized as premultiplication and postmultiplication by a chirped local oscillator. The entire process to implement (3.36) is as follows:

- *Step 1:* Multiply (mix) the input signal $f(t)$ with a chirped LO signal $\exp(j\mu t)^2$.
- *Step 2:* Convolve the result with a second chirp $\exp(j\mu - \tau)^2$ (done in the dispersive delay line).
- *Step 3:* Multiply the result by a third chirp $\exp(j\mu\tau)^2$.

Practical implementations require finite rather than infinite limits on the process. Finally, the phase information is usually not needed. Thus the final (post) multiplication via a chirp can be left out. The output of the dispersive delay line can be processed by an envelope detector as indicated in Figure 3.30.

Multiple channel microscan receivers called metascan receivers were described by Ready [26] and some design considerations are given in [27]. The use of high temperature superconductive materials to construct wideband compressive receivers is described in [28]. These units are able to achieve bandwidths of over 1 GHz and dispersive delay times of 40 ns (time-bandwidth product in excess of 100). Projections of future developments able to cover 10 GHz of bandwidth and 3.3-MHz frequency resolution were included. Such receivers must be cooled to about 77K.

3.12 System Considerations

Various receivers have different approaches to monitoring the signal space of time, frequency, and amplitude, as summarized in Table 3.11. It is clear that preserving the amplitude at every time and frequency resolution cell results in a much higher data rate than the simpler, traditional approaches, which either monitor a single narrowband channel (tuned superhet), or which monitor a broad band but can only measure the frequency of the strongest signal present in any time cell (IFM). For this reason, channelization must be introduced in a deliberate and thoughtful way. The brute force approach of preserving the entire time-frequency-amplitude space for later analysis will result in massive recording, playback, and analysis problems but very little additional knowledge will result.

Table 3.11 Summary of Receiver Advantages and Disadvantages

Type	Advantages	Disadvantages
Crystal video	Cheap	No frequency measurement
	Small	Degraded sensitivity
	Proven	Poor in dense environment
		Saturation by one strong signal
Superhet	High sensitivity	No frequency agility
	Good ECCM	Slow search
	Proven	Poor narrow pulse response
	Useful in interferometer and	Two or more channels for DF
	precision measurements	
Channelized	High sensitivity	Limited frequency agility
	High POI	Limited resolution
	Good ECCM	Poor to moderate narrow pulse
		No DF without parallel units
		Cost, weight, size, power
IFM	High POI	Same as crystal video
	Frequency agile	No simultaneous signals
	Fair frequency accuracy	No DF
	Proven	
Bragg	High POI	Limited dynamic range
	High sensitivity	Poor pulse response
	High selectivity	No frequency agility
	Narrow pulse	Two for DF
		Unproven
Microscan/Compressive	High frequency accuracy	Poor narrow pulse
	Simultaneous signals	Sweep-scan alignment
	Handles frequency agility	Wide IF required
		Delay lines state of the art

True knowledge is obtained only by *destroying* unwanted parts of the original signals. It is important to destroy unwanted information as soon as possible in the intercept-record-analyze process to avoid overloading the entire system. In an abstract sense, a high-speed A/D converter (say, a 20-GHz sampling rate with 14-bit resolution; that is, 84-dB dynamic range) coupled to a mass memory provides an ideal ELINT receiver. It has 100% probability of intercept and preserves everything in a 10-GHz bandwidth. However, recording a simple circular scan radar for a few scans (e.g., 1 minute of data) would mean preserving and processing 1.6×10^{13} bits of data. Clearly, this is an ideal system only if a high-speed processor coupled to the A/D converter can eliminate most of those bits and provide more concise outputs that are of real value, that is:

- Direction of arrival;
- Peak amplitude (for estimating ERP);
- Antenna patterns (beam width, sidelobe levels);
- Pulse shape (duration, rise and fall time);
- Pulse intervals (interval variations, average PRI);
- Radio frequency (center frequency, pulse-to-pulse and on-pulse RF variations);
- Associated signals.

All of the information is present in the 1.6×10^{13} bps data stream from the A/D converter. However, the bit stream has little value unless it can economically produce the data items above. In designing intercept receivers of the future, those that are most successful will focus on providing *only* that information that is really needed at the output, given the signal environment in which the receiver must operate.

References

[1] Frank, R. L., "Phase Shift Codes with Good Periodic Correlation Properties," *IEEE Trans. on Information Theory*, Vol. IT-8, October 1962, pp. 381–382.

[2] Friis, H. T., "Noise Figures of Radio Receivers," *IRE Trans.*, July 1944, p. 419.

[3] North, D. O., "The Absolute Sensitivity Radio Receivers," *RCA Review*, Vol. 6, January 1942, pp. 332–344.

[4] Skolnik, M. I., *Introduction to Radar Systems*, New York: McGraw-Hill, 1962, p. 24.

[5] Folsom, J. B., "That 'True-rms' Meter: Will It Be True to You?" *EDN*, November 5, 1975, pp. 91–95.

[6] Mumford W. W., and E. H. Scheibe, *Noise Performance Factors in Communications Systems*, Dedham, MA: Artech House, 1968.

[7] Barton, D. K., *Radar Systems Analysis*, Dedham, MA: Artech House, 1979.

[8] Tsui, J. B., *Microwave Receivers and Related Components*, Springfield, VA: NTIS, 1983.

[9] Urkowitz, H., "Energy Detection of Unknown Deterministic Signals," *Proc. IEEE*, Vol. 55, No. 4, April 1967, pp. 523–531.

[10] Urkowitz, H., *Signal Theory and Random Processes*, Dedham, MA: Artech House, 1983, pp. 488–465.

[11]   Torrieri, D. J., *Principles of Military Communications Systems*, Dedham, MA: Artech House, 1981, Ch. 4.

[12]   Emerson, R. C., "First Probability Densities with Square Law Detectors," *Journal of Applied Physics*, Vol. 24, No. 9, September 1953, pp. 1168–1176.

[13]   Wiley, R. G., *Electronic Intelligence: The Analysis of Radar Signals*, Dedham, MA: Artech House, 1982.

[14]   Woodward, P. M., *Probability and Information Theory, with Applications to Radar*, Dedham, MA: Artech House, 1980, Ch. 6. (Originally published by Pergammon Press, London, 1953.) See also J. R. Pierce, *Symbols, Signals and Noise*, New York: Harper and Row, 1961, Chapter X.

[15]   Skolnik, M. I., *Introduction to Radar Systems*, New York: McGraw-Hill, 1980.

[16]   Tsui, J., "Tangential Sensitivity of EW Receivers," *Microwave Journal*, October 1981, p. 99.

[17]   Harrison, A. P., Jr., "The World Versus RCA: Circumventing the Superhet," *IEEE Spectrum*, Vol. 20, No. 2, February 1983, p. 67.

[18]   Engelson, M., and F. Telewski, *Spectrum Analyzer Theory and Applications*, Dedham, MA: Artech House, 1974, Ch. 5.

[19]   Torrieri, D., J., *Principles of Military Communication Systems*, Dedham, MA: Artech House, 1981, pp. 155–160.

[20]   Tsui, J. B., *Microwave Receivers and Related Components*, Springfield, VA: NTIS, 1983, p. 59.

[21]   East, P. W., "Design Techniques and Performance of Digital IFM," *IEE Proc.*, Part F, No. 3, June 1982.

[22]   Barton, D. K., *Radar System Analysis*, Dedham, MA: Artech House, 1976, p.17.

[23]   Tsui, J. B., *Microwave Receivers and Related Components*, Springfield, VA: NTIS, 1983.

[24]   McCune, E., "SAW Filters Improve Channelized Receivers," *Microwaves and RF*, January 1984, pp. 103–110.

[25]   Torrierri, D. J., *Principles of Military Communication Systems*, Dedham, MA: Artech House, 1981.

[26]   Ready, P. J., "Multichannel Coherent Receiver," U.S. Patent No. 4,204,165; granted May 20, 1980.

[27]   Moule, G. L., "SA W Compressive Receivers for Radar Intercept," *IEEE Proc.*, Vol. 129, Part F, No. 3, June 1982, p. 180.

[28]   Lyons, W. G., et al., "High-Tc Superconductive Wideband Compressive Receivers," *The Lincoln Laboratory Journal*, Vol. 9, No. 1, 1998, pp. 33–64.

CHAPTER 4

Probability of Intercept

Arthur G. Self

4.1 Background

Probability of intercept (POI) is a key performance feature of EW surveillance and reconnaissance systems; it relates to the probability of time coincidence of two or more parametric "window functions," such as scanning antennas, sweeping or stepping receivers, and frequency agile emitters.

Many publications and examples that examine POI and its associated intercept time statistics invariably quote examples involving crystal video (wide open) receivers (CVRs), scanning superheterodyne (SSH) receivers, and the implications of scanning, directional antennae. However, over the last 20 years, significant developments in technology have occurred as have several armed conflicts. These have dictated radical changes in requirements, force modus operandi, and other issues.

In previous chapters, the impact of receiver developments, the impact of more recent conflicts (such as the Gulf War), and the emergence of a number of new threats such as the low probability of intercept (LPI) emitter have been examined. Clearly, there is a significant POI issue associated with such emitters in the presence of the littoral electromagnetic environment. Finally, as shown in the *USS Cole* incident, the new threat (now referred to as the "asymmetric threat") from fast attack craft (FAC), for example, is significant and must be accounted for in tasks, combat strategies, weapons, sensors, and so on, for fast reaction and adequate firepower.

So today's EW systems have to deal with increasing complexity and density of the electromagnetic environment; comprising multimode emitters, high and low ERPs, increased signal agilities (RF, PRI, and so forth), complex scans, and fleeting emissions.

EW system architectures are now more complex and, with more capable weapon systems (hard kill and soft kill), result in a more sophisticated modus operandi involving integration not only within their own platform or service but with other platforms and force structures. EW receiving systems have become EW suites which provide situational awareness, protection, and ELINT (fine grain analysis) capabilities. Note that some of the latest ESM systems are offering a specific emitter identification (SEI) capability—typically through amplitude and/or phase digitization on a per-pulse basis. Given the complex mission environments, and thus the required EW system architectures, ESM is now a likely, and necessary, precursor to ELINT.

In summary, there is now far less distinction between ESM and ELINT. Emphasis is turning to EW systems designs with both a tactical (ESM) and strategic (ELINT) capability. Systems designers need new tools to assist them, now that POI is now a more complex problem.

4.2 Developments in the Theory Behind POI

4.2.1 Intercept Description

Table 4.1 summarizes a range of typical kinds of intercept [1]; these can be subdivided into spatial, frequency, and time domains.

- *Spatial domain:* The beam-on-beam example where both transmitting (Tx) and receiving (Rx) systems use rotating, directional antennas and where the type of intercept is related to detectability (main beam or a percent of sidelobes).
- *Frequency domain:* Typically a scanning or sweeping receiver (e.g., a SSH, against a frequency agile transmitter).
- *Time domain:* Here, a number of distinct types of intercept are relevant. For example, the concept of a minimum number of pulses (or intercept duration) is important since it is required in ESM systems in order to differentiate between real signals and noise, as well as for subsequent signal sorting processes. Emission control policies will dictate when a transmitter is on and when it is not—thus introducing the scenario element to the intercept time arena. The time for an ESM system to process initially detected pulses will, in certain processing architectures, dictate whether or not subsequent pulses from the same emitter are tracked continuously or lost due to insufficient processing bandwidth/capability.
- *Others:* Propagation fluctuation aspects can also impact the time to intercept. These can be subdivided into two principal effects, namely, coherent (specular) scattering and tropospheric scattering. Coherent scattering is important for within-the-radio-horizon scenarios and low elevation threats (particularly sea-skimming missiles or high altitude platforms near the radio horizon). Tropospheric scattering relates to scattering of electromagnetic waves off the Earth's troposphere as witnessed by rapidly varying signal amplitude and fading. The intercept probability of the ESM system will be determined by the mean period of simultaneous overlap between the random pulse train representing such propagation fluctuations and the regular pulse trains representing the cyclic properties of the emitter and ESM systems.

4.2.2 Implications of Today's Environments/Operations on Intercept Time

The following are the types of intercepts shown in Table 4.1:

- In the spatial domain, phased array technology allows scanning space in a pseudo-random manner and such radar systems often perform multiple

Table 4.1   Types of Intercepts

	Receiver Activity	Transmitter Activity	Comments
Spatial Domain	Scanning antenna	Scanning antenna	Beam-on-beam intercept Main beam to main beam Main beam to sidelobes (variable)
Frequency Domain	Stepping or scanning Rx bandwidth across a specific operating bandwidth	Fixed RF, random jumping, or some other varying transmitted RF	Narrow instantaneous bandwidth (IBW) receiver which is stepping or sliding across a defined RF bandwidth looking for either a fixed or agile threat emitter RF. One window function if fixed RF else two.
Time Domain 1. Minimum intercept duration	Main beam	MB or SL intercept	Given a spatial intercept has occurred, how long is the coincidence; valid intercept if the number of received pulses is ≥ threshold
2. Emitter EMCON	Main beam	MB or SL intercept	For ECCM purposes, the emitter switches on for only short periods of time
3. Missing pulses			Environmental and emitter transmit effects could give rise to missing pulses in the received pulse train
4. ESM processing time	Receive and processing architecture	For nonparallel architectures and/or nonmultiple buffering data, the ESM could be blind to new data while it is processing the last received data	
Others Tropospheric scattering effects	Scanning	Scanning	Rayleigh distributed amplitude versus time Window function description: t = time above detection threshold; $(T - t)$ = time below detection threshold (nulls)

Source: [1].

functions. The electronic scanning antenna allows eliminating mechanical inertia as a design limitation; however, radar functions performed often impose a structure on the motion of the beam that can be exploited by would-be interceptors. Radar systems using mechanically scanned antennae have provided the scan characteristics of the intercept time problem. A change of mode (e.g., surveillance to tracking) or the switching on of a specific tracking radar afforded the EW designer further useful intelligence. Multifunction phased array radars reduce such possibilities, thus making intercept much more difficult.

- The time domain, including such operational aspects as emission control (EMCON), further increases the time to intercept. Operations in the littoral (i.e., close to shore) raise additional multipath and possibly missing pulse issues.
- In the frequency domain, the littoral electromagnetic environment makes signal search and identification that much more difficult in comparison to deep ocean environments. Clearly, more agility in the Tx signal parameters (be it RF or PRI) again directly impacts time to signal intercept. Additionally, low power frequency modulated continuous wave (FMCW) signals force the EW designer to long integration times, and thus, directly impacting time to intercept.

In summary, EW system designers are being pushed more and more towards architectures that deliver *single* pulse or *pulse burst* intercept capabilities.

4.2.3 Mathematical Models

The methodology of Self and Smith [2, 3] has been widely used due to its simplicity and ease of use, its applicability to scenarios involving more than two window functions, and its ability to predict intercept time accurately in a wide range of situations.

This section includes both the key formulae as well as several enhancements to the original work, including some specific, realistic examples with practical application. This is followed by a number of new and exciting theoretical developments involving some elegant mathematical theories on the subject of *synchronism*.

From [3], a convenient way of investigating the intercept problem is to represent the activities of the receiving and transmitting systems by window functions, as shown schematically in Figure 4.1. A key assumption is that the window functions are independent (or of "random-phase"). Such 'window functions' can also be described or represented as pulse trains—in this chapter, these descriptions are synonymous and interchangeable.

Key formulae are as summarized next.

For a number of coincidences, M, required, the equation for the mean period between coincidences, T_0, has the following explicit form:

$$T_0 = \frac{\prod\limits_{j=1}^{M} (T_j/\tau_j)}{\sum\limits_{j=1}^{M} (1/\tau_j)} \tag{4.1}$$

and for $M = 1, 2,$ and 3, this becomes:

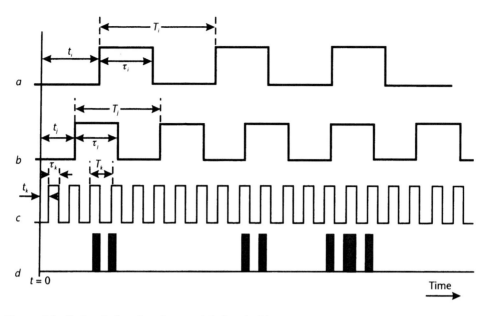

Figure 4.1 Basic window functions and their coincidences.

$$M = 1 \quad T_0 = T_1 \tag{4.2}$$

$$M = 2 \quad T_0 = \frac{T_1 T_2}{(\tau_1 + \tau_2)} \tag{4.3}$$

$$M = 3 \quad T_0 = \frac{T_1 T_2 T_3}{(\tau_1 \tau_2 + \tau_1 \tau_3 + \tau_2 \tau_3)} \tag{4.4}$$

where T_1, T_2, T_3 are the periods of the individual window functions and τ_1, τ_2, τ_3 are the durations of the individual window openings.

The probability of at least one intercept in time T is

$$P(T) = 1 - K e^{-(T/T_0)} \tag{4.5}$$

where $K = 1 - P(0)$, and $P(0)$, the probability of an intercept occurring in the first instant, is given by

$$P(0) = \prod_{j=1}^{M} \frac{\tau_j}{T_j} \tag{4.6}$$

The mean duration τ_0 of the simultaneous overlaps of all the M window functions is given by

$$\frac{1}{\tau_0} = \sum_{j=1}^{M} \frac{1}{\tau_j} \tag{4.7}$$

which, for $M = 1$, 2 and 3, becomes

$$M = 1 \qquad \tau_0 = \tau_1 \tag{4.8}$$

$$M = 2 \qquad \tau_0 = \frac{1}{\dfrac{1}{\tau_1} + \dfrac{1}{\tau_2}} \tag{4.9}$$

$$M = 3 \qquad \tau_0 = \frac{1}{\dfrac{1}{\tau_1} + \dfrac{1}{\tau_2} + \dfrac{1}{\tau_3}} \tag{4.10}$$

Turning to the issue of intercept *duration*, Figure 4.2 shows a comparison [3] of computer simulation results for examples of three-window intercept scenarios. Results are expressed as cumulative probability versus intercept duration. Analytic forms for these results show the following (in each case τ_1 denotes the smallest window and T denotes time into the actual intercept) (private communication with N. R. Burke, April 1983).

For the case of two windows:

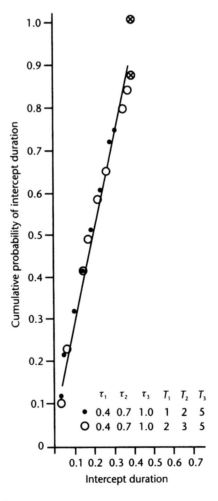

Figure 4.2 Intercept duration.

$$P(T) = \frac{2T}{(\tau_1 + \tau_2)} \qquad 0 \le T < \tau_1 \tag{4.11}$$

$$= 1 \qquad\qquad T \ge \tau_1$$

For the case of three windows:

$$P(T) = \frac{2T(\tau_1 + \tau_2 + \tau_3) - 3T^2}{(\tau_1\tau_2 + \tau_2\tau_3 + \tau_3\tau_1)} \qquad 0 \le T < \tau_1 \tag{4.12}$$

$$= 1 \qquad\qquad\qquad T \ge \tau_1$$

A number of developments of the theory of window functions with random phase have been made, and these are summarized here.

Interception of at least m *pulses*   Declaration of an adequate intercept may require at least m pulses. Thus, if the PRI of the received signal is T_2, then

$$d = mT_2 \tag{4.13}$$

$$T_0 = \frac{\displaystyle\prod_{j=1}^{M} \frac{T_j}{(\tau_j - d)}}{\displaystyle\sum_{j=1}^{M} \frac{1}{(\tau_j - d)}} \qquad \text{where } \tau_j > mT_2$$

Using (4.3), Table 4.2 demonstrates the effect on T_0 of increasing the minimum intercept duration d from 0 to 25 ms for the cases of $T_1 = 0.7$ and 1.3 seconds (two pulse trains are $\tau_1 = 0.058$, $\tau_2 = 0.078$, T_1, $T_2 = 6.0$ seconds). If a minimum of 10 pulses is required for a valid signal detection, then a minimum intercept duration of 20 ms is required for a 600-Hz PRF radar (less than 5 ms for a 4-kHz radar), which, in turn, implies a mean time between intercepts of at least 40 seconds for a receiver scan period of 0.7 second; this is in comparison to a mean time between intercepts of only 31 seconds for the purely theoretical example where receiver integration aspects are not included.

Effect of Auto Pause Mechanism   For a sweeping receiver, such a mechanism compromises intercept time. This can be deduced as an extension of the theory. If

Table 4.2 Effect on T_0 of Increasing d

	T_0		Number of Pulses	
d (ms)	$T_1 = 0.7$	$T_1 = 1.3$	PRF = 600 Hz	PRF = 4 kHz
0	31	57		
5	33	62	3	20
10	36	67	6	40
15	40	74	9	60
20	44	81	12	80
25	49	91	15	100

Source: [3].

the receiver interrupts its frequency sweep for a known amount of time every time a signal is encountered, then the *effective sweep period* of the receiver will be extended proportionately and thus the mean period between coincidences will depend both on the auto pause time and the average number of signals transmitting at any one time.

Propagation Fluctuations   The effects of propagation can be added in a simple manner to this theory [1]. Propagation fluctuations are assumed to be able to be subdivided into two types:

1. Long-term fading effects (affects the way in which average power received varies and hence will show variations in terms of hourly, daily changes);
2. Short-term fading effects (affects POI more directly).

Intercept of the radar signal from a distant transmitter will be possible when the strength of the fading signal rises above the threshold level of the receiver. The occurrence (in time) of bursts of the signal above the threshold may be represented by a random rectangular pulse train. The mean period of the pulse train $<T>$ is the reciprocal of the mean rate of upward crossings of the threshold level by the fluctuating signal amplitude. The mean duration of the pulse train $<\tau>$ is the mean duration of the signal amplitude above the threshold level.

The intercept probability will be determined by the overlaps between the random pulse train representing propagation fluctuations and the regular pulse trains representing the cyclic properties of the transmit and receive systems:

$$<T> = \left(\frac{1}{\pi a}\right)^{1/2} \left(\frac{1}{2f}\right) \cdot e^{a} \tag{4.14}$$

$$<\tau> = \left(\frac{1}{\pi a}\right)^{1/2} \left(\frac{1}{2f}\right) \tag{4.15}$$

where:

$a = L/b_0$ is a dimensionless threshold parameter

b_0 = mean power level of the fading signal

f = rms bandwidth of amplitude fluctuations

L = threshold level

For example, if $f = 1$ Hz at S-band and $a = 1$, then substituting into (4.14) and (4.15) gives $<T> = 0.77$ second and $<\tau> = 0.28$ second.

A Practical Example of Three Pulse Trains
Looking at a more realistic radar type example, let us derive the intercept time statistics [1]. Table 4.3 shows the assumed pulse train values (fading is neglected).

Table 4.3 Realistic Receiver/Transmitter Scenario

	Parameters for Example POI Calculation		
j	Mean Duration, j (sec)	Mean Period, T (sec)	Comments
1	$\tau_1 = 0.03$	$T_1 = 3.0$	Tx antenna scan
2	$\tau_2 = 2 \times 10^{-6}$	$T_2 = 5 \times 10^{-4}$	Pulse modulation
3	$\tau_3 = T_3/100$	$10^{-4} \leq T_3 \leq 10^4$	Receiver frequency sweep

The receiver's sweep is chosen to have an instantaneous bandwidth of 1%, corresponding to 20 MHz in a total sweep range of 2 GHz. The mean time between intercepts, T_0, is calculated as a function of the sweep period, T_3, which varies over a wide range, say, 10^{-4} seconds to 10^{+4} seconds.

The correct expression for T_0 is to be chosen for any particular value of T_3, from among the equations that represent the mean period of overlaps of different combinations of the three overlapping pulse trains:

$$T_0 = T_j \text{ where } j = 1, 2, 3 \tag{4.16}$$

$$T_0 = \frac{T_j T_k}{(\tau_j + \tau_k)} \text{ where } j = 1, 2, 3 \text{ and } k \neq j \tag{4.17}$$

$$T_0 = \frac{T_1 T_2 T_3}{(\tau_1 \tau_2 + \tau_2 \tau_3 + \tau_1 \tau_3)} \tag{4.18}$$

In general, the above formula for mean time between intercepts does not discriminate against the possibility of multiple overlaps, and thus, it may, in certain circumstances, underestimate the mean time. Smith analyzes such a possibility and concludes that the correct formula is given by the one which has the greater numerical value, based on there being $2^M - 1$ formulae in the case of M overlapping pulse trains (in the above case this gives seven distinct formulae) (private communication with B. G. Smith, 1983).

The curve of T_0 versus T_3 divides into four regions as shown in Figure 4.3.

- *Region #1:* Very large values of T_3. The duration of the sweep is much larger than the scan period; the pulse modulation period of the mean period of overlaps between these two trains. Therefore, for large T_3, the mean time between intercepts equals the sweep period [(4.16) with $j = 3$].
- *Region #2:* As T_3 decreases, the duration of the frequency sweep falls until it is less than the period of the scan cycle. Interception in a single sweep is no longer certain. The mean duration of the overlap between the scan and sweep cycles is long compared with the period of the pulse modulation, and so the mean time between intercepts is determined by the scan and sweep parameters in (4.17) with $j = 3$ and $k = 1$.
- *Region #3:* As T_3 falls further, the mean overlap between the scan and sweep cycles falls below the period of the pulse modulation and the mean period between intercepts is (4.18) (dependent on parameters of the three interacting pulse trains).

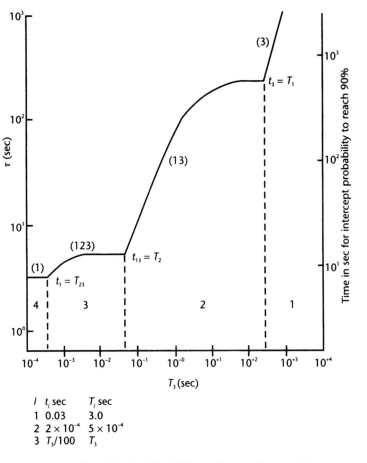

Figure 4.3   Intercept time. (*From:* [4]. © 1996 IEEE. Reprinted with permission.)

- *Region #4:* Eventually, T_3 falls sufficiently that the mean period of overlaps of the sweep and pulse modulation cycles falls below the duration of the scan cycle. Interception within a single scan is certain, and the mean time between intercepts equals (4.16) with $j = 1$.

Specific Example of a Fast Step-Scan Receiver

From [2], consider the following example: a broad-beam receiving antenna so that it is not necessary to search in azimuth, but add the use of a fast step-scan receiver which searches a 2,000-MHz frequency band in 200 steps, each step being 10 MHz. (The receiver bandwidth should be about 20 MHz to provide overlapping coverage from one step to the next.) See Table 4.4.

Hence, from (4.3), the mean time between intercepts is now 17.74 seconds, and from (4.9) the mean duration of the intercepts is 4.93×10^{-4} seconds, and the

Table 4.4   Case 1: The Result Using a Receiver Scan Time of 100 ms

τ_1	τ_2	T_1	T_2
0.0005 sec	0.0333 sec	0.1 sec	6 sec

average coincidence fraction is 2.78×10^{-5}. This leads to a time for 90% probability of intercept of 40.85 seconds. See Table 4.5.

Now the mean time between intercepts is increased to 277.13 seconds, their mean duration to 7.69 ms, and the average coincidence fraction remains at 2.78×10^{-5}. The time required for a 90% probability of intercept is increased to 638.12 seconds or 10.64 minutes. Clearly, increasing the receiver sweep time results in a much longer time to intercept the signal, with an attendant increase in the expected duration of the intercept.

Next, consider three pulse trains by adding a rotating antenna. The receiver is sweeping in frequency as well as searching in angle, and the transmitter is also circularly scanning. See Table 4.6.

From (4.4), the mean time between intercepts is increased drastically to 5,423 seconds (about 1.5 hours), and from (4.10) the average duration of an intercept is 1.266×10^{-6} seconds and to reach a probability of intercept of 90% requires 3.47 hours. See Table 4.7.

This increases the mean time between intercepts to 7.35 hours with an average intercept duration of 2.02×10^{-5} seconds and a 90% probability of intercept time of 16.92 hours.

Table 4.8 summarizes the calculations for T_0 in the above examples from [3], and shows their relationship to POI over the interval 30% to 90%.

Cases 3 and 4 illustrate why it is usually necessary to eliminate one or more of the pulse trains. A highly sensitive receiver (or strong signal level at the receiver) allows us to drop the transmitter antenna scan pulse train from the analysis by making it possible to receive the signal of interest through the emitter antenna's side or back lobes. The best sensitivity is achieved economically with a narrowband swept receiver and a narrowbeam receiving antenna. Thus, by introducing two search processes at the receiver, it may be possible to eliminate the transmitter's scan from the probability of intercept calculation.

A limitation of this approach is that it does not address detailed impacts of synchronism effects between pulse trains. The above examples clearly identify such

Table 4.5   Case 2: The Same Situation as for Case 1, with the Receiver Sweep Slowed to 2 Seconds

τ_1	τ_2	T_1	T_2
0.01 sec	0.0333 sec	2 sec	6 sec

Table 4.6   Case 3: The Result of the Faster Receiver Sweep

τ_1	τ_2	τ_3	T_1	T_2	T_3
0.00278 sec	0.0333 sec	0.0005 sec	1.0 sec	6.0 sec	0.1 sec

Table 4.7   Case 4: The Result Using the Slower Receiver Sweep

τ_1	τ_2	τ_3	T_1	T_2	T_3
0.00278 sec	0.0333 sec	0.01 sec	1.0 sec	6.0 sec	2.0 sec

Table 4.8 Predicted Results for POI and T_0 for Cases 1 to 4

Case	Parameters (sec)		Mean Period of Simultaneous Overlaps, T_0 (sec)	Probability of Intercept $P(T)$			
				0.3	0.5	0.7	0.9
1	$\tau_1 = 0.005$	$T_1 = 0.1$	18	6	12	21	41
	$\tau_2 = 0.0333$	$T_2 = 6.0$					
2	$\tau_1 = 0.01$	$T_1 = 2.0$	277	99	192	333	638
	$\tau_2 = 0.0333$	$T_2 = 6.0$					
3	$\tau_1 = 0.00278$	$T_1 = 1.0$	5,423	1,934	3,759	6,529	12,486
	$\tau_2 = 0.0333$	$T_2 = 6.0$					
	$\tau_3 = 0.0005$	$T_3 = 0.1$					
4	$\tau_1 = 0.00278$	$T_1 = 1.0$	26,450	9,434	18,334	31,846	60,905
	$\tau_2 = 0.0333$	$T_2 = 6.0$					
	$\tau_3 = 0.01$	$T_3 = 2.0$					

Source: [3].

relationships between the values of $T_{[1,2,3]}$; thus, for a two pulse train example, if the period of one pulse train is harmonically related to the period of the other, then the two pulse trains will either be in synchronism or they will not. If the latter, then an intercept will not occur (note that $\tau_i \ll T_i$ in the radar case); if the former, then there will be synchronism and intercepts will occur at known, discrete times that can be deduced. Note that for regular pulse trains, there will always be a fixed relationship between the pulse trains *once the first intercept has occurred*. In such instances, an ESM system operating with a directional antenna may improve its intercept probability if its rotation rate is variable.

4.2.4 Recent Developments on POI

More recent work on intercept time and POI has focused on the underlying theory of pulse train coincidence where the pulse train phases are *nonrandom*. In other words, earlier references [1–3, 5], and private communications with N. R. Burke (April 1983), A. G. Self (1986), and B. G. Self (1983), examined noncorrelated pulse trains (herein now referred to as the random-phase model). However, there are instances (for example, with a stepping/scanning receiver or scanning transmit and receive antennae) where the resultant intercept pulse train becomes highly correlated and dependent on the input pulse train attributes. In this section, the latest results on *synchronism* effects [6] are summarized as well as the elegant application of number theory [4, 7] to the intercept problem domain.

4.2.4.1 Synchronization Effects on POI

Historically, mechanical scanning transmit systems (especially long range surveillance systems) had rotation rates of typically 3, 6, 9, 10 (and beyond) rpm. Clearly, a receiving system with the same or sub/multiples of such rotation rates (such as 4:3 or 7:5) encounters synchronism issues, expressed as "the ratio of the periods." In the worst case, coincidence (intercept) may never occur, or occur in such long time periods that fast interception requirements could not be met. Reference [6] examines the problem of pulse synchronization with pairs of pulse trains, and derives a theoretical approach supported by Monte Carlo simulations.

The theory [6] is constrained to the intercept probability of two pulse trains with specific reference to a swept SSH Rx and a pulsed emitter. The approach is based on the concept of pulse clusters overlapping in phase space (i.e., based on the relative start times or phases between the various pulse trains). Based on this, exact expressions are developed for the POI after a given number of sweeps, averaged over all relative start phases of the transmitter's pulse train and the receiver's pulse (or sweep) train.

Given the pulse train characteristics are $[T_1, \tau_1, T_2, \tau_2]$, the analysis [6] starts by non-dimensionalizing this set by dividing by T_2, which yields $(R, \tau_1/T_2, 1, D_2)$ where $R = T_1/T_2$ is the PRI ratio and $D_2 = \tau_2/T_2$. When considering the POI, it is only the distance from one pulse to the nearest pulse of the other train that is of interest; the TOA of a pulse can then be considered as a dimensionless quantity that can be viewed as being equivalent to wrapping or folding the TOA sequence around a cylinder of circumference 1. The phase is chosen such that the initial pulse from train 1 falls at zero; all pulses from train 2 will fall at the same point given a fixed PRI (analysis considers this as a series of spikes) while pulses from train 1 are distributed on the interval [0, 1]. After n pulses from train 1, the region covered by pulses represents the set of positions of the train 2 spike, which leads to interception within n pulses. The POI is the probability that the spike generated by train 2 intersects a region covered by pulses from train 1. Regarding the folded interval concept, then, this probability is equivalent to the ratio of the length covered by pulses from train 1 to the length of the interval. As shown in [6], the POI can be considered within four distinct regions:

$$P_n = P_n^{(j)} \; j = 1, 2, 3, 4 \tag{4.19}$$

where the magnitude of j defines the region of interest dependent on the magnitude of the sweep number n relative to certain critical values which may vary with the PRI ratio and $P_1 \; [T_1/T_0]$.

These four regions can be expressed as follows, where n_c = number of clusters:

$$P_n^{(1)} = nP_1 \qquad\qquad\qquad\qquad n \le n_c \tag{4.20}$$

$$P_n^{(2)} = P_n^{(1)} + (n - n_c)\Delta \qquad\qquad n_c \le n \le n' \tag{4.21}$$

$$P_n^{(3)} = P_{n'}^{(2)} + \frac{(n - n')}{n_c}\left(1 - P_{n'}^{(2)}\right) \qquad n' \le n \le n' + n_c \tag{4.22}$$

$$P_n^{(4)} = 1 \qquad\qquad\qquad\qquad n > n' + n_c \tag{4.23}$$

and Δ represents an amount that each new pulse adds to the phase length of its cluster; for an exact coincidence, $\Delta = 1/2 - |n_c R \bmod 1 - 1/2|$. From [6], Figure 4.4 shows the POI plotted for $P_1 = 0.1$, with sweep numbers ranging from $n = 1$ to $n = 12$ as a function of the PRI ratio R. It shows a number of interesting features:

- A number of plateau regions of extended maxima—these are obtained when $n \le n_c$ and where (4.20) holds.

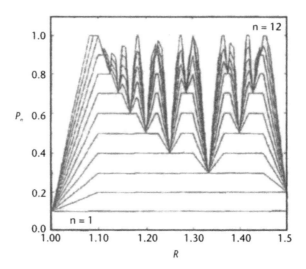

Figure 4.4   POI between uniform (periodic) pulse trains as a function of PRI ratio, R, and pulse number (or sweep number) from pulse train 1. (*From:* [6]. © 1996 IEEE. Reprinted with permission.)

- The extended maxima are separated by broad peaked minima which occur when $n > n_c$ and (4.21) holds, or if $n > n'$ and (4.22) is valid.
- The width of each minima varies in proportion to P_1.
- When $n = 1$, P_n is given by the random phase result; for $n > 1$, the random-phase theory generally underestimates the POI when it is in a plateau and overestimates the POI when it is approaching a minimum. As n increases, the density of the minima also increases.
- Other minima of $P_n(R)$ associated with $P_m(R)$ for $m < n$ are also present.

Figure 4.4 can be useful in selecting the optimal search parameters against a given transmitted PRI in order to maximize the POI.

However, this approach has limitations against unknown pulse train characteristics when using a uniform pulse train on the receiving system. Adding jitter to the search pulse train is suggested in [6] in order to smooth out regions of destructive synchronization (i.e., where no intercept will happen or not for very long periods). Adding a small amount of jitter in a cumulative fashion on each pulse is also suggested in [6]. Based on random walk theory, the position of each pulse then performs a random walk about a mean TOA position with a step distribution given by the distribution of jitter on each pulse. From [6], for a *uniform* step distribution ranging from $-J/2$ to $+J/2$, where J is defined as the jitter amplitude, the variance of the mean PRI of the pulse train is given by

$$\sigma^2 = nJ^2/3 \tag{4.24}$$

Since jitter is assumed cumulative, the variance on the TOA of the nth pulse increases linearly with n, and so the pair of pulse trains are more likely to satisfy the requirements of the random phase assumption as the sweep number increases.

From [6] and using Monte Carlo simulations, Figure 4.5 shows POI as a function of the mean PRI ratio [R] and the jitter amplitude on T_1, measured in units of T_2, for $n = 4$ sweeps and $P_1 = 0.1$. Jitter was added to T so that the PRI varied over the range $T_1 (1 - J/2)$ to $T_1 (1 + J/2)$. Each point on the graph shows the mean POI in 4 sweeps. When the jitter amplitude is zero, it is seen that Figure 4.4 is derived; as the jitter amplitude increases, $P_n(R)$ approaches the value given by the random-phase theory earlier. As [6] points out, quite small values of J can move the POI as derived from this cluster theory to that of the random-phase model as a function of P_1. Also, as expected, the required value for J decreases as the number of sweeps, n, increases.

In summary, there now exists a more complete and accurate picture of POI for cases where the phases of pulse trains are nonrandom. As shown in [6], such occurrences give rise to sharp maxima and minima in POI space, and this can aid an EW system designer in optimizing search strategies. However, in real life, synchronization effects can work against the performance of receiving systems, and the authors propose introducing a jitter mechanism to smooth out such cases. This, in turn, is shown to link directly back to the random-phase case of earlier POI publications.

4.2.4.2   POI and Number Theory Application

In [4, 7] the mathematical background is developed even further for the two periodic pulse train cases and beyond. The applicability of number theory to a range of POI problems involving pulse trains is demonstrated in [7], where (1) phases are known and equal; (2) known and unequal; and (3) one or both are random variables. The problem becomes finding an algorithm for computing when the first intercept occurs and then when subsequent intercepts will occur. Key in this analysis is that the ratio of the PRIs (principal focus is for two pulse trains only) is not restricted to rational numbers. Through formulating the problem as a Diophantine approximation (i.e., based on only integer solutions), [7] demonstrates how number theory can directly assist in the intercept problem.

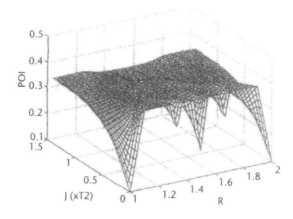

Figure 4.5   Effect of adding jitter. POI as a function of PRI ratio and jitter amplitude for sweep number $n = 4$ on pulse train 1. (*From:* [6]. © 1996 IEEE. Reprinted with permission.)

While mathematically intensive, [7] contains an elegant theoretical basis for pulse train synchronism. Indeed, a recursive algorithm is outlined showing how POI is related to the Farey point notation, where the Farey series F_N is the set of all fractions in lowest terms between 0 and 1 whose denominators do not exceed N, arranged in order of magnitude. For example, F_6 is given by [8]:

$$\frac{0}{6}, \frac{1}{6}, \frac{1}{5}, \frac{1}{4}, \frac{1}{3}, \frac{2}{5}, \frac{1}{2}, \frac{3}{5}, \frac{2}{3}, \frac{3}{4}, \frac{4}{5}, \frac{5}{6}, \frac{6}{6}$$

The mean time to intercept for the discrete time case can be expressed as [7]:

$$E(N) = \{T_2 + q(\tau)A(\tau)^2 + a(\tau)Q(\tau)^2 + 2[a(\tau) + q(\tau) - \tau]A(\tau)Q(\tau)\}/2T_2$$
$$(4.25)$$

Using the above, plus a range of other theorems and definitions contained in detail in [4, 7], the POI can be determined over the interval

$$T_1 \in [p'T_2/p, q'T_2/q] \tag{4.26}$$

other parameters being held constant. Based on being given two adjacent Farey points and their left/right parents, Figure 4.6 shows the POI as a function of the number of pulses N from the first pulse train, its PRI T_1 with $T_2 = 1$ and $\tau = 0.1$. The diagram shows the rise of POI from small values of N to unity; also, at various PRIs, there are deep troughs which are centered around the Farey points (conversely, this is where the mean time to intercept approaches infinity). Figure 4.7 shows how Figure 4.6 can be interpreted as *regions* [7]. For any given N and T_1, it shows

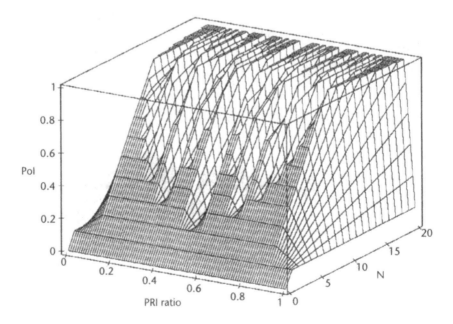

Figure 4.6   Probability of intercept as a function of PRI ratio. (*From:* [7]. © 2004 Association of the Old Crows. Reprinted with permission.)

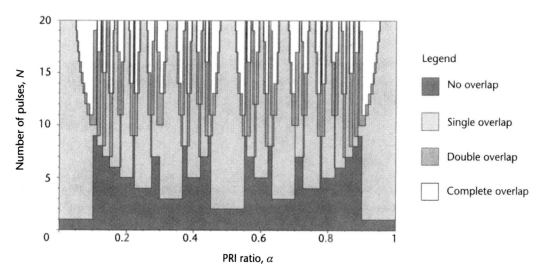

Figure 4.7 POI regions as a function of PRI ratio. (*From:* [4]. © 1996 IEEE. Reprinted with permission.)

how the POI lies in the region of initial growth, single or double overlap or probability 1.

Taking a specific example [4], Figure 4.8 shows the intercept time between two pulse trains—the duty cycles for pulse trains 1 and 2 are 0.13/0.26, respectively. The PRI of train 1, T_1, is held constant with $T_1 = 1$ and T_2 allowed to vary between 0.1 and 4. Intercept time is shown as the solid line, and it can be seen how it goes to infinity at the synchronization ratios. The theoretical lower limit of intercept time given by

$$T_0 = \frac{T_1 T_2}{(\tau_1 + \tau_2)}$$

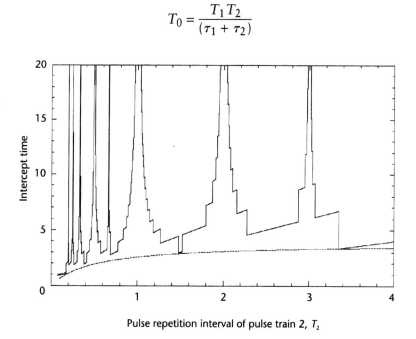

Figure 4.8 Intercept time as a function of PRI 2. (*From:* [4]. © 1996 IEEE. Reprinted with permission.)

is also shown as the dotted line where the predicted intercept time approaches T_0 at several points. Using such a formalism, it is possible to assist in deriving receiver characteristics such as sweep time that will minimize T_0 against known radar transmitter scan characteristics. Ongoing work into three window intercept cases is suggested at the conclusion of [4].

4.3 Summary

Since [2], there have been some significant developments on the important subject of POI. Several new publications have appeared indicating both developments of the (random-phase) window functions model and its application across a variety of Rx architectures. Tools now exist to support EW system designers beyond the earlier two or three window function scenarios such as scan-on-scan and stepping or scanning Rx. This wider understanding of POI has been critical not just to improve system designs but also to acknowledge that certain conventional window functions such as scan may diminish in today's POI analyses (due to the proliferation of phased array radars), thus requiring more in-depth review and appreciation.

More recently, there have been a number of new publications addressing the real-life issue of synchronization; this has been described through some elegant mathematics. Current theory appears well matched to describing the two window case (such as with a scanning receiver system against a rotating transmitter antenna); future work is already examining three windows and beyond, together with a variety of techniques to aid EW system designers. EW practitioners are already considering practical solutions to correlation; for example, [8] looks at POI issues for a naval EW system against radar threats incorporating a wide instantaneous field of view (IFOV) receiver behind a high rotation speed, directional antenna. Detailed calculations are shown for a scan-on-scan scenario where synchronism between receive and transmit window functions is removed (thus enabling intercept) through a variable receiver scan rotation rate.

Overall, it is clear that there now exists strong and clear linkage between uncorrelated and correlated aspects of POI.

A scenario of a frequency agile emitter and a sweeping narrowband receiver is analyzed in some detail in Appendix C. The random frequency agility allows the use of conventional probability concepts to characterize the data provided at the receiver output.

References

[1] Self, A. G., *EW Design Engineers' Handbook*, Norwood, MA: Horizon House, 1990.

[2] Wiley, R. G., *Electronic Intelligence: The Interception of Radar Signals*, Dedham, MA: Artech House, 1985.

[3] Self, A. G., and B. G. Smith, "Intercept Time and Its Prediction," *IEE Proceedings*, Vol. 132, Pt. F, No. 4, July 1985.

[4] Vaughn, I., et al., "Number Theoretic Solutions to Intercept Time Problems," *IEEE Trans. on Information Theory*, Vol. 42, No. 3, May 1996.

[5]   Self, A. G., "Intercept Time and Its Prediction," *Journal of Electronic Defense*, August 1983.

[6]   Kelly, S. W., G. P. Noone, and J. E. Perkins, "Synchronization Effects on Probability of Pulse Train Interception," *IEEE Trans. on Aerospace and Electronic Systems*, Vol. 32, No. 1, January 1996.

[7]   Vaughn, I., and L. Clarkson, "The Farey Series in Synchronization and Intercept Time Analysis for Electronic Support," *Trans. of the AOC*, Vol. 1, No. 1, October 2004.

[8]   Sullivan, W. B., "A Simple Tune: Single-Channel Naval ESM Receivers Remain Popular and Effective," *Journal of Electronic Defense*, November 2003.

CHAPTER 5

Antennas and Direction Finders

The antenna is a vital link in the reception process. It is the transducer to convert electromagnetic radiation propagating through space into a microwave signal that can be processed by a receiver. It is also a signal processor ahead of the receiver. The antenna can be used to discriminate against some signals and enhance others based on angle of arrival and polarization. It is also the key element in the process of measuring angle of arrival or polarization.

The window functions described in Chapter 4 are often a result of directional antennas being used to search for signals over an angular region. Furthermore, the power available at the intercept receiver input is a function of the antenna gain.

5.1 Omni-Directional Antennas

An idealized isotropic antenna responds equally well to a signal from any direction of arrival over a full sphere of coverage. Such an antenna does not actually exist, but the performance of real antennas is often measured with respect to such an idealized isotropic antenna. The gain of an antenna is the ratio of the signal power output of the antenna to the signal power from an ideal isotropic antenna. This quantity is expressed in decibels with respect to an isotropic antenna as dBi.

Since no actual isotropic antenna exists, standard antennas with calibrated gain are used to provide the reference levels. "Omni-directional" antennas are often used in signal interception applications. Omni-directional in this context means approximately equal response to signals arriving from any direction in one plane. For example, a vertical monopole antenna above a ground plane will respond equally to signals from any azimuth angle at a fixed elevation angle, with the peak response at a 0° elevation angle (i.e., toward the horizon) and no response at all to a signal arriving from straight above the antenna. If the isotropic antenna's response is thought of as a sphere, then any other antenna pattern can be thought of as deforming the sphere. A high gain antenna pattern can be thought of as being formed by squeezing a spherical isotropic pattern until the entire volume of the unit sphere extends outward only in a narrow angular region.

5.1.1 Omni-Directional Antenna Applications

While the low gain of omni-directional antennas reduces the received power, this is not a significant drawback in some applications. For radar warning receiver applications, the signal intensity from the radar is generally quite high, and the characteristic of most interest is the ability to provide a warning regardless of the direction from which the signal arrives, and also to give approximate indication of the direction of the threat. A typical example system uses four broad beam (~90°) antennas. By comparing the signal amplitudes, the system can determine direction of arrival.[1]

Of more direct interest for a nontactical application of broad beam antennas is to cover a large angular sector so that no signals are missed (i.e., to achieve 100% probability of intercept). A circular array of broad beam elements with appropriate RF signal processing can determine the direction of arrival of a single radar pulse over a broad frequency band. (See Section 5.4.3 for more details.) Finally, there is the combination of an omni-directional antenna with a directional antenna for purposes of eliminating those radar pulses intercepted through the sidelobes of the directional antenna. In such a system, the amplitude of the received signal from both the omni-directional antenna and the high gain direction-finding antenna are compared. If the signal from the omni-directional antenna is larger, the conclusion is that the pulse was received from other than the direction in which the directional antenna is pointed. This is illustrated in Figure 5.1.

Yet another application of broad beam antennas is to use them on the same antenna pedestal with a narrow beam antenna. The narrow beam antenna can be used to track a moving target via radar, while the associated intercept antennas with broad beams receives any signals from the general vicinity of the target; as, for example, when an aircraft fires a missile, the guidance signals (if any) can be intercepted for only a short time. The tracking antenna ensures that the missile is in the field of view of the intercept antennas at the time of launch.

5.1.2   Parameters for Omni-Directional Antennas

The significant parameters that define the performance of broad beam antennas are the gain and the capture area. The directive gain of an antenna is defined as the ratio of the maximum radiation intensity to the average radiation intensity

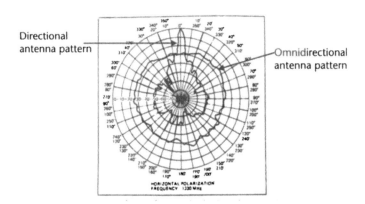

Directional antenna pattern

Omnidirectional antenna pattern

Figure 5.1   Omni-directional antenna used to control the field of view.

1.   The AN/ALR-56 is an example of a warning receiver with such an arrangement of antennas [1].

over a full sphere. The power gain is defined as the ratio of the maximum radiation intensity to the radiation intensity of an ideal (no loss) isotropic antenna. If the antenna has no loss, the directional gain is equal to the power gain. Antenna theory makes use of the concept of a current element, which consists of a very short (relative to one wavelength), very thin conductor carrying a uniform current over its short length. The gain of such an antenna (which cannot be realized in practice) is 1.5 [2, p. 20]. Since neither the isotropic radiator nor the current element is realizable, it is convenient to use another type of antenna as a standard. One such standard is a small dipole (or doublet), which also has a gain of 1.5. If the dipole is one-half wavelength long, the gain is 1.64. Using the method of images, a monopole above a ground plane has the same pattern (in one hemisphere) as a dipole but no radiation in the other hemisphere. Hence, its gain is twice that of a dipole. These gains and other parameters are given in Table 5.1 [3].

The radiation resistance is the real part of the ratio of the voltage to the current at the antenna terminals. The impedance of a short dipole is capacitively reactive (the current leads the voltage) and its radiation resistance decreases as the square of the length. As the length increases to $\lambda/2$, the capacitive reactance vanishes and the radiation resistance becomes 73Ω.

The capture area of an antenna is easier to visualize when the antenna consists of a large reflector or array (dimensions are large compared to a wavelength), when the electrical area and physical area are approximately the same. The capture area is found by measuring the power density (watts/meter squared) at the antenna and

Table 5.1 Omni-Directional Antenna Parameters

| | Gain | | Radiation | Capture |
Antenna Type[a]	Numeric	dB	Resistance (Ω)	Area (λ^2)
Isotropic	1.0	0.0	—	0.08
Current element	1.5	1.76	$80\pi^2 (l/\lambda)^2$	0.12
Short dipole	1.5	1.76	$20\pi^2 (l/\lambda)^2$	0.12
Short dipole using local ground reflection[b]	6.0	7.78	c	0.48
$\lambda/2$ dipole	1.64	2.15	73	0.13
$\lambda/2$ dipole using local ground reflection	6.56	8.17	$\approx 73°$	0.52^d
Short monopole: Transmitting—all cases Receiving—space wave	3.0	4.77	$40\pi^2 (h/\lambda)^2$	0.24^d
Short monopole receiving a single ray (surface wave)	0.75	−1.25	$80\pi^2 (h/\lambda)^2$	0.06^d
$\lambda/4$ monopole Transmitting—all cases	3.28	5.16	36.5	0.26^d
Receiving—space waves $\lambda/4$ monopole receiving a single ray (surface wave)	0.82	−0.86	36.5	0.07^d

[a]Unless noted otherwise, all antennas, except monopoles, are assumed situated in free space. Monopoles must be located on a ground plane of substantial area.
[b]Gain values for antennas that employ a ground reflection apply to the maximum of the resulting elevation angle radiation pattern.
[c]Depends on height above ground and ground characteristics [2, p. 303].
[d]These values of capture area and effective height apply to the space wave impinging on the antenna-ground combination on which gain is based.
Source: [3].

then measuring the received power delivered to a matched load. Since the total power received is equal to the power density multiplied by the capture area, the capture area is defined as the ratio of the received power to the incident power density.[2] Using this idea, the capture area for any antenna can be computed. The capture area of an isotropic antenna is

$$A_i = \frac{\lambda^2}{4\pi} \cong 0.08\lambda^2 \qquad (5.1)$$

and the capture area of the other antennas in Table 5.1 is simply the gain times the capture area of the isotropic radiator.

Effective height (h_{eff}) is a concept that is applied at lower frequencies (<10 MHz). In this case, the incident power density is often expressed in terms of the electric field strength (volts/meter). Then the voltage at the antenna terminals (assuming a matched load) will be

$$V = E \cdot h_{eff} \qquad (5.2)$$

The incident power density in a plane wave is given by the cross product of ExH, which in free space is $|E|^2/120\pi$ in the MKS system of units. The voltage in (5.2) appears across the radiation resistance if the antenna is connected to a matched load. Since half the received power is delivered to the load, the total received power is $2V^2/R$,

$$\frac{2(Eh_{eff})^2}{R} = \frac{E^2A}{120\pi} \qquad (5.3)$$

or

$$A = \frac{240\pi h_{eff}^2}{R} \qquad (5.4)$$

Substituting the radiation resistance for the short monopole from Table 5.1 of $40\pi^2 (h/\lambda)^2$ verifies that the effective area of the short monopole is

$$A = 1.5\frac{\lambda^2}{4\pi} \qquad (5.5)$$

A surface wave is one traveling from one monopole to another (i.e., both monopoles are above the same ground plane, possibly the Earth). Hence the distinction in Table 5.1 between "space waves" and "surface waves." The presence of the infinite ground plane assumed for the monopole makes it impossible to also assume free-space conditions of propagation [3].

In the development of intercept systems, especially for aircraft, it has been important to obtain broadband performance over 360° in azimuth with small

2.   This assumes the polarization of the antenna is matched to that of the incident wave.

antennas, preferably flush-mounted to avoid drag. Such antennas can achieve multiple octave bandwidth, 60° to 90° of elevation coverage, and nominal gains of −3 to +5 dBi using a cavity-backed spiral design. Other flush antennas are of the microstrip type [4].

A typical nonflush mounting omni-directional antenna is the conical spiral. The patterns for such an antenna at two frequencies, 2 and 12 GHz, are shown in Figure 5.2.

5.2 Directional Intercept Antennas

In many microwave applications, directional antennas are the general rule. Certainly most of the radar signals to be intercepted are transmitted by high gain, very directional antennas for at least two reasons: (1) the radar needs the gain to increase the target return, and (2) the directionality is needed to find the bearing to the radar's target. Electronic intelligence operations may need high gain antennas for the same reasons. From the interceptor's viewpoint, another reason to use a directional antenna is to eliminate signals arriving from directions other than the sector of interest. In an active radar environment, a directional antenna may be a very important signal processing aid to reduce interference. Suppose the weakest radar signal at the horizon is well above the threshold level of the intercept receiver. Then an antenna beamwidth of 3.6° can reduce the number of signal pulses being received by a factor of approximately 100 compared to an omni-directional antenna.

Compared to a radar system, an intercept system will generally use a lower antenna gain. While the radar designer may choose antenna gain in excess of 30 dB,

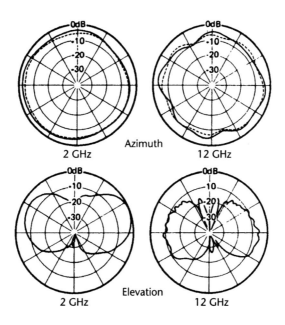

Figure 5.2 Antenna pattern for a broadband conical spiral.

the typical intercept antenna "high gain" may be 15 to 30 dB, depending on the frequency of interest. Since the interceptor designer is concerned with frequencies over a much wider band than is a radar designer, the intercept antenna may cover several octaves (e.g., 1 to 12 GHz). Broadband directional antennas can be made using a broadband omni-directional antenna as the feed for a parabolic dish. This means that the gain increases in proportion to the square of the RF since the aperture area is constant. Hence, a 10 to 1 frequency band implies a 20-dB gain variation and a 10 to 1 beamwidth change.

Broadband coverage can also be achieved using horn antennas with internal ridges or a dielectric lens over the mouth of the horn to provide phase correction across the aperture. A typical horn of this type, covering 1 to 11 GHz and having an aperture of about 5×7 inches, has gain ranging from 2 to 20 dBi. Figure 5.3 shows the pattern expected from such a horn antenna.

These directional antennas, when used for angular search, can be rotated at much higher rates than their search radar counterparts. First, their broader beamwidth means that a given dwell time in a given direction can be achieved with a higher rotation rate. Second, the dwell time in a given direction can probably be less than that used by the radar for the simple reason that there is often a much higher power density at the intercept antenna compared to the weak target echo at the radar antenna. For this reason, antenna scan rates of hundreds of revolutions per minute may be used by the intercept system designer, whereas the radar designer (for search systems) typically chooses rotation rates on the order of 10 rpm (or even less).

The natural increase of antenna gain with frequency serves a useful purpose in matching the intercept receiver to the signal environment. The one-way range equation gives the signal power available at the ELINT receiver as

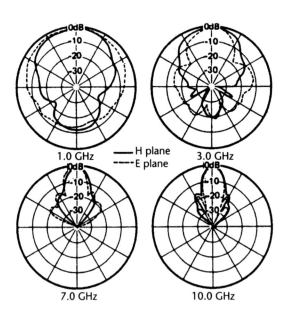

Figure 5.3 Antenna patterns for a broadband horn antenna.

$$S_E = \frac{P_T G_{TE} G_E \lambda^2}{(4\pi)^2 R_E^2 L_T L_E} \tag{5.6}$$

where:

S_E = signal level at ELINT receiver (W)

G_{TE} = gain of radar transmitting antenna in the direction of the ELINT receiver

G_E = gain of the ELINT antenna

R_E = range from radar to ELINT receiver

L_E = loss from ELINT antenna to receiver

P_T = radar transmitter power (W)

λ = wavelength

L_R = radar transmission losses

If the aperture of the ELINT antenna is fixed, G_E is given by

$$G_E = \frac{4\pi A_E}{\lambda^2} \tag{5.7}$$

Combining (5.6) and (5.7) gives

$$S = \frac{P_T G_{TE} A_E}{(4\pi) R_E^2 L_T L_E} \tag{5.8}$$

Thus, the frequency (wavelength) dependence of the received power level is effectively eliminated. There is still a tendency for a reduced received power level at higher frequencies because these frequencies are generally used for shorter range (and hence lower ERP) radar systems.

The higher gain of broadband antenna systems at higher frequencies affects the probability of intercept since the window duration at a given search (scan) rate is reduced in proportion to the reduction in the beamwidth.

The discussion of antenna beam patterns in Chapters 8 and 10 concerning radar antennas applies equally well for ELINT antennas, as does the polarization discussion in Chapter 8. In particular, ELINT antennas should be equipped to mechanically rotate or to switch polarizations, both to aid in determining the polarization of the intercepted signals and maximize the received signal level.

Sidelobe levels are not as critical to intercept system designers as they are to radar system designers. The ELINT station can be sure that a particular signal is entering only the main beam by attenuating the received signal until it can be received only at a particular angle. Likewise, the typical ELINT system is not subject to jamming, spoofing, or angle deception signals, all of which cause radar designers to try reducing the sidelobe levels. The combination of an omni-directional antenna whose gain is above the sidelobe levels of the directional antenna can be used to eliminate errors due to sidelobe reception. Of course, the ELINT antenna should be located so that there are no severe nulls or lobes caused by the antenna's surroundings (e.g., adjacent antennas, masts, buildings, and so on).

The use of radomes to cover the ELINT antenna apertures can significantly affect performance. The fact that broadband radomes are needed for ELINT systems means that radome loss can vary significantly across the band. Such losses and gain variations must be thoroughly documented if accurate measurements of the received power level are to be made. As noted in Chapter 7, estimating the radar's ERP is a part of the interceptor's task, which requires very good calibration of the antenna gain versus frequency, as well as accurate pointing of the antenna toward the emitter of interest. As shown by (5.6), if S_E is measured, and if the ELINT antenna gain and system losses are known, and if the range to the emitter is known, the radar's ERP can be determined.

5.3   Direction Finding

Determining the direction or angle of arrival of a radar signal is fundamental to electronic intelligence. The AOA is particularly important in an era of radar parameter modulation and random variation: the AOA is not subject to control by the radar designer. Direction finding is an extensive subject in its own right. The brief discussion here points out the fundamental ideas and principles involved. Also, some examples are provided.

Historically, direction finding goes back to the early 1900s when long-range communication in the HF band was already possible. Even earlier, Marconi had used copper parabolic mirrors to increase range to 2 miles. Before 1893, Hertz used cylindrical parabolic mirrors to focus energy from his transmitter. (These early experiments actually used a frequency of about 200 MHz. Later, the HF band was used because of the long ranges made possible by ionospheric propagation [5].) Since the majority of ELINT signals are in the microwave region of the spectrum, the large body of knowledge concerning direction finding for HF signals is of limited interest. The most common technique from the ELINT viewpoint uses a rapidly spinning reflector, with "instantaneous" direction finding techniques used in more modern systems.

A third approach, using time difference of arrival over a relatively long baseline, may also be used (for emitter location purposes). This is included in Chapter 6 because the role of the antenna in this process is not crucial.

The traditional direction finding unit used in the past on U.S. ships is the AS-899F, a part of the WLR-1 receiving system [6]. The key specifications of this antenna are given in Table 5.2. Figure 5.4 is a photograph of the antenna system with the radome removed. Note that only the reflector rotates. It is mounted on a slant above the antenna feed, which is aimed vertically at the reflector. This avoids the need for rotary joints. The window functions (see Chapter 4) that such a system can achieve at the highest scan rate of 300-rpm range from a window duration of 22 ms at the low end of the band to 2.2 ms at the high end of the band (40° beamwidth at 1 GHz, 4° beamwidth at 20 GHz), with a period of 200 ms. The 200 ms required for 360° rotation is quite fast compared to the scan of

Table 5.2 Specifications of a Typical Direction Finding Antenna for Electronic Intelligence Applications

ANTENNA AS-899F /SLR

TYPE/USE: The AS-899F is a rotating, very broadband microwave direction finding antenna designed for use with such counter measures receiving equipment as the AN/WLR-I where it is responsive to signals in bands 6, 7, 8, 9, and 10 (1 to 20 GHz). The antenna elements consist of a single, stationary spiral feed, a multiplexer, and a rotating, shaped reflector. (Earlier models of the AS-8OO employed a more complex system of four feed horns, two mounted coaxially at the top center and two coaxially centered at the bottom, with a dual-surface parabolic reflector oriented at 45° in between. These antennas, which were limited in frequency response to bands 6, 7, 8, and 9 of the WLR-I, were subsequently modified to be equivalent to the extended range AS-899F.) The antenna is capable of being operated in an automatic, continuous rotation mode with speeds variable from 0 to 250 rpm, or manually positioned to any desired bearing. To protect it from a sea environment, the antenna is enclosed within a radome.

DIMENSIONS: Height: 82 inches (with radome); Diameter: 34 inches (radome); Weight: 268 pounds (total).

FREQUENCY RANGE: 1 to 20 GHz.

INPUTIMPEDANCEIVSWR: 50 ohms. VSWR not greater than 4:1 over entire frequency range.

RF POWER RATING: N/A (receive only).

POLARIZATION: Right-hand circular or any sense of linear polarization.

TYPE OF FEED: Unbalanced, three coaxial cables and one waveguide for RF.

PRIMARY POWER REQUIRED: Servo power and control voltages supplied from control unit; 115 vac, 60 Hz, one-phase required for space heaters.

INPUT CONNECTOR: Coaxial Type N. (Waveguide WR-62 for highest band.)

BEAMWIDTH: 40° on lowest band to 4° on highest band.

AZIMUTH SCAN: 360° mechanical.

SCAN RATE: 300 rpm maximum, CW or CCW.

INSTALLATION REQUIREMENTS: The antenna assembly should be mounted as high as possible on the ship's superstructure in order to obtain maximum omni-directional coverage. The assembly base casting has four 1-inch bolt holes on 19.25-inch centers for mounting purposes. A clearance of 12 inches is required beneath the base for cover removal.

SPECIAL CONSIDERATIONS: In order to reduce the possibility of receiver overload damage, the AS-899F /SLR should not be located in the strong fields of transmitting antennas.

REFERENCES: Technical Manual for Antenna Assembly AS899E/SLR and AS-899F/SLR, NAVELEX 0967-550.3010.
Source: [6].

most radars. Thus, the DF antenna rotation essentially measures the radiation from such a radar as if the radar antenna had stopped its scan. As a result, if the receiver sensitivity allows receiving the radar's sidelobes, an operator can determine direction of arrival at all times by noting the angle at which the strongest signal occurs, whether or not the main beam of the radar scan coincides with the scanning of the DF antenna. (Of course, the idea of a snapshot of the radar signal while its antenna has stopped scanning is only approximately true. Since the radar's own antenna pattern will have a complex structure of sidelobes and nulls, the signal level will also be changing during the 200-ms rotation period of the DF antenna.)

The problem of signal sorting is handled in the normal ways at the receiver (i.e., either a narrowband receiver is used or time domain PRI gating is used, or the two may be combined). Also, an omni-directional antenna can be used to eliminate pulses arriving through the sidelobes of the DF antenna. The DF display

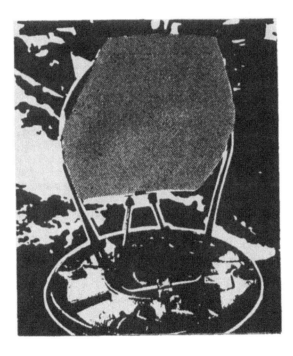

Figure 5.4 Photograph of rotating DF antenna (radome removed).

can consist of a polar plot of the signal amplitude versus the bearing angle to which the spinning reflector is aimed at each instant. A particular signal may be studied by stopping the antenna scan. Its beam can then be pointed toward a particular bearing. Also, a sector search mode can be used to make it easier to read the absolute bearing angle from an expanded angular scale.

Notice that the specifications in Table 5.2 include the capability for right-hand circular polarization and any linear polarization. Clearly, the polarization of each signal will need to be investigated along with its direction.

5.4 Instantaneous Direction Finding

Rapid rotation of high gain antenna apertures is possible in ground and ship-borne receiving systems. However, this type of direction finder is not suited for use on aircraft since it occupies too much space. In addition, the complexity of the radar environment may make it very important to be able to measure the AOA of a single radar pulse.[3]

The AOA of a single pulse can be measured in several ways. Multilobe amplitude comparison makes use of antennas with their beam peaks (boresights) aimed in different directions (squinted). The amplitude is measured independently in each antenna, and the AOA is calculated as being between the boresight angles of the

---

3.   The main drawback to all instantaneous AOA techniques is the error caused if more than one signal is present during the measurement.

pair of antennas having the largest amplitudes. The actual angle is computed based on the ratio of the amplitudes.

Another technique is to use phase comparison. This is called an interferometer. One advantage to the interferometer is that the use of nondirectional antennas means that off-axis AOA measurements do not suffer from a reduced SNR as they do in amplitude comparison system. In practice, similar broad beam antennas are often used for either amplitude comparison or an interferometer, and thus the interferometer may also suffer an SNR reduction for off-axis AOAs.

5.4.1 Amplitude Comparison AOA Measurement

The multilobe amplitude comparison system concept is shown in Figure 5.5(a) for four receiver channels. (Any number of antennas can be used, but four or eight are common choices for 360° azimuth coverage.) The individual channel amplitudes are measured, as are the difference shown in Figure 5.5(a) to allow selection of the channel pair having the largest and next-to-largest amplitudes. Figure 5.5(b)

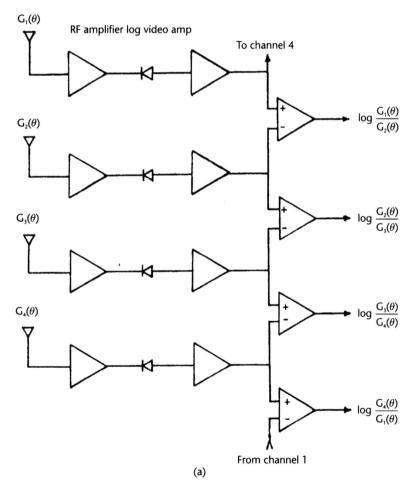

Figure 5.5 (a) Four-channel multilobe AOA system concept: (b) antenna patterns superimposed; and (c) amplitude comparison system angles defined.

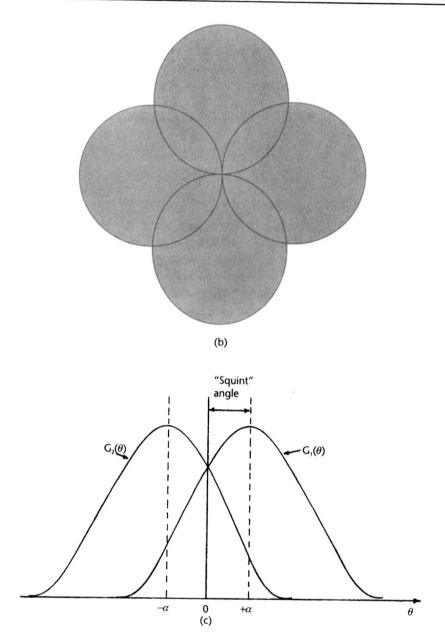

Figure 5.5 (Continued.)

shows the type of antenna patterns needed, with the boresight direction of each antenna offset by 90°. Figure 5.5(c) shows one pair of patterns with definitions for the angles needed in the subsequent analysis.

The angular error in the AOA is a function of the slope of the ratio of the antenna patterns to the AOA. Assuming channels 1 and 2 are selected, we have

$$R(\theta) = \frac{G_1(\theta)}{G_2(\theta)} \tag{5.9}$$

or, in decibels,

$$R_{dB} = G_{1\,dB} - G_{2\,dB} \tag{5.10}$$

The variation in the measured AOA is related to the variation in the ratio measurement by

$$\sigma_\theta = \frac{\sigma_{R_{dB}}}{\partial R_{dB}/\partial\theta} \tag{5.11}$$

where:

σ_θ = rms bearing or AOA error (degrees)

σ_R, dB = rms error in measuring R_{dB} (decibels)

$\partial R_{dB}/\partial\theta$ = rate of change of R_{dB} with respect to the AOA (decibel/degree)

To relate bearing errors to amplitude errors, it is convenient to assume (voltage) antenna patterns that are Gaussian in shape [7, 8]; that is,

$$G_1(\theta) = A_1^2 \exp\left[-\frac{k^2(\theta - \alpha)^2}{\theta_B^2}\right]$$

$$G_2(\theta) = A_2^2 \exp\left[-\frac{k^2(\theta + \alpha)^2}{\theta_B^2}\right]$$

where:

G_1 and G_2 = offset antenna gain patterns

A_2 and A_1 = square root of antenna gain at peak or boresight (relative to isotropic)

α = squint angle, (i.e., one-half of the angle between the boresight directions of the two antennas)

θ_B = 3-dB beamwidth

k^2 = (2 ln 4), so that $G(\theta_B) = 0.5$ (so that the crossover point of the patterns occurs at the −3-dB point)

While the assumptions that the antenna patterns are Gaussian is clearly not realistic for the sidelobe regions of the patterns, the interest in this case is in the main beam patterns not many decibels down from the peak. In this case, the Gaussian assumption is not too far from reality.

Then, R_{dB} is given by

$$R_{dB} = 10 \log \frac{G_1(\theta)}{G_2(\theta)} \tag{5.12}$$

$$= 20 \log \frac{A_1}{A_2} - 10 \, (\log e) \left(\frac{k^2}{\theta_B^2}\right) [(\theta + \alpha)^2 - (\theta + \alpha)^2]$$

The partial derivative is then

$$\frac{\partial R_{dB}}{\partial \theta} = 10 \, (\log e) \left(\frac{4k^2 \alpha}{\theta_B^2}\right)$$

$$= 10 \, (\log e)(8 \ln 4) \left(\frac{\alpha}{\theta_B^2}\right)$$

For the case where $\alpha = 0.5\theta_B$, we have

$$\frac{\partial R_{dB}}{\partial \theta} = +10 \, (\log e)(4 \ln 4)/\theta_B$$

$$\approx 24 \text{ dB/beamwidth}$$

The rms bearing can now be related to the SNR using (5.1). Suppose the amplitude in each channel is subject to a measurement error ΔA due to noise. In (5.12), the term $\log (A_1/A_2)$ does not affect the partial derivative. However, it contains the errors due to noise. Even with perfectly matched channels (or a technique to correct for any mismatch), thermal noise causes the measured values of the channel outputs to fluctuate. The change in the ratio caused by amplitude variations due to noise is

$$\Delta R_{dB} = (20 \log e) \left(\frac{\Delta A_1}{A_1} - \frac{\Delta A_2}{A_2}\right)$$

Since A_1 and A_2 are the signal voltages at the peaks of the antenna patterns in channels 1 and 2, respectively, the error in R_{dB} is related to the SNR in each channel.

$$\frac{\Delta A_1}{A_1} = \frac{1}{\sqrt{2\,SNR_1}} \quad \text{and} \quad \frac{\Delta A_2}{A_2} = \frac{1}{\sqrt{2\,SNR_2}} \tag{5.13}$$

Therefore, assuming the noise in the two channels is uncorrelated,

$$\sigma_{R_{dB}} = 20 \, (\log e) \left(\frac{1}{\sqrt{2\,SNR_1}} + \frac{1}{\sqrt{2\,SNR_2}}\right) \tag{5.14}$$

Combining (5.14) with (5.11) gives

$$\sigma_\theta = \frac{20 \log e \left(\dfrac{1}{2\,\mathrm{SNR}_1} + \dfrac{1}{2\,\mathrm{SNR}_2} \right)^{1/2}}{\partial R_{\mathrm{dB}}/\partial \theta} \tag{5.15}$$

For the Gaussian antenna patterns, the SNR is a function of the angle θ. At the crossover point where $G_2(\theta) = G_1(\theta) = 0.5$, the SNR in each channel is the same. If the SNR at the peak of either beam is SNR_0, then

$$\sigma_\alpha = \frac{20\,(\log e) \left(\dfrac{2}{\mathrm{SNR}_0} \right)^{1/2} \theta_B}{10\,(\log e)\,[8\,\ln(4)]} \ \text{degrees}$$

or

$$\sigma_\alpha = 0.510 \left(\frac{\theta_B}{\mathrm{SNR}_0^{1/2}} \right) \ \text{degrees}$$

For $\theta_B = 90°$ and $\mathrm{SNR}_0 = 13$ dB, $\sigma_\alpha = 10.26°$.

Substituting the expression for the SNR, as a function of angle into (5.15) along with the value of $\partial R_{\mathrm{dB}}/\partial \theta$ from (5.13), allows a calculation of the rms bearing error as a function of the squint angle. For Gaussian patterns, the optimum squint angle is

$$\alpha = \frac{\theta_B}{2\sqrt{\ln 2}} = 0.6\,\theta_B$$

For this case, to allow $\alpha = 45°$ (90° between antenna boresights), the 3-dB beamwidth can be approximately 75°. For an eight-antenna system, the beamwidth can be cut in half, which would also reduce the angular error due to noise by a factor of two at the cost of twice as many antennas (and possibly twice as many receiver channels).

The performance degrades very slowly for squint angle values other than 0.6 times the beamwidth. Squint angles from 0.4 to 0.8 times the beamwidth give nearly optimum performance for Gaussian antenna patterns.

Errors are also caused by variation in the antenna boresight as a function of frequency. This can be a significant problem in wideband systems.

5.4.2 Phase Interferometers

In phase comparison systems (interferometers), the phase difference in two channels is measured. First consider the interferometer as shown in Figure 5.6. The phase front of the plane wave arriving at an angle θ from the boresight direction must travel the additional distance $1 = d \sin \theta$, where d is the distance separating the two antennas. The resulting phase shift can be expressed as [7]

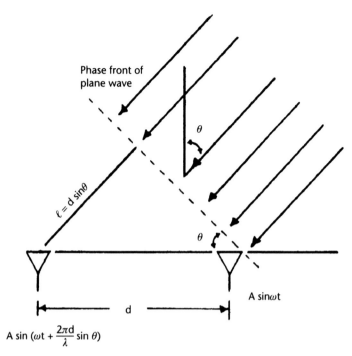

θ is the *angle of arrival* measured from the boresight direction,
which is perpendicular to the base line d

Figure 5.6 Two-channel interferometer.

$$\phi = \frac{2\pi d}{\lambda} \sin \theta = \frac{2\pi d}{c} f \sin \theta \qquad (5.16)$$

The maximum separation for a total phase change of no more than 2π (for unambiguous angle measurement) is half of the wavelength, for which a phase change of $\pm\pi$ occurs over the angular region $\pm90°$. Equation (5.16) shows that errors in the phase measurement are directly proportional to errors in the baseline, d, to errors in the measurement of the carrier frequency, f, as well as to thermal noise.

To take advantage of baselines longer then $\lambda/2$, it is possible to use several baselines, with the shorter ones resolving the angular ambiguities of the longer ones. For example, a four-antenna configuration using baselines of $\lambda/2$, λ, and 2λ is a possibility. The phase shift over the longest baseline can then be up to $\pm4\pi$ and the angular resolution has been improved by a factor of 4. The size of the longest baseline is restricted by the resolving power of the shorter ones. For example, if the $(-\pi$ to $+\pi)$ range of the $\lambda/2$ baseline pair can be resolved into $\pi/8$ sectors, the next longer baseline can be up to 8 times as long as the first. In such a system the problems of matched phase shifts over broad bands in the components becomes a critical problem.

The effect of thermal noise on interferometric bearing measurement can be found with the aid of (5.16) and a derivation of phase variation due to noise. First, note that

$$\delta\theta = \frac{\partial\theta}{\partial\phi}\Delta\phi$$

Using (5.16), the rate of change of the bearing measurement with respect to a change in phase is

$$\frac{\partial\theta}{\partial\phi} = \left(\frac{\partial\phi}{d\theta}\right)^{-1} = \left(\frac{2\pi d \cos\theta}{\lambda}\right)^{-1}$$

A calculation of the rms bearing error σ_θ in terms of the rms phase error yields

$$\sigma_\theta = \sigma_\phi\left(\frac{\lambda}{2\pi d \cos(\theta)}\right) \tag{5.17}$$

The rms phase variation due to noise can be computed using the phasor diagram in Figure 5.7. The quantity of interest is the incremental phase change, $\Delta\phi$, resulting from a noise disturbance, ΔA. Using the law of cosines,

$$\cos\Delta\phi = \frac{|A|^2 + |A^1|^2 - |\Delta A|^2}{2|A||A^1|}$$

For relatively high signal-to-noise ratios, $\Delta\phi$ will be small so that $\cos\Delta\phi \cong \Delta\phi/2$ and $|A^1| \cong |A|$. This yields

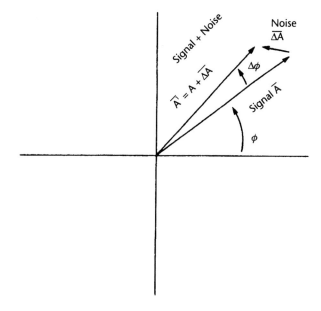

Figure 5.7   Phasor diagram to compute phase variation due to noise.

$$1 - \frac{\Delta\phi^2}{2} \cong 1 - \frac{1}{2}\frac{|\Delta A|^2}{|A|^2}$$

or

$$\Delta\phi^2 \cong \frac{|\Delta A|^2}{|A|^2} \tag{5.18}$$

Since ΔA is the instantaneous noise phasor and A is the signal phasor, the right-hand side of (5.18) averaged over time is the reciprocal of the SNR and the left-hand side is the variance of the phase measurement:

$$\sigma_\phi^2 = \frac{1}{\text{SNR}}$$

Combining this result with (5.17) yields the rms variation in AOA measurement due to noise as

$$\sigma_\theta = \frac{\lambda}{2\pi d \cos\theta \sqrt{\text{SNR}}} \tag{5.19}$$

The implementation of a wideband phase interferometer presents the very difficult task of maintaining knowledge of the phase shift at any frequency of interest in two receiving channels. In practice, this requires injection of test signals at the frequency of interest for calibration purposes or some other technique (such as multiplexing a single receiver among two or more antennas). The variation of phase with component aging, temperature changes, and maintenance operations means that mere calibration across the band for looking up in a table will not allow for AOA measurements of sufficient accuracy in microwave applications.

The need for bearing measurements on a single-pulse basis leads to the consideration of the monopulse angle measurement antenna configurations used in radar systems. Such systems can be used to obtain accurate bearing information while remaining open to reception over broad sectors [9]. In these systems, two or four broadband antennas (e.g., spiral or log periodic) are used, as well as two or three matched receiving channels. A beam-forming network processes the antenna outputs to form one *sum beam* and a *difference beam* in azimuth and a *difference beam* in elevation. The sum beam is used as a reference to which the other beams are compared, and the AOA in azimuth and elevation is computed by comparing the signals from the difference channels to that in the sum channel. Either the signal amplitude or phase (or both) can be used. Achieving broadband operation means that the same phase or amplitude difference must occur at the same angle of arrival. This means that the phase centers of the antenna elements must be the same number of wavelengths apart. This can be accomplished with conical log spiral antennas by placing the outer (high frequency) end closer together than the base, as shown in Figure 5.8.

Figure 5.8 A two-element array of conical log spirals (frequency independent).

The signal processing needed is similar to that used in monopulse radar systems, as shown in Figure 5.9. Typical antenna patterns at the sum and difference ports are shown in Figure 5.10. The antenna spacing is 0.6λ in one case shown and 0.8λ in the other. To achieve frequency independent performance, it is necessary to maintain antenna spacing that is a fixed fraction of a wavelength at over the frequency band to be covered. The variation in the indicated bearing angle with frequency is one of several sources of error in monopulse angular measurement systems. The error in the phase and amplitude of the microwave beam-forming hybrids, the imbalance in the receiver channels, and the misalignment of the mechanical components contribute to the AOA error, as does the available SNR. These effects are summarized in Figure 5.11 [8]. According to the calculations summarized in Figure 5.11, a phase comparison system with a baseline of three wavelengths can achieve an rms accuracy of $0.63°$, which is generally sufficient for ELINT purposes. Note that linear arrays of antennas generally cover a sector of less than $90°$.

An rms bearing accuracy of $2°$ means that 99.7% of the measurements will fall inside a $12°$ sector. This is sufficiently small to allow a jammer to be directed towards the emitter or for launching an antiradiation missile using only the information provided by a single receiver. ELINT, which is generally not concerned with responding in real time, can make use of less accurate AOA measurements if there are a large number of bearing measurements that can collectively narrow the geographic region in which the emitter is located.

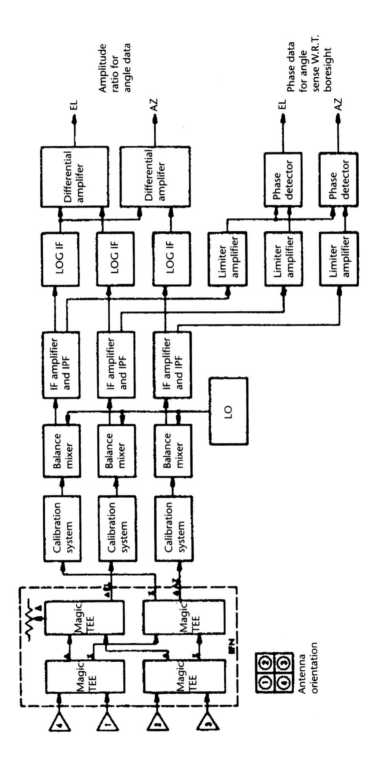

Figure 5.9 Block diagram of three-channel monopulse system. (*From:* [8]. © 1971 IEEE. Reprinted with permission.)

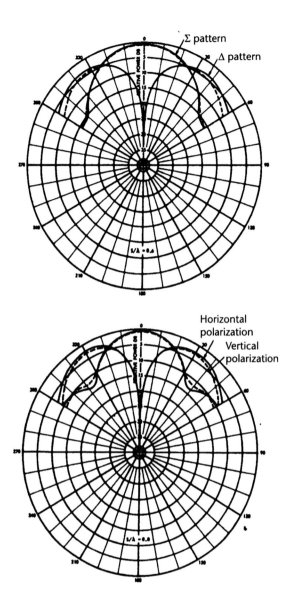

Figure 5.10   Sum and difference beams from DF system for two-element spacing. (*From:* [8].
© 1971 IEEE. Reprinted with permission.)

Multiple Channel Interferometers[4]

Some of the systems described in Figure 5.11 use antennas spaced too far apart
for unambiguous measurement of angles by an interferometer. The desire for greater
accuracy leads to the use of longer baselines. However, the angle indication becomes
ambiguous when the phase shift exceeds ±180°. This leads to the use of additional
antennas and receiving channels. A closely spaced pair of antennas can indicate
the approximate angle, and a larger spacing can indicate the more accurate but

---

4.   The author is grateful to Brian Moore for providing this material.

System Type	2-Channel Amplitude and Phase Comparison			3-Channel Amplitude and Phase Comparison			Amplitude Comparison			Phase Comparison					
										0.85-Wavelength Baseline			3-Wavelength Baseline		
Average Error Gradient Over a ±30° Field of View	0.79 dB/deg			0.85 dB/deg (0.8-wavelength element spacing)			0.44 dB/deg (70° 3-dB BW and 45° element offset)			5.1 deg/deg			18 deg/deg		
Error Parameter	dB	spatial deg	(deg)²	dB	spatial deg	(deg)²	dB	spatial deg	(deg)²	electrical deg	spatial deg	(deg)²	electrical deg	spatial deg	(deg)²
Polarization and Antenna Pattern Errors	–	2.45	6.0	1.3	1.53	2.35	2.2	5.00	25.0	4.0	0.79	0.62	1.3	0.83	0.07
RF Beam-Forming Network Errors	0.5	1.39	1.92	0.5	0.59	0.35	–	–	–	–	–	–	–	–	–
Thermal Noise Error (Signal-to-Noise Ratio of 30 dB and 1 Pulse)	–	1.17	1.36	–	0.75	0.56	–	2.73	7.47	0.05	0.01	0.0001	0.05	0.0028	–
Calibrated Receiver Imbalance Error (Includes 0.25 dB Quantization Error)	1.0	2.76	7.63	1.0	1.18	1.39	1.0	2.27	5.15	10.0	2.5	6.25	10.0	0.55	0.31
Mechanical Alignment Errors	–	0.1	0.01		0.1	0.01	–	0.1	0.01	·	0.1	0.01	–	0.1	0.01
Sum of the Variances			16.92			4.66	–	–	37.63	–	–	6.88	–	–	0.39
Rms (1σ) Bearing	4.12°			2.16°			6.12°			2.62°			0.63°		

Figure 5.11 Comparison of rms bearing accuracy (1σ) for a ±30° field of view. (*From:* [8]. © 1971 IEEE. Reprinted with permission.)

ambiguous angle. The two measurements together resolve the ambiguity. A block diagram for a dual baseline system (three channels) is shown in Figure 5.12.

Performance is expressed in terms of the ratio of the wide baseline, d_w, to the narrow baseline, d_n. This ratio is set to be an integer.

$$m/n = d_w/d_n \qquad m \text{ and } n \text{ are integers}$$

The unambiguous angular field of view for such a system is given by

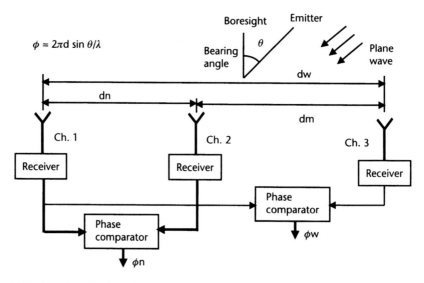

Figure 5.12 Dual baseline interferometer.

$$\text{FOV}_{\text{SYS}} = 2 \arcsin\left(\frac{m\lambda}{2d_w}\right), \text{ for } \lambda < 2d_w/m \tag{5.20}$$

$$= 180° \text{ for } \lambda \geq 2d_w/m$$

or, equivalently,

$$\text{FOV}_{\text{SYS}} = 2 \arcsin\left(\frac{n\lambda}{2d_n}\right), \text{ for } \lambda < 2d_n/n \tag{5.21}$$

$$= 180° \text{ for } \lambda \geq 2d_n/n$$

Of course, the accuracy is determined by the wide baseline, d_w:

$$\sigma_\theta = \frac{\sigma_\varphi \lambda}{2\pi d_w \cos \theta} \tag{5.22}$$

The maximum phase error allowed to properly resolve the ambiguity is given by

$$|\Delta\varphi| < \frac{\pi}{\sqrt{2}} \frac{1}{\sqrt{(n^2 + m^2)}} \text{ radians} \tag{5.23}$$

Some deployed systems make use of four antennas with three different baselines and further divide the band covered into high, medium, and low frequency ranges for a total of 12 antennas.

5.4.3 Bearing Discriminators

The need for 360° azimuth coverage for a direction finding system combined with the need for instantaneous (monopulse) AOA measurement leads to the idea of a circular array of antennas. This is a natural extension of the four-antenna system shown in Figure 5.5(b). We have seen that comparing the phase of the signals from two separated antennas can be used to indicate the AOA of a signal (Figure 5.6) and that the phase difference can be increased in the signal processing (Figure 5.7) to provide an increased electrical phase change for a given angular change if the SNR is sufficient. We have monopulse angle measurements. This sets the stage for a discussion of a circular array of antennas, which can achieve accurate, instantaneous bearing measurements over wide bandwidths for a full 360°. This bearing discriminator makes use of high-resolution phase measurements using the same kind of phase discriminator, which is also the heart of the instantaneous frequency measuring receivers (IFMs). The only drawback of this approach (as with all of the instantaneous AOA techniques) is the error caused by the presence of more than one signal at the time the angle is measured. In this situation, if one signal is much stronger (e.g., >6 dB) than any other, the angle of the strongest signal will be measured; however, with increase error due to the presence of the interfering signal.

To start, consider a phase comparison version of a four-antenna system (similar to Figure 5.6). As shown in Figure 5.13, the output port (labeled $n = 0$) is the

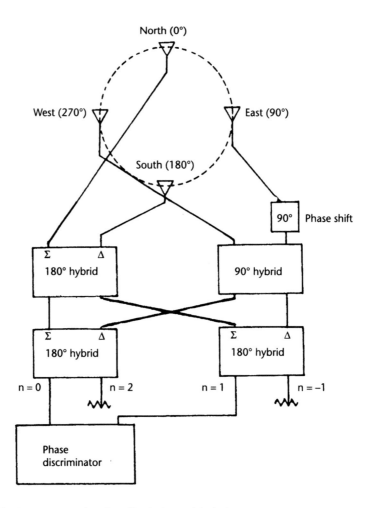

Figure 5.13 Four-antenna bearing discriminator block diagram.

reference signal. The north antenna output passes through the two left 180° hybrids with no phase shift to port zero. The east antenna output passes through the +90° shifter, is shifted −90° through the 90° hybrid, and passes through the bottom left hybrid with no phase shift for a net total of 0°. The south antenna's signal passes through both the left hybrids with no phase shift. The west antenna's signal reaches output port zero by passing through both the 90° hybrid and the bottom left hybrid with no phase shift. Thus, no matter which antenna receives a signal, it reaches the $n = 0$ port with 0° of phase shift. Next consider the phase at the $n = 1$ output port. The signal from the north antenna reaches this port with no phase shift, passing directly through the two left hybrids. The signal from the east antenna reaches the $n = 1$ output port by passing through the 90° phase shift and then directly through the two right hybrids for a net +90° phase shift. The south antenna's signal reaches the $n = 1$ port via 180° phase shift in the top left hybrid and then directly through the bottom right hybrid for a net 180° phase shift. The west antenna's signal reaches the $n = 1$ output port with 90° phase shift through the

90° hybrid and 180° phase shift through the bottom right hybrid for a total of 270°. The results are shown in Table 5.3.

As shown, the phase difference between the $n = 0$ and $n = 1$ ports is equal to the spatial angle around the circular array. Note that only one-fourth of the power from one antenna is available at a given output port due to the power dividing properties of the hybrids. Also note that the phase difference between the $n = 0$ and the $n = 2$ port is exactly twice the phase difference between the $n = 0$ and $n = 1$ ports (modulo 360°).

By measuring the $n = 0$ to $n = 1$ phase difference, the bearing angle is obtained. This is true even if the signal is, say, at a 45° bearing angle. Then the north and east antennas have equal signals and the result is a 45° phase difference at the output (i.e., equal components at 0° and 90° produce a 45° phase shift). As long as the antennas produce equal amplitudes, the proper bearing angle can be found regardless of AOA. If the hybrids and antenna elements are broadband, the AOA measurement can be made over a broad band.

A network with the properties of the one shown in Figure 5.13 is called a Butler matrix [10]. The power at any of N input ports divides equally among N output ports. Furthermore, a signal at one input port results in a signal at each output with a constant phase difference between them. The output ports are numbered 0, ±1, ±2, . . . , up to ±($N - 1$) and N. For a 32-port Butler matrix, the output ports would be numbered 0, ±1, ±2, ±3, . . . , ±14, ±15, ±16. The phase progression at any output would be its number times the spatial angle (bearing) of the signal arriving at a circular array of N antennas [9]. This concept is the heart of a digital bearing discriminator using a circular array of 32 broadband antennas to achieve a broadband bearing accuracy of 2° rms over a full 360° in azimuth and −10° to +40° in elevation [11].

A block diagram of this receiver is shown in Figure 5.14. The Butler matrix outputs 1, 2, 4, and 8 are compared to output 0 using digital phase comparators to achieve a bearing resolution of 8 bits (or 1.4°). This means the net phase resolution required from the port 8 measurement is 11.25 electrical degrees since the phase changes 8 times as much as the spatial angle changes. The bearing accuracy is maintained over the 2- to 18-GHz band down to −55 dBm relative to an isotropic antenna and for pulse duration as short as 200 ns.

The antenna array is a stripline configuration with conical reflectors mounted above and below to shape the beam in elevation. The circular array is shown in Figure 5.15 and the entire unit is shown in Figure 5.16. This unit has two antenna arrays. The upper one covers the 7.5- to 18-GHz band and the lower one covers the 2- to 7.5-GHz band.

Table 5.3 Bearing Discriminator Using Four Antennas

Antenna	Spatial Angle	n = 0	n = 1	n = 2	n = −1
			Electrical Phase Shift at Port		
N	0°	0°	0°	0°	0°
E	90°	0°	90°	180°	270° (−90°)
S	180°	0°	180°	0° (360°)	180° (−180°)
W	270°	0°	270°	180° (540°)	90° (−270°)

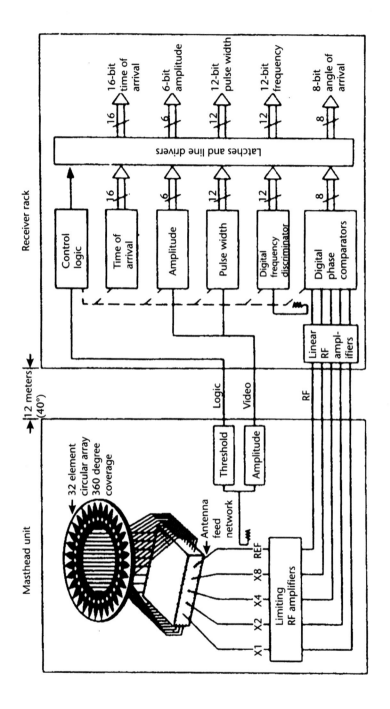

Figure 5.14 Block diagram of digital ESM receiver incorporating a bearing discriminator. (*From:* [9]. Courtesy of Anaren Microwave.)

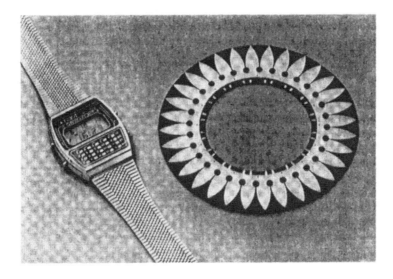

Figure 5.15 Circular antenna array (7.5 to 18 GHz). (*From:* [9]. Courtesy of Anaren Microwave.)

Figure 5.16 Antenna unit (2 to 18 GHz) (radome removed). (*From:* [9]. Courtesy of Anaren Microwave.)

5.5   Arrays, Lenses, and Subspace DF Methods

As the electromagnetic environment grows in complexity, the need for high quality AOA measurements on a single-pulse basis increases. Certainly, single-pulse AOA is critical to automated systems. As the likelihood of multiple received signals overlapping in time increases, the need for finding the AOA for multiple signals also grows. One method for doing this is analogous to the channelized receiver. Instead of multiple channels across the frequency band of interest, multiple antenna beams can be formed to cover the angular region of interest. Then, independent receivers and processors can be used to deal with the signals arriving in each beam. Such a system can be created using a microwave lens. An example is shown in Figure 5.17.

The use of an array of antenna elements includes the amplitude comparison and phase comparison (interferometer) systems described above. In the general case, there are many antenna elements and their outputs are combined in various ways. One way is to form beams indifferent directions, as is done in the system shown in Figure 5.17. A general way of looking at this is shown in Figure 5.18 [13]. Here the array outputs (arrayports) are shown feeding into a Lens Transfer Matrix to produce the lens outputs (lensports). These are processed by CVRs to provide an output of the power from each lensport. These outputs are the beamports. The Butler matrix is one example of a lens transfer matrix. Another is the beamformer used to form the beams in the multiple beam array above.

The use coherent processing of the arrayport outputs is the basis of subspace methods such as the Multiple Signal Classification (MUSIC) algorithm. In this algorithm, it is necessary to calibrate the system using a variety of angles of arrival to form what is called the array manifold. Using the eigenvalues and eigenvectors obtained from the data correlation matrix, one can express the signals in terms of the larger eigenvalues and their eigenvectors, as indicated in Figure 5.19.

An example of the performance of the MUSIC algorithm is shown in Figure 5.20. In [13] a method is derived for using the outputs of the CVRs in Figure 5.17 to obtain the correlation matrix and hence accurate AOA values. The method depends on ergodic signals (time average the same as the ensemble average).

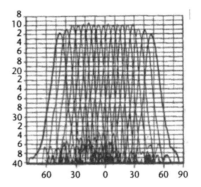

Figure 5.17   Lens fed constant beamwidth array antenna. (*From:* [12]. © 1984 Horizon House, Inc. Reprinted with permission.)

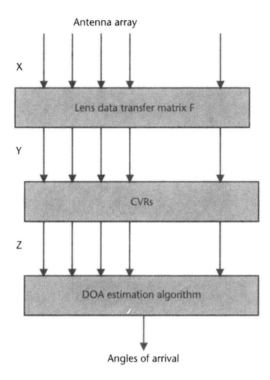

Figure 5.18 Lens-based array system with CVRs. (*After:* [12, 13].)

5.6   Short Baseline TDOA for AOA

Although time difference of arrival (TDOA) methods are usually thought of as long baseline systems with independent platforms (see Chapter 6), it is also possible to obtain AOA information by measuring the time difference instead of the phase difference. (Refer to Figure 5.6.) The extra distance traveled by the signal is $d \sin \theta$. At the speed of light, this produces a time difference of $\tau = d\,(\sin \theta)/c$. Then the AOA can be found from

$$\theta = \arcsin(\tau \cdot c/d) \text{ radians} \tag{5.24}$$

The error in AOA due to an error in time difference is given by the derivative

$$\Delta\theta = \frac{c[1 - \sin^2(\theta)]^{-1/2}}{d} \Delta\tau \text{ radians} \tag{5.25}$$

For a time difference error of 1 ns and for a 30° angle of arrival and $d = 3$m, the AOA error is 6.6°. It is clear that very precise time measurements are required, which in turn, requires a high SNR and/or a pulse with a steep rise time. Of course, increasing the baseline improves the accuracy directly. For 30-m baseline, the AOA accuracy is 0.66°. There is no need to measure the frequency or wavelength with this technique.

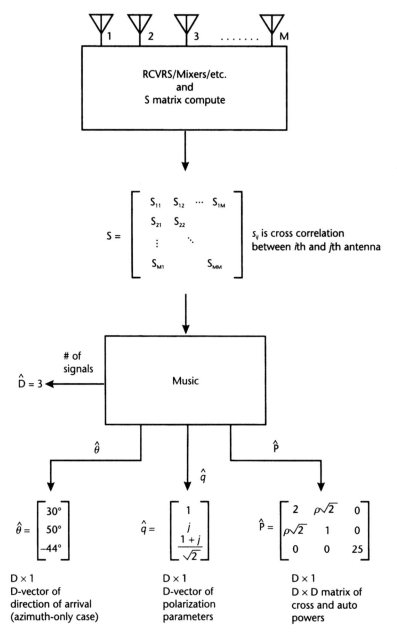

Figure 5.19 MUSIC block diagram. (*From:* [14]. © 1986 IEEE. Reprinted with permission.)

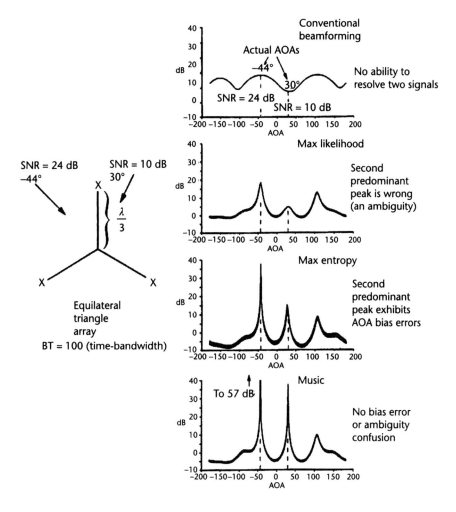

Figure 5.20 Example of MUSIC algorithm performance.

References

[1] Braun, A. E., "RWRs Face New Threats," *MSN*, November 1983, p. 53.

[2] Kraus, J. D., *Antennas*, New York: McGraw-Hill, 1950.

[3] Ames, J. W., and W. A. Edson, "Gain, Capture Area, and Transmission Loss for Grounded Monopulse and Elevated Dipoles," *RF Design*, November/December 1983.

[4] Bahl, I. J., and P. Bhartia, *Microstrip Antennas*, Dedham, MA: Artech House, 1980.

[5] Boyd, J. A., et al., *Electronic Countermeasures*, Los Altos, CA: Peninsula Publishing, 1978, Chapter 5; originally published in 1961 as a classified book.

[6] Law, P. E., *Shipboard Antennas*, Dedham, MA: Artech House, 1983, p. 426.

[7] Tsui, J. B., *Microwave Receivers and Related Components*, Springfield, VA: National Technical Information Service, 1983, (PB84-1O8711), pp. 33–37.

[8] Bullock, L. G., G. R. Och, and J. J. Sparagna, "An Analysis of Wideband Microwave Monopulse Direction Finding Techniques," *IEEE Trans. on Aerospace and Electronic Systems*, Vol. AES-7, No. 1, January 1971, pp. 188–203.

[9] Anaren Microwave, Inc., Publication MI826-I8, East Syracuse, New York.

[10] Moody, H. J., "The Systematic Design of the Butler Matrix," *IEEE Trans. on Antennas and Propagation*, November 1964, pp. 786–788.

[11] Rehnmark, S., "2–18 GHz Digital ESM Receivers with 2 Degree RMS Bearing Accuracy," *Military Microwaves '82*, London, 1982.

[12] Archer, D. H., "Lens Fed Multiple Beam Arrays," *Microwave Journal*, September 1984, p. 194.

[13] Diggavi, D., J. Shynk, and A. Laub, "Direction-of-Arrival Estimation for a Lens Based Array," *IEEE Trans. on Antennas and Propagation*, Vol. 42, No. 5, May 1994.

[14] Schmidt, R. O., "Multiple Emitter Location and Signal Parameter Estimation," *Proc. RADC Spectrum Estimation Workshop*, Griffiss Air Force Base, New York, 1979, pp. 243–258. Reprinted in *IEEE Trans. Antennas and Propagation*, Vol. AP-34, March 1986, pp. 276–280.

CHAPTER 6

Emitter Location

Charles Estrella

6.1 Introduction

The capability of performing accurate emitter location has become a primary requirement for today's reconnaissance platforms. In practice, one or more platforms can be used to perform geolocation using emitter bearing, time, and/or frequency information. A single platform can use pulse angle of arrival information to form lines of bearing (LOBs) along which the emitter lies. The intersections of the LOBs provide a location estimate, as shown in Figure 6.1.

With multiple platforms, measurements of an emitter's time of arrival from one platform can be used with TOA measurements from another to compute a time difference of arrival. Each TDOA forms a hyperbola, or isochrone, upon the surface of the Earth. An intersection of TDOA isochrones provides a potential location of the emitter. An example of isochrones for various values of TDOA between three aircraft is shown in Figure 6.2.

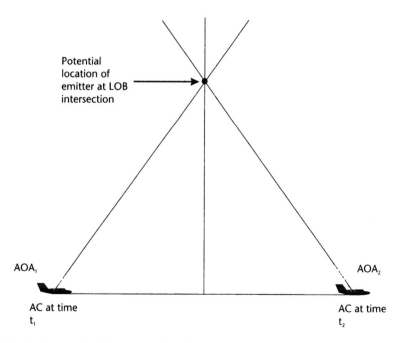

Figure 6.1 Single aircraft AOA using emitter bearing measurements.

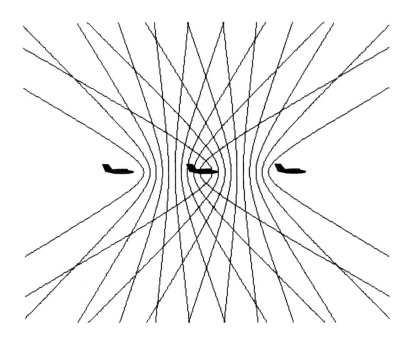

Figure 6.2 Various isochrone lines; three aircraft TDOA.

Frequency difference of arrival (FDOA) measurements can be made in conjunction with TDOA measurements by determining the peak of the complex ambiguity function (CAF) between two intercepts from moving platforms [1]. In this case, the intersection of isochrones and iso-Doppler lines as shown in Figure 6.3 provide location estimates of the emitter.

6.2 Emitter Location Estimation

Because of uncertainties in TDOA, FDOA, and AOA measurements, uncertainties exist in the estimated emitter location. An estimate of uncertainties in emitter location (x_{Em}, y_{Em}) and the associated location covariance matrix $C_{x_{Em}y_{Em}}$ are ultimately desired in geolocation analysis. The location covariance matrix $C_{x_{Em}y_{Em}}$ provides uncertainties in terms of the variance $\sigma^2_{x_{Em}}$, $\sigma^2_{y_{Em}}$ and correlation $\rho_{x_{Em}}$, $\rho_{y_{Em}}$ between emitter x-y location coordinates, and is given as

$$C_{x_{Em}y_{Em}} = \begin{bmatrix} \sigma^2_{x_{Em}} & \rho_{x_{Em}y_{Em}}\sigma_{x_{Em}}\sigma_{y_{Em}} \\ \rho_{y_{Em}x_{Em}}\sigma_{y_{Em}}\sigma_{x_{Em}} & \sigma^2_{y_{Em}} \end{bmatrix} \quad (6.1)$$

For a given emitter location, the resulting eigenvalues (λ_1, λ_2) of the computed location covariance matrix are used to compute the semimajor/minor axes (a, b) of an uncertainty ellipse:

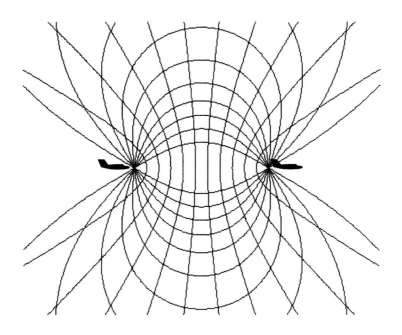

Figure 6.3   Isochrone/iso-Doppler lines; two aircraft time/frequency difference of arrival.

$$a = \sqrt{\lambda_1} \atop b = \sqrt{\lambda_2} \Big\}_{\lambda_1 > \lambda_2} \tag{6.2}$$

The uncertainty ellipse has an orientation angle $\theta_{x_{Em}y_{Em}}$ obtained from the associated location covariance matrix computed as

$$\theta_{x_{Em}y_{Em}} = \frac{1}{2} \cdot \tan^{-1}\left(\frac{\sigma_{x_{Em}}^2 - \sigma_{y_{Em}}^2}{2 \cdot \rho_{x_{Em}y_{Em}} \sigma_{x_{Em}} \sigma_{y_{Em}}}\right) \tag{6.3}$$

For two dimensions, the circular error probable (CEP) is often used as a measure of accuracy, defined to contain a percentage of the values within the error ellipse. A CEP calculation containing 50% of values within the location ellipse can be estimated from [2] using

$$CEP = 0.75 \cdot \sqrt{\lambda_1 + \lambda_2} \tag{6.4}$$

An example error ellipse and associated CEP is shown in Figure 6.4.

This ellipse is typically centered at the emitter's true location (x_{Em}, y_{Em}) when uncertainties in TDOA, FDOA, and AOA measurements are unbiased; however, the relative position of the error ellipse can incur a biased total location error (TLE) even when the individual non-signal-dependent errors are unbiased. A

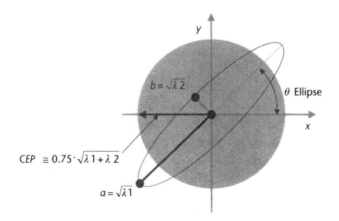

Figure 6.4 Location ellipse example.

significant TLE can occur because of the nonlinear relationship between the error sources and the resulting location covariance matrix. Examples for the AOA case are given in Section 6.4.

6.3 Deriving the Location Covariance Matrix

The analysis presented herein derives theoretical expressions for the covariance matrix $C_{x_{Em}y_{Em}}$ in terms of various errors (e.g., collector position, heading, velocity). These errors are treated as zero-mean random variables. Given two functions f_1 and f_2 consisting of zero-mean random variables X and Y, having variances $[\sigma_X^2, \sigma_Y^2]$ and correlation ρ_{XY}, the resulting uncertainty in f_1 and f_2 can be estimated using

$$\text{Cov}_{f_1 f_2} = J_{\underset{xy}{f_1 f_2}} \cdot \text{Cov}_{xy} \cdot J_{\underset{xy}{f_1 f_2}}^T \tag{6.5}$$

where

$$\text{Cov}_{f_1 f_2} = \begin{bmatrix} \sigma_{f_1}^2 & \rho_{f_1 f_2} \sigma_{f_1} \sigma_{f_2} \\ \rho_{f_2 f_1} \sigma_{f_2} \sigma_{f_1} & \sigma_{f_2}^2 \end{bmatrix}$$

$$J_{\underset{xy}{f_1 f_2}} = \begin{bmatrix} \dfrac{\partial f_1}{\partial x} & \dfrac{\partial f_1}{\partial y} \\ \dfrac{\partial f_2}{\partial x} & \dfrac{\partial f_2}{\partial y} \end{bmatrix}$$

$$\text{Cov}_{xy} = \begin{bmatrix} \sigma_x^2 & \rho_{xy} \sigma_x \sigma_y \\ \rho_{yx} \sigma_y \sigma_x & \sigma_y^2 \end{bmatrix}$$

In employing (6.5), the relationship between the uncertainties in collector position, heading, velocity, and emitter AOA and the resulting overall uncertainty in AOA, TDOA, and FDOA uncertainty is assumed to be linear. Because in practice these relationships are not linear, the expressions can only provide an approximation of the overall location accuracy. To validate the derived expressions using the approach above, Monte Carlo simulation is used to generate results for comparison and verification. The resulting analysis can then be used to define platform geodetic requirements and used to estimate overall location performance when combined with signal dependent errors. Derivations for the associated location covariance matrix using AOA, TDOA, and TDOA/FDOA location methods are presented in the following sections.

6.4 Angle of Arrival Location Analysis

To derive the associated location covariance matrix $C_{x_{Em}y_{Em}}$ for single aircraft AOA location in terms of nonsignal-dependent errors in aircraft position and estimated emitter AOA, the expression for AOA for a stationary emitter at (x_{Em}, y_{Em}) and an aircraft located at (x_n, y_n) is used:

$$AOA_n = \tan^{-1}\left(\frac{x_{Em} - x_n}{y_{Em} - y_n}\right) \tag{6.6}$$

The emitter position (x_{Em}, y_{Em}) is computed from two AOA measurements (also known as LOBs) from aircraft positions (x_1, y_1), (x_2, y_2) as

$$x_{Em} = \frac{x_1 - x_2 \cdot \dfrac{\tan(AOA_1)}{\tan(AOA_2)} - y_1 \cdot \tan(AOA_1) + y_2 \cdot \tan(AOA_1)}{1 - \dfrac{\tan(AOA_1)}{\tan(AOA_2)}} \tag{6.7}$$

$$y_{Em} = \frac{x_1 - x_2 \cdot \dfrac{\tan(AOA_1)}{\tan(AOA_2)} - y_1 \cdot \tan(AOA_1) + y_2 \cdot \tan(AOA_1)}{\tan(AOA_2) - \tan(AOA_1)} - \frac{x_2}{\tan(AOA_2)} + y_2 \tag{6.8}$$

Treating each uncertainty in aircraft position as independent random variables $\Delta x_n, \Delta y_n$ having zero mean with standard deviation $\sigma_{x_n}, \sigma_{y_n}$, the associated aircraft position covariance matrix can be written as

$$C_{x_n y_n} = \begin{bmatrix} \sigma_{x_n}^2 & \overbrace{\rho_{x_n y_n} \sigma_{x_n} \sigma_{y_n}}^{0} \\ \underbrace{\rho_{y_n x_n} \sigma_{y_n} \sigma_{x_n}}_{0} & \sigma_{y_n}^2 \end{bmatrix} \tag{6.9}$$

As with aircraft position, uncertainty in measured AOA is treated as independent random variables ΔAOA_n having zero mean with standard deviation σ_{AOA_n}. The associated AOA covariance matrix $C_{AOA_{12}}$ can be written as

$$C_{AOA_{12}} = \begin{bmatrix} \sigma^2_{AOA_1} & \overbrace{\rho_{AOA_1 AOA_2} \sigma_{AOA_1} \sigma_{AOA_2}}^{0} \\ \underbrace{\rho_{AOA_2 AOA_1} \sigma_{AOA_2} \sigma_{AOA_1}}_{0} & \sigma^2_{AOA_2} \end{bmatrix} \quad (6.10)$$

The emitter position covariance matrix $C_{x_{Em} y_{Em}}$ due to uncertainties in aircraft positions (x_1, y_1), (x_2, y_2) and respective emitter angles of arrival (AOA_1, AOA_2) can be computed as

$$C_{x_{Em} y_{Em}} = J_{x_{Em} y_{Em} | x_n y_n} C_{x_n y_n} J^T_{x_{Em} y_{Em} | x_n y_n} + J_{x_{Em} y_{Em} | AOA_1 AOA_2} C_{AOA_1 AOA_2} J^T_{x_{Em} y_{Em} AOA_1 AOA_2} \quad (6.11)$$

The variables $J_{x_{Em} y_{Em} | x_1 x_2 y_1 y_2}$ and $J_{x_{Em} y_{Em} | AOA_1 AOA_2}$ are the partial derivatives of the location estimate with respect to the aircraft position and AOA, given as

$$J_{x_{Em} y_{Em} | x_1 x_2 y_1 y_2} = \begin{bmatrix} \dfrac{\partial x_{Em}}{\partial x_1} & \dfrac{\partial x_{Em}}{\partial x_2} & \dfrac{\partial x_{Em}}{\partial y_1} & \dfrac{\partial x_{Em}}{\partial y_2} \\ \dfrac{\partial y_{Em}}{\partial x_1} & \dfrac{\partial y_{Em}}{\partial x_2} & \dfrac{\partial y_{Em}}{\partial y_1} & \dfrac{\partial y_{Em}}{\partial y_2} \end{bmatrix} \quad (6.12)$$

$$J_{x_{Em} y_{Em} | AOA_1 AOA_2} = \begin{bmatrix} \dfrac{\partial x_{Em}}{\partial AOA_1} & \dfrac{\partial x_{Em}}{\partial AOA_2} \\ \dfrac{\partial y_{Em}}{\partial AOA_1} & \dfrac{\partial y_{Em}}{\partial AOA_2} \end{bmatrix} \quad (6.13)$$

An example is given in Figure 6.5 to compare AOA error ellipse parameters generated via Monte Carlo simulation to those computed using (6.11) for an emitter located at [0, 150]. An AOA and position error of 1° and 1m, respectively, were used for aircraft positioned at [−100, 0] and [100, 0].

The results for this case have negligible bias; however, when the AOA error is increased to 10°, results given in Figure 6.6 show that the bias becomes significant.

The resulting bias in the error estimate illustrates the limitations in assuming a linear relationship between the nonsignal-dependent error sources and their resulting impact on the geolocation estimate. A more thorough investigation of the bias in the estimate as a function of collection geometry is provided in [3], using multiple LOBs computed and combined over time. Work done by Brown provides three algorithms for the combining of LOBs along a flight track using distance least squares, quadratic, and asymptotic algorithms described in [4]. A comparison of location performance for these algorithms is shown in Figure 6.7 for several AOA standard deviation values using 200 LOBs collected as a collector moves from [−100, 0] to [100, 0].

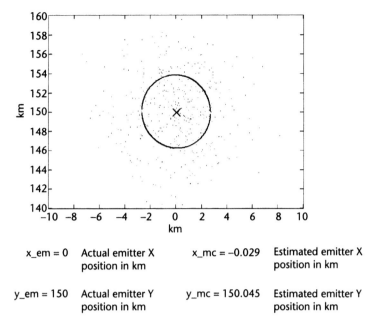

x_em = 0   Actual emitter X   x_mc = –0.029   Estimated emitter X
            position in km   position in km

y_em = 150 Actual emitter Y   y_mc = 150.045   Estimated emitter Y
            position in km   position in km

TLE = 0.053 *True location error, in km*

Theoretical ellipse	Monte Carlo ellipse
a_theo = 4.011 km	a_mc = 3.819 km
b_theo = 2.674 km	b_mc = 2.7 km
θ_theo = –90 degrees	q_mc = –87.699 degrees

Figure 6.5   AOA location ellipse example, 1° AOA, 1-m position error.

Figure 6.7 shows significant improvement in the reduction of the TLE when employing either the quadratic or asymptotic algorithm over the distance least squares for larger AOA errors. The Stansfield algorithm is similar to the Brown least squares algorithm except for a weighting matrix. The bias of the Stansfield algorithm was analyzed extensively in [5].

6.5   Time Difference of Arrival Location Analysis

The derivation of the associated location covariance matrix $C_{x_{Em}y_{Em}}$ for three aircraft TDOA location is now derived in terms of errors in collector position. The TOA for a stationary emitter at (x_{Em}, y_{Em}) and an aircraft located at (x_n, y_n) can be computed as

$$\text{TOA}_n = \frac{1}{c}\sqrt{(x_{Em} - x_n)^2 + (y_{Em} - y_n)^2} \qquad (6.14)$$

Associated errors in each TOA measurement will occur due to uncertainties in collector position:

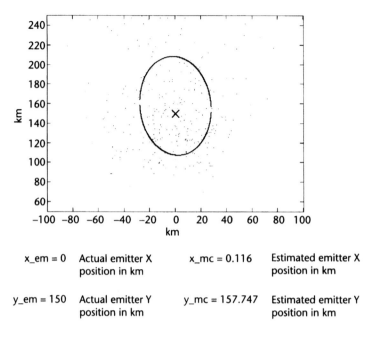

x_em = 0	Actual emitter X position in km	x_mc = 0.116	Estimated emitter X position in km
y_em = 150	Actual emitter Y position in km	y_mc = 157.747	Estimated emitter Y position in km

TLE = 7.748 *True location error, in km*

Theoretical ellipse	Monte Carlo ellipse
a_theo = 40.109 km	a_mc = 50.728 km
b_theo = 26.74 km	b_mc = 27.847 km
θ_theo = −90 degrees	θ_mc = −86.674 degrees

Figure 6.6 AOA location ellipse example, 10° AOA error.

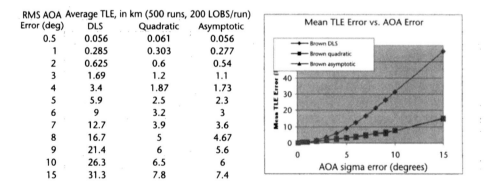

RMS AOA Error (deg)	Average TLE, in km (500 runs, 200 LOBS/run)		
	DLS	Quadratic	Asymptotic
0.5	0.056	0.061	0.056
1	0.285	0.303	0.277
2	0.625	0.6	0.54
3	1.69	1.2	1.1
4	3.4	1.87	1.73
5	5.9	2.5	2.3
6	9	3.2	3
7	12.7	3.9	3.6
8	16.7	5	4.67
9	21.4	6	5.6
10	26.3	6.5	6
15	31.3	7.8	7.4

Figure 6.7 Comparison of performance: Brown distance least squares, quadratic, and asymptotic algorithms.

$$\text{TOA}_n + \Delta\text{TOA}_n = \frac{1}{c}\sqrt{[x_{Em} - (x_n + \Delta x_n)]^2 + [y_{Em} - (y_n + \Delta y_n)]^2} \quad (6.15)$$

Treating each uncertainty in collector position as independent random variables Δx_n, Δy_n having zero mean with standard deviation σ_{x_n}, σ_{y_n}, the associated collector position covariance matrix can be written as

$$C_{x_n y_n} = \begin{bmatrix} \sigma_{x_n}^2 & \overbrace{\rho_{x_n y_n} \sigma_{x_n} \sigma_{y_n}}^{0} \\ \underbrace{0}_{\rho_{y_n x_n} \sigma_{y_n} \sigma_{x_n}} & \sigma_{y_n}^2 \end{bmatrix} \qquad (6.16)$$

The associated errors in individual TOA measurements can be written in terms of the TOA covariance matrix C_{TOA}, derived from known errors in collector position as

$$C_{\text{TOA}} = J_{\text{TOA}|x_n y_n} \cdot C_{x_n y_n} \cdot J_{\text{TOA}|x_n y_n}^T \qquad (6.17)$$

where

$$C_{\text{TOA}} = \begin{bmatrix} \sigma_{\text{TOA}_1}^2 & \rho_{\text{TOA}_1 \text{TOA}_2} \sigma_{\text{TOA}_1} \sigma_{\text{TOA}_2} & \rho_{\text{TOA}_1 \text{TOA}_3} \sigma_{\text{TOA}_1} \sigma_{\text{TOA}_3} \\ \rho_{\text{TOA}_2 \text{TOA}_1} \sigma_{\text{TOA}_2} \sigma_{\text{TOA}_1} & \sigma_{\text{TOA}_2}^2 & \rho_{\text{TOA}_2 \text{TOA}_3} \sigma_{\text{TOA}_2} \sigma_{\text{TOA}_3} \\ \rho_{\text{TOA}_3 \text{TOA}_1} \sigma_{\text{TOA}_3} \sigma_{\text{TOA}_1} & \rho_{\text{TOA}_3 \text{TOA}_2} \sigma_{\text{TOA}_3} \sigma_{\text{TOA}_2} & \sigma_{\text{TOA}_3}^2 \end{bmatrix}$$

$$(6.18)$$

The variable $J_{\text{TOA}|x_n y_n}$ is a 3×2 matrix of partial derivatives of the TOA measurements with respect to collector position, written as

$$J_{\text{TOA}|x_n y_n} = \begin{bmatrix} \dfrac{\partial \text{TOA}_1}{\partial x_n} & \dfrac{\partial \text{TOA}_1}{\partial y_n} \\ \dfrac{\partial \text{TOA}_2}{\partial x_n} & \dfrac{\partial \text{TOA}_2}{\partial y_n} \\ \dfrac{\partial \text{TOA}_3}{\partial x_n} & \dfrac{\partial \text{TOA}_3}{\partial y_n} \end{bmatrix} \qquad (6.19)$$

For TDOA position estimation in two dimensions, three TOA measurements are required. Using three collectors with one as a common reference, the two TDOA measurements resulting from three independent TOA measurements can be written as

$$\begin{bmatrix} \text{TDOA}_{12} \\ \text{TDOA}_{32} \end{bmatrix} = \begin{bmatrix} 1 \cdot \text{TOA}_1 & -1 \cdot \text{TOA}_2 & +0 \cdot \text{TOA}_3 \\ 0 \cdot \text{TOA}_1 & -1 \cdot \text{TOA}_2 & +1 \cdot \text{TOA}_3 \end{bmatrix} \qquad (6.20)$$

Associated errors in each TDOA measurement will occur due to resulting uncertainties in TOA:

$$\begin{bmatrix} \text{TDOA}_{12} + \Delta\text{TDOA}_{12} \\ \text{TDOA}_{32} + \Delta\text{TDOA}_{32} \end{bmatrix} \qquad (6.21)$$

$$= \begin{bmatrix} 1 \cdot (\text{TOA}_1 + \Delta\text{TOA}_1) & -1 \cdot (\text{TOA}_2 + \Delta\text{TOA}_2) & +0 \cdot (\text{TOA}_3 + \Delta\text{TOA}_3) \\ 0 \cdot (\text{TOA}_1 + \Delta\text{TOA}_1) & -1 \cdot (\text{TOA}_2 + \Delta\text{TOA}_2) & +1 \cdot (\text{TOA}_3 + \Delta\text{TOA}_3) \end{bmatrix}$$

The associated errors in TDOA measurement pairs can be written in the form of the TDOA covariance matrix C_{TDOA}, derived from known errors in TOA measurements as

$$C_{TDOA} = J_{TDOA|TOA} \cdot C_{TOA} \cdot J_{TDOA|TOA}^T \qquad (6.22)$$

$$C_{TDOA} = \begin{bmatrix} \sigma_{TDOA_{12}}^2 & \rho_{TDOA_{12}TDOA_{32}}\sigma_{TDOA_{12}}\sigma_{TDOA_{32}} \\ \rho_{TDOA_{32}TDOA_{12}}\sigma_{TDOA_{32}}\sigma_{TDOA_{12}} & \sigma_{TDOA_{32}}^2 \end{bmatrix}$$
$$(6.23)$$

The variable $J_{TDOA|TOA}$ is a 2×3 matrix of partial derivatives of the TDOA measurement pairs with respect to the individual platform measurements, written as

$$J_{TDOA|TOA} = \begin{bmatrix} \dfrac{\partial TDOA_{12}}{\partial TOA_1} & \dfrac{\partial TDOA_{12}}{\partial TOA_2} & \dfrac{\partial TDOA_{12}}{\partial TOA_3} \\ \dfrac{\partial TDOA_{32}}{\partial TOA_1} & \dfrac{\partial TDOA_{32}}{\partial TOA_2} & \dfrac{\partial TDOA_{32}}{\partial TOA_3} \end{bmatrix} = \begin{bmatrix} 1 & -1 & 0 \\ 0 & -1 & 1 \end{bmatrix}$$
$$(6.24)$$

The expressions for TDOA in terms of collector position uncertainty can be written as

$$TDOA_{12} + \Delta TDOA_{12} = \frac{1}{c} \left\{ \sqrt{[x_{Em} - (x_1 + \Delta x_1)]^2 + [y_{Em} - (y_1 + \Delta y_1)]^2} \right. \qquad (6.25)$$
$$\left. - \sqrt{[x_{Em} - (x_2 + \Delta x_2)]^2 + [y_{Em} - (y_2 + \Delta y_2)]^2} \right\}$$

$$TDOA_{32} + \Delta TDOA_{13} = \frac{1}{c} \left\{ \sqrt{[x_{Em} - (x_3 + \Delta x_3)]^2 + [y_{Em} - (y_3 + \Delta y_3)]^2} \right. \qquad (6.26)$$
$$\left. - \sqrt{[x_{Em} - (x_2 + \Delta x_2)]^2 + [y_{Em} - (y_2 + \Delta y_2)]^2} \right\}$$

With uncertainties in the TDOA measurement pairs known, the resulting location covariance matrix $C_{x_{Em}y_{Em}}$ can be written in terms of TDOA uncertainties as

$$C_{x_{Em}y_{Em}} = J_{x_{Em}y_{Em}|TDOA_{12}TDOA_{32}} \cdot C_{TDOA} \cdot J_{x_{Em}y_{Em}|TDOA_{12}TDOA_{32}}^T \qquad (6.27)$$

Because the expressions for emitter location in terms of $(TDOA_{12}, TDOA_{32})$ are not readily found, the following properties are used:

$$J \cdot C \cdot J^T = \{(J^T)^{-1} \cdot C^{-1} \cdot J^{-1}\}^{-1} \tag{6.28}$$

$$J_{\mathrm{TDOA}_{12}\mathrm{TDOA}_{32}|x_{Em}y_{Em}} = (J_{x_{Em}y_{Em}|\mathrm{TDOA}_{12}\mathrm{TDOA}_{32}})^{-1} \tag{6.29}$$

The variable $J_{x_{Em}y_{Em}|\mathrm{TDOA}_{12}\mathrm{TDOA}_{32}}$ is the Jacobian matrix of the partial derivatives of (x_{Em}, y_{Em}) with respect to $(\mathrm{TDOA}_{12}, \mathrm{TDOA}_{32})$, written as

$$J_{\mathrm{TDOA}_{12}\mathrm{TDOA}_{32}|x_{Em}y_{Em}} = \begin{bmatrix} \dfrac{\partial \mathrm{TDOA}_{12}}{\partial x_{Em}} & \dfrac{\partial \mathrm{TDOA}_{12}}{\partial y_{Em}} \\[2ex] \dfrac{\partial \mathrm{TDOA}_{32}}{\partial x_{Em}} & \dfrac{\partial \mathrm{TDOA}_{32}}{\partial y_{Em}} \end{bmatrix} \tag{6.30}$$

Using (6.28) and (6.29), the three aircraft TDOA location covariance matrix can then be written as

$$C_{x_{Em}y_{Em}} = \left\{ J^T_{\mathrm{TDOA}_{12}\mathrm{TDOA}_{32}|x_{Em}y_{Em}} \cdot C^{-1}_{\mathrm{TDOA}} \cdot J_{\mathrm{TDOA}_{12}\mathrm{TDOA}_{32}|x_{Em}y_{Em}} \right\}^{-1} \tag{6.31}$$

$$
\begin{aligned}
C_{x_{Em}y_{Em}} = \Big\{ & J^T_{\mathrm{TDOA}_{12}\mathrm{TDOA}_{32}|x_{Em}y_{Em}} \\
& \cdot \left\{ J_{\mathrm{TDOA}_{12}\mathrm{TDOA}_{32}|\mathrm{TOA}} \cdot J_{\mathrm{TOA}|x_n y_n} \cdot C_{x_n y_n} \cdot J^T_{\mathrm{TOA}|x_n y_n} \cdot J^T_{\mathrm{TDOA}_{12}\mathrm{TDOA}_{32}|\mathrm{TOA}} \right\}^{-1} \\
& \cdot J_{\mathrm{TDOA}_{12}\mathrm{TDOA}_{32}|x_{Em}y_{Em}} \Big\}^{-1}
\end{aligned} \tag{6.32}
$$

An example given in Figure 6.8 compares expected TDOA error ellipse parameters to those computed using (6.32) for an emitter located at [0, 150]. A position error sigma of 1m was used for three collector positioned at [−100, 0], [0, 0], and [100, 0].

6.6   Time/Frequency Difference of Arrival Location Analysis

The derivation of the associated location covariance matrix $C_{x_{Em}y_{Em}}$ for two collector TDOA/FDOA location is derived in terms of errors in aircraft position, heading, and velocity. The TOA for a stationary emitter at (x_{Em}, y_{Em}) and a collector located at (x_n, y_n) as given in (6.14) is used with a corresponding FOA measurement, given as

$$\mathrm{FOA}_n = \frac{1}{\lambda_{Em}} \cdot \frac{V_n \cdot \cos \theta_n \cdot (x_{Em} - x_n) + V_n \cdot \sin \theta_n \cdot (y_{Em} - y_n)}{\sqrt{(x_{Em} - x_n)^2 + (y_{Em} - y_n)^2}} \tag{6.33}$$

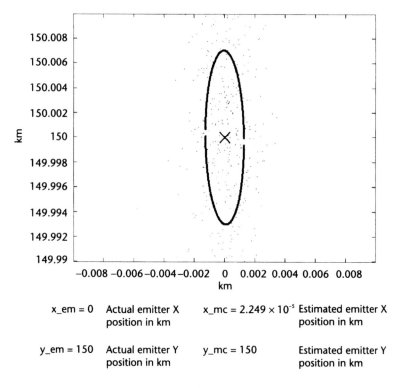

x_em = 0   Actual emitter X   x_mc = 2.249 × 10⁻⁵   Estimated emitter X
           position in km   position in km

y_em = 150   Actual emitter Y   y_mc = 150   Estimated emitter Y
             position in km   position in km

TLE = 1.936 × 10⁻⁴ *True location error, in km*

Theoretical results	Monte Carlo results
a_theo = 7.292 × 10⁻³ km	a_mc = 7.014 × 10⁻³ km
b_theo = 1.275 × 10⁻³ km	b_mc = 1.273 × 10⁻³ km
θ_theo = 90 degrees	θ_mc = −89.41 degrees

Figure 6.8   Comparison of three collector TDOA ellipse parameters, simulated versus theoretical.

Associated errors in each FOA measurement will occur due to uncertainties in collector position, heading, and velocity:

$$\text{FOA}_n + \Delta\text{FOA}_n = \frac{1}{\lambda_{Em}} \cdot \qquad (6.34)$$

$$\frac{(V_n + \Delta V_n) \cdot \{\cos(\theta_n + \Delta\theta_n) \cdot [x_{Em} - (x_n + \Delta x_n)] + \sin(\theta_n + \Delta\theta_n) \cdot [y_{Em} - (y_n + \Delta y_n)]\}}{\sqrt{[x_{Em} - (x_n + \Delta x_n)]^2 + [y_{Em} - (y_n + \Delta y_n)]^2}}$$

Treating each uncertainty in aircraft position, heading, and velocity as independent random variables Δx_n, Δy_n, $\Delta\theta_n$, and ΔV_n having zero mean with standard deviation σ_{x_n}, σ_{y_n}, σ_{θ_n}, and σ_{V_n}, the associated collector position, heading, and velocity covariance matrix $C_{\text{PosHeadVel}}$ for two collectors results in an 8×8 matrix, written as

$C_{\text{PosHeadVel}} =$

$$
\begin{bmatrix}
\sigma^2_{x_1} & \rho_{x_1y_1}\sigma_{x_1}\sigma_{y_1} & \rho_{x_1\theta_1}\sigma_{x_1}\sigma_{\theta_1} & \rho_{x_1V_1}\sigma_{x_1}\sigma_{V_1} & \rho_{x_1x_2}\sigma_{x_1}\sigma_{x_2} & \rho_{x_1y_2}\sigma_{x_1}\sigma_{y_2} & \rho_{x_1\theta_2}\sigma_{x_1}\sigma_{\theta_2} & \rho_{x_1V_2}\sigma_{x_1}\sigma_{V_2} \\
\rho_{y_1x_1}\sigma_{y_1}\sigma_{x_1} & \sigma^2_{y_1} & \rho_{y_1\theta_1}\sigma_{y_1}\sigma_{\theta_1} & \rho_{y_1V_1}\sigma_{y_1}\sigma_{V_1} & \rho_{y_1x_2}\sigma_{y_1}\sigma_{x_2} & \rho_{y_1y_2}\sigma_{y_1}\sigma_{y_2} & \rho_{y_1\theta_2}\sigma_{y_1}\sigma_{\theta_2} & \rho_{y_1V_2}\sigma_{y_1}\sigma_{V_2} \\
\rho_{\theta_1x_1}\sigma_{\theta_1}\sigma_{x_1} & \rho_{\theta_1y_1}\sigma_{\theta_1}\sigma_{y_1} & \sigma^2_{\theta_1} & \rho_{\theta_1V_1}\sigma_{\theta_1}\sigma_{V_1} & \rho_{\theta_1x_2}\sigma_{\theta_1}\sigma_{x_2} & \rho_{\theta_1y_2}\sigma_{\theta_1}\sigma_{y_2} & \rho_{\theta_1\theta_2}\sigma_{\theta_1}\sigma_{\theta_2} & \rho_{\theta_1V_2}\sigma_{\theta_1}\sigma_{V_2} \\
\rho_{V_1x_1}\sigma_{V_1}\sigma_{x_1} & \rho_{V_1y_1}\sigma_{V_1}\sigma_{y_1} & \rho_{V_1\theta_1}\sigma_{V_1}\sigma_{\theta_1} & \sigma^2_{V_1} & \rho_{V_1x_2}\sigma_{V_1}\sigma_{x_2} & \rho_{V_1y_2}\sigma_{V_1}\sigma_{y_2} & \rho_{V_1\theta_2}\sigma_{V_1}\sigma_{\theta_2} & \rho_{V_1V_2}\sigma_{V_1}\sigma_{V_2} \\
\rho_{x_2x_1}\sigma_{x_2}\sigma_{x_1} & \rho_{x_2y_1}\sigma_{x_2}\sigma_{y_1} & \rho_{x_2\theta_1}\sigma_{x_2}\sigma_{\theta_1} & \rho_{x_2V_1}\sigma_{x_2}\sigma_{V_1} & \sigma^2_{x_2} & \rho_{x_2y_2}\sigma_{x_2}\sigma_{y_2} & \rho_{x_2\theta_2}\sigma_{x_2}\sigma_{\theta_2} & \rho_{x_2V_2}\sigma_{x_2}\sigma_{V_2} \\
\rho_{y_2x_1}\sigma_{y_2}\sigma_{x_1} & \rho_{y_2y_1}\sigma_{y_2}\sigma_{y_1} & \rho_{y_2\theta_1}\sigma_{y_2}\sigma_{\theta_1} & \rho_{y_2V_1}\sigma_{y_2}\sigma_{V_1} & \rho_{y_2x_2}\sigma_{y_2}\sigma_{x_2} & \sigma^2_{y_2} & \rho_{y_2\theta_2}\sigma_{y_2}\sigma_{\theta_2} & \rho_{y_2V_2}\sigma_{y_2}\sigma_{V_2} \\
\rho_{\theta_2x_1}\sigma_{\theta_2}\sigma_{x_1} & \rho_{\theta_2y_1}\sigma_{\theta_2}\sigma_{y_1} & \rho_{\theta_2\theta_1}\sigma_{\theta_2}\sigma_{\theta_1} & \rho_{\theta_2V_1}\sigma_{\theta_2}\sigma_{V_1} & \rho_{\theta_2x_2}\sigma_{\theta_2}\sigma_{x_2} & \rho_{\theta_2y_2}\sigma_{\theta_2}\sigma_{y_2} & \sigma^2_{\theta_2} & \rho_{\theta_2V_2}\sigma_{\theta_2}\sigma_{V_2} \\
\rho_{V_2x_1}\sigma_{V_2}\sigma_{x_1} & \rho_{V_2y_1}\sigma_{V_2}\sigma_{y_1} & \rho_{V_2\theta_1}\sigma_{V_2}\sigma_{\theta_1} & \rho_{V_2V_1}\sigma_{V_2}\sigma_{V_1} & \rho_{V_2x_2}\sigma_{V_2}\sigma_{x_2} & \rho_{V_2y_2}\sigma_{V_2}\sigma_{y_2} & \rho_{V_2\theta_2}\sigma_{V_2}\sigma_{\theta_2} & \sigma^2_{V_2}
\end{bmatrix}
$$

$$(6.35)$$

The associated errors in individual TOA/FOA measurements can be written in the form of the TOA/FOA covariance matrix C_{TOAFOA}, and derived from known errors in collector position, heading, and velocity as

$C_{\text{TOAFOA}} = J_{\text{TOAFOA}|\text{PosHeadVel}} \cdot C_{\text{PosHeadVel}} \cdot J^T_{\text{TOA}|\text{PosHeadVel}}$

$$
=
\begin{bmatrix}
\sigma^2_{\text{TOA}_1} & \rho_{\text{TOA}_1\text{TOA}_2}\sigma_{\text{TOA}_1}\sigma_{\text{TOA}_2} & \rho_{\text{TOA}_1\text{FOA}_1}\sigma_{\text{TOA}_1}\sigma_{\text{FOA}_1} & \rho_{\text{TOA}_1\text{FOA}_2}\sigma_{\text{TOA}_1}\sigma_{\text{FOA}_2} \\
\rho_{\text{TOA}_2\text{TOA}_1}\sigma_{\text{TOA}_2}\sigma_{\text{TOA}_1} & \sigma^2_{\text{TOA}_2} & \rho_{\text{TOA}_2\text{FOA}_1}\sigma_{\text{TOA}_2}\sigma_{\text{FOA}_1} & \rho_{\text{TOA}_2\text{FOA}_2}\sigma_{\text{TOA}_2}\sigma_{\text{FOA}_2} \\
\rho_{\text{FOA}_1\text{TOA}_1}\sigma_{\text{FOA}_1}\sigma_{\text{TOA}_1} & \rho_{\text{FOA}_1\text{TOA}_2}\sigma_{\text{FOA}_1}\sigma_{\text{TOA}_2} & \sigma^2_{\text{FOA}_1} & \rho_{\text{FOA}_1\text{FOA}_2}\sigma_{\text{FOA}_1}\sigma_{\text{FOA}_2} \\
\rho_{\text{FOA}_2\text{TOA}_1}\sigma_{\text{FOA}_2}\sigma_{\text{TOA}_1} & \rho_{\text{FOA}_2\text{TOA}_2}\sigma_{\text{FOA}_2}\sigma_{\text{TOA}_2} & \rho_{\text{FOA}_2\text{FOA}_1}\sigma_{\text{FOA}_2}\sigma_{\text{FOA}_1} & \sigma^2_{\text{FOA}_2}
\end{bmatrix}
$$

$$(6.36)$$

The variable $J_{\text{TOAFOA}|\text{PosHeadVel}}$ is a 4×8 matrix of partial derivatives of the TOA/FOA measurements with respect to aircraft position, heading, and velocity, written as

$J_{\text{TOAFOA}|\text{PosHeadVel}} =$

$$
\begin{bmatrix}
\dfrac{\partial \text{TOA}_1}{\partial x_1} & \dfrac{\partial \text{TOA}_1}{\partial y_1} & \dfrac{\partial \text{TOA}_1}{\partial \theta_1} & \dfrac{\partial \text{TOA}_1}{\partial V_1} & \dfrac{\partial \text{TOA}_1}{\partial x_2} & \dfrac{\partial \text{TOA}_1}{\partial y_2} & \dfrac{\partial \text{TOA}_1}{\partial \theta_2} & \dfrac{\partial \text{TOA}_1}{\partial V_2} \\[2mm]
\dfrac{\partial \text{TOA}_2}{\partial x_1} & \dfrac{\partial \text{TOA}_2}{\partial y_1} & \dfrac{\partial \text{TOA}_2}{\partial \theta_1} & \dfrac{\partial \text{TOA}_2}{\partial V_1} & \dfrac{\partial \text{TOA}_2}{\partial x_2} & \dfrac{\partial \text{TOA}_2}{\partial y_2} & \dfrac{\partial \text{TOA}_2}{\partial \theta_2} & \dfrac{\partial \text{TOA}_2}{\partial V_2} \\[2mm]
\dfrac{\partial \text{FOA}_1}{\partial x_1} & \dfrac{\partial \text{FOA}_1}{\partial y_1} & \dfrac{\partial \text{FOA}_1}{\partial \theta_1} & \dfrac{\partial \text{FOA}_1}{\partial V_1} & \dfrac{\partial \text{FOA}_1}{\partial x_2} & \dfrac{\partial \text{FOA}_1}{\partial y_2} & \dfrac{\partial \text{FOA}_1}{\partial \theta_2} & \dfrac{\partial \text{FOA}_1}{\partial V_2} \\[2mm]
\dfrac{\partial \text{FOA}_2}{\partial x_1} & \dfrac{\partial \text{FOA}_2}{\partial y_1} & \dfrac{\partial \text{FOA}_2}{\partial \theta_1} & \dfrac{\partial \text{FOA}_2}{\partial V_1} & \dfrac{\partial \text{FOA}_2}{\partial x_2} & \dfrac{\partial \text{FOA}_2}{\partial y_2} & \dfrac{\partial \text{FOA}_2}{\partial \theta_2} & \dfrac{\partial \text{FOA}_2}{\partial V_2}
\end{bmatrix}
$$

$$(6.37)$$

For position estimation in two dimensions, two TOA and two FOA measurements are required. The TDOA/FDOA measurements resulting from two TOA/FOA measurements can be written as

$$
\begin{bmatrix} \text{TDOA}_{12} \\ \text{FDOA}_{12} \end{bmatrix} = \begin{bmatrix} 1 \cdot \text{TOA}_1 & -1 \cdot \text{TOA}_2 & +0 \cdot \text{FOA}_1 & +0 \cdot \text{FOA}_2 \\ 0 \cdot \text{TOA}_1 & +0 \cdot \text{TOA}_2 & +1 \cdot \text{FOA}_1 & -1 \cdot \text{FOA}_2 \end{bmatrix}
$$

$$(6.38)$$

Substituting associated errors in collector position, heading, and velocity into (6.38) yields

$$
\text{TDOA}_{12} + \Delta\text{TOA}_{12} = \frac{1}{c} \left\{ \sqrt{[x_{Em} - (x_1 + \Delta x_1)]^2 + [y_{Em} - (y_1 + \Delta y_1)]^2} \right.
$$

$$
\left. - \sqrt{[x_{Em} - (x_2 + \Delta x_2)]^2 + [y_{Em} - (y_2 + \Delta y_2)]^2} \right\}
$$

$$(6.39)$$

$$
\text{FDOA}_{12} + \Delta\text{FOA}_{12} = \frac{1}{\lambda_{Em}} \cdot
$$

$$
\left\{ \frac{(V_1 + \Delta V_1) \cdot \{\cos(\theta_1 + \Delta\theta_1) \cdot [x_{Em} - (x_1 + \Delta x_1)] + \sin(\theta_1 + \Delta\theta_1) \cdot [y_{Em} - (y_1 + \Delta y_1)]\}}{\sqrt{[x_{Em} - (x_1 + \Delta x_1)]^2 + [y_{Em} - (y_1 + \Delta y_1)]^2}} \right.
$$
$$
\left. - \frac{(V_2 + \Delta V_2) \cdot \{\cos(\theta_2 + \Delta\theta_2) \cdot [x_{Em} - (x_2 + \Delta x_2)] + \sin(\theta_2 + \Delta\theta_2) \cdot [y_{Em} - (y_2 + \Delta y_2)]\}}{\sqrt{[x_{Em} - (x_2 + \Delta x_2)]^2 + [y_{Em} - (y_2 + \Delta y_2)]^2}} \right\}
$$

The TOA/FOA covariance matrix C_{TOAFOA} can then be used to compute uncertainty in TDOA/FDOA measurement pairs, given in the TDOA/FDOA covariance matrix C_{TDOAFDOA} as

$$
C_{\text{TDOAFDOA}} = J_{\text{TDOAFDOA}|\text{TOAFOA}} \cdot C_{\text{TOAFOA}} \cdot J_{\text{TDOAFDOA}|\text{TOAFOA}}^T
$$

$$
= \begin{bmatrix} \sigma_{\text{TDOA}_{12}}^2 & \rho_{\text{TDOA}_{12}\text{FDOA}_{12}} \sigma_{\text{TDOA}_{12}} \sigma_{\text{FDOA}_{12}} \\ \rho_{\text{FDOA}_{12}\text{TDOA}_{12}} \sigma_{\text{FDOA}_{12}} \sigma_{\text{TDOA}_{12}} & \sigma_{\text{FDOA}_{12}}^2 \end{bmatrix}
$$

$$(6.40)$$

The variable $J_{\text{TDOAFDOA}|\text{TOAFOA}}$ is the partial derivative of TDOA/FDOA measurements with respect to individual TOA/FOA measurements, derived from (6.38) as

$$
J_{\text{TDOAFDOA}|\text{TOAFOA}} = \begin{bmatrix} 1 & -1 & 0 & 0 \\ 0 & 0 & 1 & -1 \end{bmatrix}
$$

$$(6.41)$$

With uncertainties in the TDOA/FDOA measurement pairs known, the resulting location covariance matrix $C_{x_{Em}y_{Em}}$ can be written in terms of TDOA/FDOA uncertainties as

$$C_{x_{Em}y_{Em}} = J_{x_{Em}y_{Em}|TDOA_{12}FDOA_{12}} \cdot C_{TDOAFDOA} \cdot J^T_{x_{Em}y_{Em}|TDOA_{12}FDOA_{12}}$$

$$(6.42)$$

The variable $J_{x_{Em}y_{Em}|TDOA_{12}FDOA_{12}}$ is the Jacobian matrix of the partial derivatives of (x_{Em}, y_{Em}) with respect to $(TDOA_{12}, FDOA_{12})$, written as

$$J_{TDOA_{12}FDOA_{12}|x_{Em}y_{Em}} = \begin{bmatrix} \dfrac{\partial TDOA_{12}}{\partial x_{Em}} & \dfrac{\partial TDOA_{12}}{\partial y_{Em}} \\[2ex] \dfrac{\partial FDOA_{12}}{\partial x_{Em}} & \dfrac{\partial FDOA_{12}}{\partial y_{Em}} \end{bmatrix}$$

$$(6.43)$$

Because the expressions for emitter location in terms of $(TDOA_{12}, FDOA_{12})$ are not readily found (as the case with three aircraft TDOA), the following relationship is again employed:

$$J_{TDOA_{12}FDOA_{12}|x_{Em}y_{Em}} = (J_{x_{Em}y_{Em}|TDOA_{12}FDOA_{12}})^{-1}$$

$$(6.44)$$

The two aircraft TDOA/FDOA location covariance matrix can then be written as

$$C_{x_{Em}y_{Em}} = \left\{ J^T_{TDOA_{12}FDOA_{12}|x_{Em}y_{Em}} \cdot C^{-1}_{TDOAFDOA} \cdot J_{TDOA_{12}FDOA_{12}|x_{Em}y_{Em}} \right\}^{-1}$$

$$(6.45)$$

An example is given in Figure 6.9 to compare expected TDOA/FDOA error ellipse parameters to those computed using (6.45). An emitter located at [1, 150] with a frequency of 900 MHz was used. A position error sigma of 1m was used for two collectors positioned at [−100, 0] and [100, 0], traveling in an easterly direction at 200 m/s with heading and velocity uncertainties of 1° and 1 m/s, respectively.

6.7   Geometric Dilution of Precision

Other errors occur with geometry. Poorer location performance is observed when the angle between intersecting LOBs, TDOA, and/or TDOA/FDOA lines approaches zero. This phenomenon is known as geometric dilution of precision (GDOP) and is typically observed at cross- and down-range regions outside of the baseline, as shown in Figure 6.10 for TDOA using three collectors.

The effects of GDOP can be readily seen in observing CEP values for emitters at various cross- and down-range locations. CEP values for the three collector TDOA scenario with 1° position error are given in Figure 6.11 to illustrate the poorer performance away from the center of the baseline.

To minimize the unwanted effects of GDOP, different collector paths can be used. Typically, nearly perpendicular intersecting lines of bearing and constant

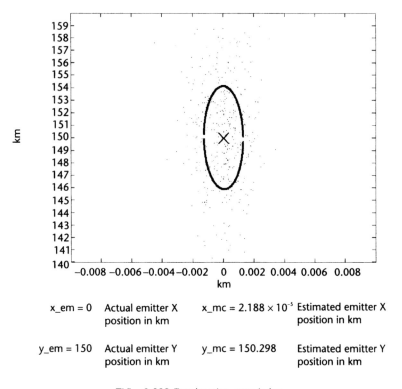

km

-0.008 -0.006 -0.004 -0.002 0 0.002 0.004 0.006 0.008
km

x_em = 0   Actual emitter X   x_mc = 2.188 × 10⁻⁵ Estimated emitter X
              position in km   position in km

y_em = 150   Actual emitter Y   y_mc = 150.298   Estimated emitter Y
              position in km   position in km

TLE = 0.298 *True location error, in km*

Theoretical results		Monte Carlo results	
a_theo = 4.083	km	a_mc = 4.116	km
b_theo = 1.275 × 10⁻³	km	b_mc = 1.273 × 10⁻³ km	
θ_theo = 90	degrees	θ_mc = −90 degrees	

Figure 6.9 Comparison of two collector TDOA/FDOA ellipse parameters, simulated versus theoretical.

TDOA or FDOA are desired, especially when errors at each platform are equal. In such a case with two collectors measuring AOA, the minimum CEP occurs when the emitter is down-range from the midpoint of the line joining the collectors at a distance of $R/\sqrt{2}$ when the collectors are separated by a distance of R, as shown in Figure 6.12.

6.8   Incorporation of Measurement Error

The derived AOA, TDOA, and TDOA/FDOA covariance matrices for non-signal-dependent errors can be summed with associated covariance matrices computed from signal-dependent error sources to provide an overall estimate of system performance. Signal-dependent error is largely determined by the SNR. As indicated in Chapter 5, the standard deviation of the AOA error function of SNR for an interferometer is given by

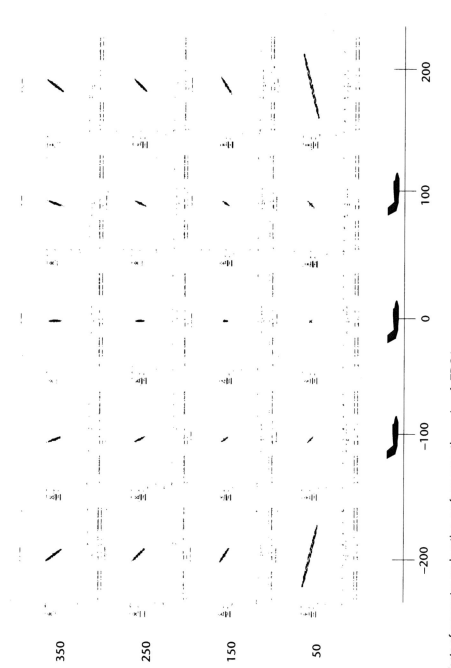

Figure 6.10 Effects of geometry on location performance; three aircraft TDOA.

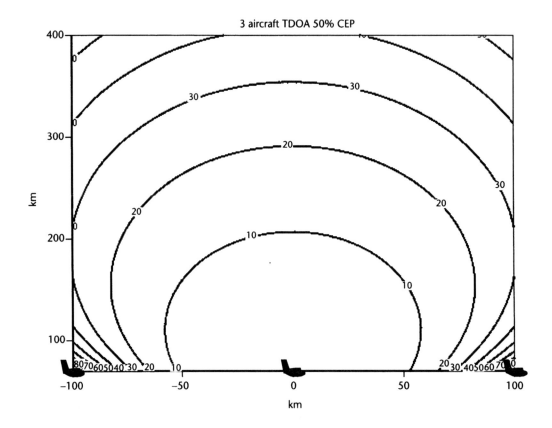

(xCEP, yCEP, CEP)

Figure 6.11 Effects of geometry on geolocation performance; three collector TDOA.

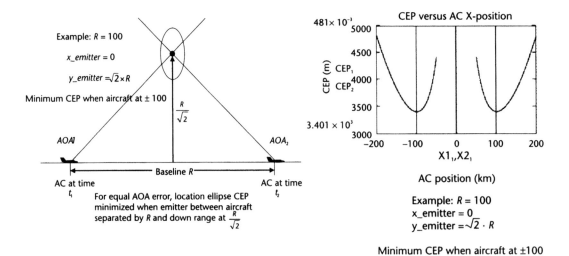

Figure 6.12 Minimum CEP; two aircraft AOA.

$$\sigma_{AOA} = \frac{\lambda_{Emitter}}{2\pi \cdot d \cdot \cos(AOA) \cdot \sqrt{SNR}} \text{ (deg)} \tag{6.46}$$

The variables d and $\lambda_{Emitter}$ are the antenna separation and emitter wavelength, respectively. The AOA value is given with respect to the perpendicular off boresight; SNR is given in as a magnitude ratio (not in decibels). Errors in TOA as a function of SNR are typically computed as a function of the pulses' leading edge rise time, t_R, given in Chapter 12 as

$$\delta_T = \frac{1.25 t_R}{\sqrt{2\,SNR}} \text{ (sec)} \tag{6.47}$$

When TDOA/FDOA measurements are made with complex ambiguity function processing, errors in terms of signal bandwidth B_s, noise bandwidth B, integration time T, and receiver SNR values γ_1 and γ_2 can be estimated using [1]

$$\sigma_{TDOA} \approx \frac{0.55}{B_s \sqrt{BT\gamma}} \text{ (sec)} \tag{6.48}$$

$$\sigma_{FDOA} \approx \frac{0.55}{T \sqrt{BT\gamma}} \text{ (Hz)} \tag{6.49}$$

The variable γ is the effective SNR computed from the SNR at each platform as

$$\gamma = \frac{1}{2} \left[\frac{1}{\gamma_1} + \frac{1}{\gamma_2} + \frac{1}{\gamma_1 \gamma_2} \right] \tag{6.50}$$

6.9 Summary

In this chapter the effect of various errors on estimates of emitter location was addressed. Theoretical expressions for location uncertainty in terms of errors in collection platform position, heading, and velocity were derived and used to predict error ellipse parameters for single collector AOA and multiple collector TDOA and TDOA/FDOA techniques. The error ellipse parameters predicted using theoretical expressions were compared with computer simulation results. Geometric dilution of precision was also discussed. Several expressions for errors in AOA, TDOA, and FDOA measurements due to noise were also mentioned. Combining all of these expressions provides estimates for total location system performance. The detailed location analysis in [6] is a comprehensive summary of the subject for the interested reader.

Instantaneous phase or frequency tracking using a single moving platform can also be used to locate an emitter. If the emitter is coherent, one measures the pulse-to-pulse phase change as the platform moves. These phase changes may be converted to instantaneous frequency by differentiating the instantaneous phase. By assuming a location for the emitter, one can plot the instantaneous frequency that would have been observed by a collector moving along the known path. This is compared to the observed frequency variation and the assumed location is adjusted to minimize the difference and thus estimate the emitter location. This process can only be used if the emitter is coherent. If the emitter is not coherent, two antennas and receivers are placed on the collector separated by some distance. The phase difference now depends on the angle of arrival of the pulses from the emitter. (The two antennas form an interferometer.) The observed phase variations along the path are compared to what would have been observed from an emitter at a trial location, and then the differences are minimized to find the best location estimate. Generally the best location performance is obtained using an S-shaped collector path. A similar location process is possible using very accurate TOA measurements if the emitter has a stable PRI. Then as the collector moves, the TOA changes slightly and the emitter location is estimated by determining which location gives the observed TOA variations.

References

[1]   Stein, S., "Algorithms for Ambiguity Processing," *IEEE Trans. on Acoustics, Speech, and Signal Processing*, June 1981.

[2]   Brown, R., *Emitter Location Using Bearing Measurements from a Moving Platform*, NRL Report 8483, June 1981.

[3]   Torrieri, D. J., "Statistical Theory of Passive Location Systems," *IEEE Trans. on Aerospace and Electronic Systems*, March 1984.

[4]   Wiley, R. G., *Electronic Intelligence: The Interception of Radar Signals*, Dedham, MA: Artech House, 1985.

[5]   Gavish, M., and A. Weiss, "Performance Analysis of Bearing-Only Target Location Algorithms," *IEEE Trans. on Aerospace and Electronic Systems*, July 1992.

[6]   Poisel, R., *Electronic Warfare Target Location Methods*, Norwood, MA: Artech House, 2005.

CHAPTER 7

Estimating Power at the Transmitter

7.1 Power Estimation Through ELINT

The radar's transmitted power is a basic quantity that defines its range performance. Unfortunately, transmit power is difficult to estimate from the ELINT point of view. Historically, transmitter power was one of the last aspects of a signal to be examined by signal analysis. In ELINT work, the quantity that can be most readily estimated is the peak effective radiated power (ERP), which is the product of the peak transmit power, the antenna gain, and any loss between the transmitter and transmit antenna. (The radar's antenna gain can be estimated from its beam pattern, and then the peak transmit power can also be estimated once the ERP is determined.) Consider the one-way or ELINT range equation. Solving this equation for the ERP gives the following result:

$$\text{ERP} = \frac{P_T G_{TE}}{L_T} = \frac{(4\pi)^2 S_E R_E^2 L_E}{G_E \lambda} \tag{7.1}$$

where:

P_T = transmitted power

G_{TE} = transmit antenna gain in the direction of the ELINT site

L_T = losses at the transmitter

S_E = received signal power

R_E = distance from the transmitter to the ELINT site

L_E = losses at the ELINT receiver

G_E = gain of the ELINT site antenna in the direction of the transmitter

λ = wavelength = speed of light divided by the RF

In radar usage, the term "peak power" means the transmitted power averaged over the duration of one pulse. The term "average power" means the power averaged over one PRI. The average power is related to the peak power by

$$P_{AV} = \frac{P_T \tau}{T} \tag{7.2}$$

169

where:

P_{AV} = average transmitted power

P_T = peak transmitter power

τ = pulse duration

T = pulse repetition interval

The ratio τ/T is referred to as the duty cycle of the radar transmitter.

Many radar transmitters are designed for large peak power output. They are typically operated as saturated amplifiers or oscillators. At present, it is still unusual for a radar transmitter to have the capability of varying its power level; however, some systems can select from among several discrete power-output levels on a pulse-by-pulse basis.

One of the key terms in (7.1) is the distance from the transmitter to the ELINT site. Generally speaking the transmitter location is not known. Emitter location is a difficult task. It is the subject of Chapter 6, but could the subject of an entire book. Some of the basic ideas useful for estimation of transmit power are included below.

7.2 Distance to the Horizon

Generally speaking, the emitter must be located within line of sight of the ELINT station. If the interceptor is in an aircraft and the emitter is on the ground (or vice versa), the line-of-sight range can be rather large—several hundreds of kilometers. Key parameters for determining the maximum line of sight are the height of the transmit and receive antennas above the Earth. The line of sight to the horizon, assuming a spherical Earth, can be calculated from plane geometry using Figure 7.1.

$$d^2 = h^2 + 2rh \tag{7.3}$$

Because usually $r \gg h$, this is often approximated by

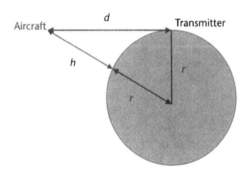

Figure 7.1 Distance to the horizon.

$$d \approx \sqrt{2rh}$$

With propagation at microwave frequencies, the effective radio horizon is observed to be greater than the geometrical horizon. This is caused by refraction by the atmosphere. Refractive effects vary widely as a function of the water vapor in the atmosphere; however, it is common practice to approximate this by multiplying the Earth's radius by a factor greater than 1. The usual factor is 4/3, making the effective Earth radius, $r_{eff} = 4r/3$. This yields the distance to the radio horizon as

$$d \approx (2r_{eff}h)^{1/2} \approx \left(\frac{8rh}{3}\right)^{1/2}$$

For Earth radius $r = 3,900$ miles, where height (h) is in feet, this reduces to

$$d_{miles} \approx \sqrt{2h_{ft}}$$

For $r = 3,900$ miles, where height (h) is in meters and distance (d) is in kilometers

$$d_{km} \approx \sqrt{17.3h_m}$$

Figure 7.2 illustrates the distance to the horizon as a function of altitude.

One of the methods to find an upper bound of the ERP is to insert the distance to the horizon into (7.1). This gives the maximum ERP value possible.

$$\mathrm{ERP}_{max} = \frac{P_T G_{TE}}{L_T} = \frac{(4\pi)^2 S_E r_{eff} h L_E}{G_E \lambda} \tag{7.4}$$

As (7.4) shows, for a given received signal level S_E, the ERP estimate increases in proportion to the collector altitude for an emitter at the horizon.

If the height of the ELINT antenna and the height of the transmitter are known, simply use Figure 7.2 to compute the distance to the horizon for each antenna height and add the values together. For example, if the ELINT antenna is on an aircraft flying at an altitude of 10,000m, the distance to the horizon is 416 km. If the emitter antenna is on a hilltop 1,000m high, the distance to its horizon is 131 km. The total distance possible for line-of-sight propagation is then 547 km.

In much of what follows, the discussion is concerned with determining what conditions must be met to limit the ERP error caused by the individual factors in (7.1) to 0.5 dB.

7.3   ERP Errors Due to Antenna Pointing Errors

In principle, the gain of the ELINT antenna G_E is known. But, it may be difficult to accurately point the ELINT antenna toward the radar. The power received is fluctuating due to the scanning of the radar beam and the collection platform

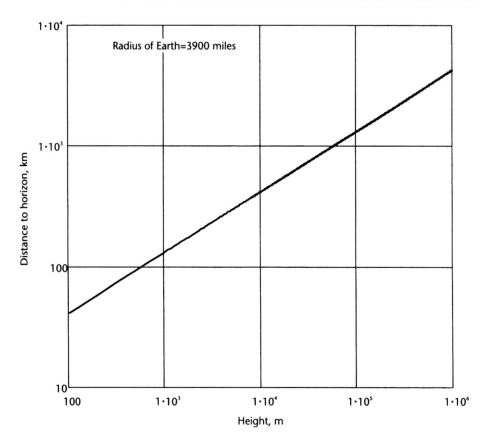

Figure 7.2 Distance to the horizon (for R multiplied by 4/3).

may be moving. Thus, maximizing the signal amplitude at the ELINT site means performing scan-to-scan amplitude comparisons. The ELINT site would prefer a relatively broad beam for its antenna so that pointing accuracy is not a problem for power measurement. On the other hand, the ELINT station may need a narrow beam so that the bearing (AOA) to the radar can be determined. This bearing should be accurately determined in order to find the range to the radar by means of triangulation (this requires the cooperation of at least one other ELINT site or else requires that the ELINT station be moving). (Triangulation is discussed in Chapter 6.) These contradictory requirements mean that some stations may use one antenna for direction finding and another for power measurement. Another approach is to use monopulse techniques to aid in pointing the antenna to within a fraction of the 3-dB beamwidth. The pointing accuracy required to achieve a given ERP accuracy can be computed if the antenna pattern is known. Consider the case of a uniformly illuminated rectangular aperture. This produces one-way power gain that varies as [1]

$$\frac{G(\theta, \phi)}{G(0, 0)} = \frac{\sin^2\left(\dfrac{\pi L}{\lambda} \sin \theta \cos \phi\right) \cdot \sin^2\left(\dfrac{\pi W}{\lambda} \sin \phi \cos \theta\right)}{\left(\dfrac{\pi L \sin \theta \cos \phi}{\lambda}\right)^2 \cdot \left(\dfrac{\pi W \sin \phi \cos \theta}{\lambda}\right)^2} \qquad (7.5)$$

where:

$G(\theta, \phi)$ = gain in direction (θ, ϕ)

L = length of antenna in azimuth direction

W = width of antenna in elevation direction

θ = azimuth angle relative to normal to the aperture

ϕ = elevation angle relative to normal to the aperture

$G(0, 0)$ = boresight for $(0, 0)$

λ = wavelength

Suppose the receive antenna gain used to compute the ERP is to be accurate to within 0.5 dB. Then the antenna must be pointed toward the transmitter to within about 20% of the 3-dB beamwidth if the pointing error is only in one dimension (azimuth or elevation). If the antenna must be pointed in both dimensions (azimuth and elevation), the pointing error components must each be less than about 15% of the corresponding beamwidth, corresponding to a vector error of 21% if the beamwidths are equal. (See Figure 7.3.)

It is unlikely that the antenna gain of the ELINT antenna will be known to within 0.5 dB. In addition to the pointing of the antenna, other factors include the fact that the antenna pattern is known only at certain frequencies and for specific

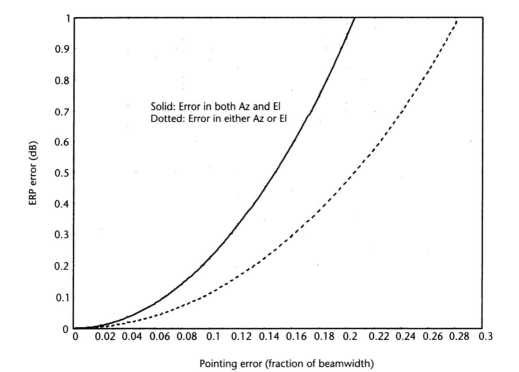

Figure 7.3   Antenna gain as a function of pointing angle error.

polarization conditions. It is likely that the polarization of the received signal will be matched to the polarization of the receiving antenna or that the frequency of the received signal will be the same as the frequencies at which the antenna gain has been measured.

Polarization errors also make determining the received power difficult. It is necessary to try different receiving antenna polarizations at the ELINT site to determine which polarization results in maximum received power when the center of the main beam is directed toward the ELINT site. For a linearly polarized radar transmit antenna, the loss at the ELINT site due to improper polarization orientation is given by $[\cos(\epsilon)]^{-2}$, where ϵ is the error in polarization [2]. Achieving an error of less than 0.5 dB requires that the error in matching the polarization of the receiver to that of the transmitting antenna be less than 19.25°.

Another problem is that when the collector is moving, it will not be at the peak of the main beam indefinitely. Consider a situation where the collector is flying toward a ground-based radar at constant altitude. As the collector approaches, the elevation angle relative to the transmitter increases. Therefore, both the receive antenna gain and the transmit antenna gain change slightly as the collector approaches.

7.4   Estimating the Distance to the Radar

Consider the effect of range errors on determining the radar power. Equation (7.1) shows that the received signal level is proportional to $(\lambda/R_E)^2$. Measuring the wavelength (or radio frequency) can be accomplished quite accurately; the main problem is finding the range, R_E. The power is related to the range by

$$P \alpha \frac{1}{R_E^2} \tag{7.6a}$$

Differentiating and rearranging terms gives

$$\frac{\Delta P}{P} = \frac{-2\Delta R_E}{R_E} \tag{7.6b}$$

This means that a 10% range error produces a power error of about 20%. To maintain power accuracy to 0.5 dB requires

$$\frac{\Delta P}{P} = 1 - 10^{-0.05} = 0.109$$

or

$$\frac{\Delta R_E}{R_E} = \frac{\Delta P}{2P} = 0.054$$

This means that the range must be known to within about 5%. Using triangulation, the range error can be related to the pointing error. As shown in Chapter 6, the error in location is a function of the ratio of the baseline between the two interception points to the distance to the emitter. Consider the simple case with two stations and an accurately known baseline distance, D, between them, as shown in Figure 7.4.

The following relationships are clear:

$$R_{E1} = \frac{x}{\cos \theta_1} = \frac{y}{\sin \theta_1} \tag{7.7}$$

$$R_{E2} = \frac{D - x}{\cos \theta_2} = \frac{y}{\sin \theta_2} \tag{7.8}$$

Equation (7.7) may be solved to give

$$R_{E1} = \frac{D \sin \theta_2}{\sin(\theta_1 + \theta_2)} \tag{7.9}$$

Differentiating gives

$$\frac{\Delta R_{E1}}{R_{E1}} = \frac{\Delta \theta_2}{\tan \theta_2} - \frac{\Delta \theta_1 + \Delta \theta_2}{\tan(\theta_1 + \theta_2)} \tag{7.10}$$

To choose a simple case, take $\theta_1 = \theta_2 = 45°$. Then,

$$\frac{\Delta R_{E1}}{R_{E1}} = \Delta \theta_2$$

For a power accuracy of less than 0.5 dB, the range error must be less than 0.054, so that the direction-finding accuracy of station 2 must be 0.054 rad or 3.1°. For this example, the pointing error due to station 1 does not appear because, for the angles chosen, the vector from station 2 to the emitter is perpendicular to the vector from station 1 to the emitter.

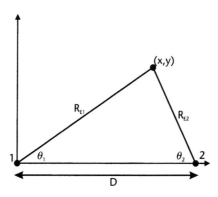

Figure 7.4   Emitter location using two station triangulation.

Consider the case $\theta_1 = 90°$, $\theta_2 = 45°$. If the two ELINT stations have the same angular accuracy $(\Delta\theta = \Delta\theta_1 = \Delta\theta_2)$, then

$$\frac{\Delta R_{E1}}{R_{E1}} = 3\Delta\theta$$

For $\Delta R_{E1}/R_{E1} = 0.054$, $\Delta\theta = 0.018$ rad $= 1.04°$. In other words, for a power measurement accuracy of 0.5 dB, the angular bearing error (for triangulation) must be no more than 1.04° in this case.

There are other methods of range estimation, including TDOA and FDOA as discussed in Chapter 6.

7.5 Multiple Signal and Multipath Problems

Significant errors in measuring the received signal power may occur due to the inadvertent reception of more than one signal at one time or through multipath. In the case of a pulsed signal, it is usually possible to isolate the desired signals from others in the receiver pass band through the time separation of the pulses even if the carrier frequencies of the two signals are quite close. Usually, the two signals will differ considerably in power, and it is possible to trace the amplitude envelope of one signal as it scans by in spite of interference from another pulsed (or CW) signal. Figure 7.5 shows the pulse amplitude variations over time of two overlapping radar scans; separation of these signal pulses, at least manually, is clearly possible using their amplitude versus time functions.

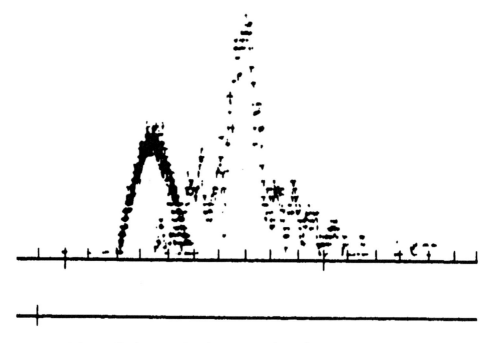

Figure 7.5 Pulse amplitude versus time for two scanning radars.

In the case of CW signals, it is necessary to rely on frequency separation. However, it may be possible to use much narrower receiver bandwidths, depending on the modulation in use.

Multipath presents a difficult problem in that the interfering signal is a precise replica of the signal of interest, except that it is delayed in time, its amplitude is different, and its phase relative to that of the direct signal is unpredictable. The best protection in this situation is to emphasize the leading edge of pulsed signals. If the rise time of the pulse is larger than the delay introduced by the greater length of the indirect path, this method also fails. The path length difference produces a delay given by

$$\Delta\tau = \frac{\Delta L}{c}$$

where:

$\Delta\tau$ = multipath time delay

c = speed of light (3×10^8 m/s)

ΔL = path length difference (meters)

Many rise times are less than 300 ns, so that multipath differences of 100m or more produce multipath interference that can be discriminated from the direct path by time differences. Very large path length differences reduce the amplitude of the multipath. Also, if the arrival angle of the multipath differs from that of the direct path (as when reflections occur from objects near the ELINT station), the gain of the ELINT antenna may be used to help reduce their amplitude.

Because multipath signals are exact coherent replicas of the direct path signal (except for an amplitude change and a phase shift), the interference signal can either add or subtract from the direct path signal. If the multipath and direct signals were equal in amplitude, the amplitude of their vector sum could increase by as much as 6 dB or decrease to 0 depending on the phase shift. The length squared of the sum of two vectors is

$$|\vec{A}_1 + \vec{A}_2|^2 = A_1^2 + A_2^2 + 2A_1 A_2 \cos\theta \qquad (7.11)$$

where A_1 and A_2 are the two signal amplitudes, and θ is the phase angle between the two signals.

If the ratio of the multipath power to the direct power, A_2^2/A_1^2 is denoted as PR, the error in the power estimate is

$$\frac{|\vec{A}_1 + \vec{A}_2|^2}{|A_1|^2} = 1 + \text{PR} + 2\cos(\theta)\sqrt{\text{PR}} \qquad (7.12)$$

Figure 7.6 shows the power measurement error due to the multipath as a function of the power ratio (PR) of the indirect signal to the direct signal for in-phase multipath and for 45°, 90°, and 180° phase differences. For 180° phase

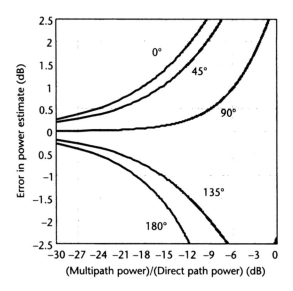

Figure 7.6   Power measurement error due to multipath.

difference, there is cancellation and the power estimate is too low. The multipath power ratio range is from −25 dB for a 0° phase difference to about −9.9 dB for a 90° phase difference if the error in the power estimate is to be held to 0.5 dB. For a 180° phase difference, the power error is −0.5 dB for a power ratio of −25 dB. It turns out that the rms error in ERP for uniformly distributed θ exceeds 0.5 dB when the multipath reflection exceeds −21.8 dB. Because the phase is unpredictable, maintaining an ERP error of 0.5 dB rms requires that the multipath power be 21.8 dB lower than the direct path power.

7.6   Summary of Power Measurement Requirements

This chapter has shown what is necessary to measure the ERP of the radar to within 0.5 dB for each of the following factors:

1. Pointing the ELINT antenna (within 15% of its 3-dB beamwidth);
2. Polarization mismatch (for linear polarization must be less than 19.25°);
3. Bearing accuracy for triangulation (better than 1.04° or 3.1° for the two cases considered).
4. Ratio of multipath power to direct path power (must be less than −21.8 dB).

   If all four of these errors are maintained to produce a peak error of less than 0.5 dB, combining them on an rms basis gives the total expected ERP measurement error as about ±1 dB and a worst case error of ±2 dB. Considering the many other sources of error (e.g., calibration of the antenna, receiver gain, and uncertainty about propagation path attenuation), ERP measurements more accurate than ±2 dB are rarely possible.

7.7 Sample ERP Calculations

To illustrate the points of this chapter, it is helpful to consider a sample ERP calculation. Table 7.1 shows the results of two ELINT stations located 100 km apart intercepting the same signal. To find the distance from each emitter, (7.9) can be used, or plane trigonometry may be applied to yield the results shown in Table 7.2.

The other parameter needed is the wavelength. This is found by dividing the speed of light by the RF, which gives a wavelength of 0.1m. Then, using the values from station 1, the ERP in dBm is found using Table 7.3.

It is essential to make sure the units are correct, namely, that the range and wavelength have the same units. The next step is to decide if this value is reasonable for the type of signal reported. If the antenna gain is about 25 dB (a typical value), then the transmit power would be in the range of 51.2 to 65 dBW—about 130 to 3,000 kW. If this represents the peak of the main beam, then it is clear that this high power radar could be expected to have a long PRI. The uncertainty in the ERP of nearly 14 dB is rather typical of the type of errors inherent in ERP measurements using only a single pair of reports. Improved accuracy can be achieved, especially by a more exact emitter location or by averaging many reports from many stations. Nevertheless ERP measurements are subject to many uncertainties.

Another way to check the reasonableness of the distance to the emitter is to compare the difference in the received power levels (in decibels) to that expected

Table 7.1 Two ELINT Stations Intercepting the Same Signal

	Station 1	Station 2
ELINT antenna gain (dBi)	10 ± 2	9 ± 1
Signal bearing (degrees)	90 ± 4	60 ± 5
Peak received power (dBm)	-24 ± 1	-26 ± 1
RF (MHz)	$3,000 \pm 1$	$3,000 \pm 1$

Table 7.2 Determining the Distance from Each Emitter

Angle A	Angle B	Angle C	Side a	Side b	Side c
30	90	60	100	200	173.2
21	94	65	100	278.4	252.9
39	86	55	100	158.5	130.2

Table 7.3 Values from Station 1

Parameter Name	Parameter Value	Parameter dB
Constant	4π	22
Received power		-24 ± 2 dBm
Maximum range	253 km	$+108$ dB (m^2)
Minimum range	130 km	$+102.3$ dB (m^2)
ELINT antenna gain		-10 ± 2 dBi
Wavelength	0.1m	$+20$ dB(m^2)
Total (ERP)		120 dBm maximum
		106.2 dBm minimum

from the difference in range to the two stations. In this case, if the longest ranges are used, the power level difference would be 0.83 dB. If the shortest ranges are used, the difference would be 1.71 dB. From the reported power and antenna gain values, the difference could be anywhere in the range of –2 to +8 dB. In this case, there is no inconsistency with the difference expected from geometrical considerations.

It can be shown that if the receive sites are separated by a unit distance, then the contours of the constant received power ratio, K, are circles centered on the baseline extended a distance of $1/(K − 1)$ times the distance between the receive sites, with radius of the square root of K divided by $(K − 1)$. In this formulation, K is larger than 1 (i.e., it is the ratio of the power from the stronger station divided by the power from the weaker station after compensating for any differences in receive antenna gain). For a power ratio of 3 dB or 2:1, $K = 2$, the radius of the circle is 1.414 times the distance between the stations, and the center of the circle is along the baseline extended a distance equal to the distance between the stations. These circles are illustrated in Figure 7.7.

Suppose the TOA of each pulse can be measured accurately at each station. The time difference is given by the difference in the distance to the emitter from each station divided by the speed of light. In this example, the maximum difference in distance is 28.3 km and the minimum difference is 25.5 km. These correspond to time differences of 94.3 to 85 μs. Because time differences can be easily measured to an accuracy of 0.1 μs, the difference in the distance to the two stations can sometimes be measured to an accuracy of 30m. Determining the complete emitter location requires a pair of stations for each dimension to be found—three-dimen-

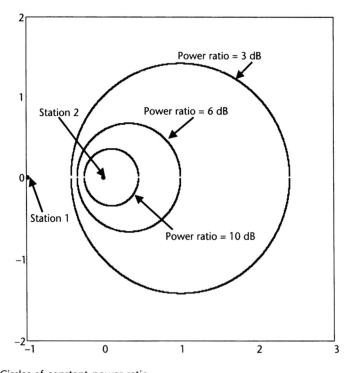

Figure 7.7 Circles of constant power ratio.

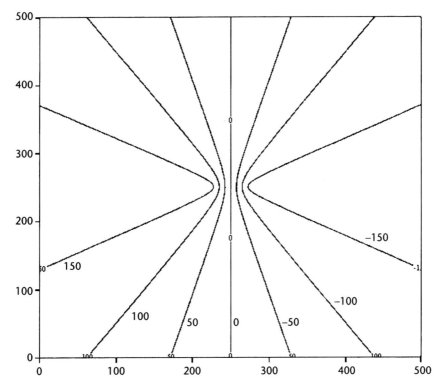

Figure 7.8 Lines of constant time difference. Station 1 is at (225,0) and station 2 is at (275,0) (TDOA in microseconds; location coordinates in kilometers).

sional emitter locations require at least four stations and they must not all be in a single geometric plane. Note that there is a problem deciding which pulse at one station is the same pulse at another station. This is difficult if the environment is so dense that many pulses occur in a time comparable to the time difference. Also, if the time difference exceeds the radar PRI, an ambiguous situation exists that must be resolved. Lines of constant time difference are illustrated in Figure 7.8 for two stations 50 km apart. The vertical line is the line of zero time difference and is the perpendicular bisector of the line between the stations or baseline. The hyperbolic curves are lines of 50, 100 and 150 μs of time difference.

References

[1]   Skolnik, M. I., *Radar Handbook*, New York: McGraw-Hill, 1970.

[2]   Skolnik, M. I., *Radar Handbook*, New York: McGraw-Hill, 1970, p. 2–61.

CHAPTER 8

Antenna Parameters

The antenna is an extremely important part of radar. It must couple the energy from the radar to and from space; it must radiate in the desired direction and discriminate against receiving radiation from other directions. The ELINT-observable antenna parameters are chiefly its polarization characteristics, the shape of its main beam, and its sidelobe levels. This chapter begins by considering aspects of polarization and continues with consideration of beam shape and sidelobe levels.

8.1 Polarization Defined[1]

The polarization of an electromagnetic wave is described by the geometric figure traced by the electric field vector as the wave travels through space.[2] The polarization of the transmitted radar signal can be determined at an ELINT station by specialized measurement equipment or simply by manually determining which polarization of the ELINT antenna provides the greatest signal strength from the transmitter. For electronic warfare purposes, it is important to determine the polarization because the effectiveness of a jammer having improper polarization is reduced.

8.2 Elliptical Polarization [1][3]

The geometric figure traced by the electric field vector is, in general, an ellipse. Under some conditions the ellipse may collapse into a straight line, in which case the polarization is called linear. In the other extreme, the ellipse may become circular. Thus, linear and circular polarizations are special cases of elliptical polarization. The parameters used to describe polarization are either those appropriate to describing an ellipse or, for a more general description, the Stokes' parameters may be used. Elliptical polarization can be thought of as the result of two linearly polarized waves at the same frequency. Polarization is determined by the electric field vector, E. If the electric field vector lies in a vertical plane, the wave is vertically polarized. For elliptically polarized waves, the tip of the electric field vector, as a

---

1. The author is grateful for the help of Mr. Grover M. Boose in preparing the polarization sections.
2. This is the convention in electrical engineering. In physics and optics, polarization refers to the magnetic field vector.
3. Some of this material is a condensation of Section 15.10 in [1].

function of time, traces out an ellipse. Circular polarization is a special case of an ellipse whose major and minor axes are equal.

Following Kraus [1, 2], let the components of the instantaneous electric field of the horizontally and vertically polarized wave be designated E_x and E_y. Then, as a function of time, t, and distance, z,

$$E_x = E_1 \sin(\omega t - \beta z) \tag{8.1}$$

$$E_y = E_2 \sin(\omega t - \beta z + \delta) \tag{8.2}$$

where δ is the phase angle between E_x and E_y. In general, the tip of E describes a locus that is an ellipse. This may be shown by proving that (8.1) and (8.2) with $z = 0$ are the parametric equations of an ellipse. With $z = 0$,

$$E_x = E_1 \sin(\omega t) \tag{8.3}$$

and

$$E_y = E_2 \sin(\omega t + \delta) \tag{8.4}$$

Expanding (8.4), eliminating ωt, and rearranging terms, leads to

$$a E_x^2 - b E_x E_y + c E_y^2 = 1 \tag{8.5}$$

where:

$a = 1/E_1^2 \sin^2 \delta$

$b = 2 \cos \delta/E_1 E_2 \sin^2 \delta$

$c = 1/E_2^2 \sin^2 \delta$

Equation (8.5) is the equation for an ellipse in its most general form. Figure 8.1 shows an illustration of a polarization ellipse where the axes of the ellipse do

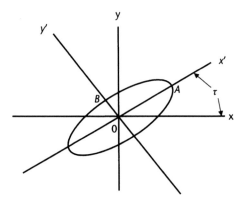

Figure 8.1 Polarization ellipse.

not correspond to the x- and y-axes. The ratio of the semimajor axis, OA, to the semiminor axis, OB, is called the axial ratio, ar:

$$ar = \frac{OA}{OB} \tag{8.6}$$

The axial ratio can also be determined from the amplitudes of the horizontally and vertically polarized waves, E_1 and E_2, and the phase angle, δ, between them:

$$ar = \frac{\sqrt{E_1^2 + E_2^2 + 2E_1E_2 \sin \delta} + \sqrt{E_1^2 + E_2^2 - 2E_1E_2 \sin \delta}}{\sqrt{E_1^2 + E_2^2 + 2E_1E_2 \sin \delta} - \sqrt{E_1^2 + E_2^2 - 2E_1E_2 \sin \delta}} \tag{8.7}$$

Kraus [1] derives an expression for the tilt angle, τ, where:

$$\tau = 0.5 \tan^{-1} \frac{2E_1E_2 \cos \delta}{E_1^2 E_2^2} \tag{8.8}$$

The angle, τ, as given by (8.8) is the angle between the x-axis and either the major or minor axis of the ellipse. The same result is obtained from (8.8) if the ellipse is as shown in Figure 8.1 or if the ellipse is rotated 90° so that the major axis is in the y-direction.

Elliptical polarization is completely characterized by the axial ratio, ar, of the polarization ellipse; the sense of rotation; and the spatial orientation of the ellipse as determined by the tilt angle, τ. Linear polarization has a tilt angle of 0° because either E_1 or E_2 is zero. Circular polarization has an axial ratio of 1.

The sense of rotation of E that describes either a circle or an ellipse in the plane of polarization (perpendicular to the direction of propagation) is called the sense of polarization or handedness [3]. For an observer looking in the direction of propagation, the sense is called right handed if the direction of rotation is clockwise and left handed if the direction is counterclockwise. In ELINT applications, the sense of the polarization of the radar transmitter is determined when facing the ELINT antenna. Figure 8.2 illustrates the polarization for different horizontal and vertical component amplitudes and phase angles.

8.3 Stokes' Parameters [2]

Another way to describe polarization is through the use of Stokes' parameters. These are often used to describe radiation from celestial radio sources. Such radiation extends over a wide frequency range and within any bandwidth, Δf, consists of the superposition of a large number of statistically independent waves of a variety of polarizations. The resultant wave is said to be unpolarized. For such a wave, (8.3) and (8.4) are still useful, but E_1, E_2, and δ are now random variables.

A wave of this type could be generated by connecting one noise generator to a vertically polarized antenna and another noise generator to a horizontally polarized

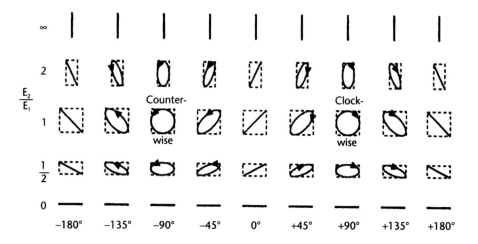

Figure 8.2 Polarization ellipse as a function of E_1/E_2 and the phase angle. (*From:* [1]. © 1950 McGraw-Hill, Inc. Used with permission of McGraw-Hill.)

antenna having the same phase center. If the waves appear at an observation point having the same average power, the total wave will be completely unpolarized. A partially polarized wave is one where one part may be considered completely polarized and the other part completely unpolarized.

Stokes' parameters may be used to characterize all waves, including partially polarized ones. Stokes' parameters are defined as follows:

$$I = S = S_x + S_y = \frac{\langle E_1^2 \rangle}{Z} + \frac{\langle E_2^2 \rangle}{Z} \tag{8.9}$$

$$Q = S = S_x - S_y = \frac{\langle E_1^2 \rangle}{Z} - \frac{\langle E_2^2 \rangle}{Z} \tag{8.10}$$

$$U = \frac{2}{Z} \langle E_1 E_2 \cos \delta \rangle \tag{8.11}$$

$$V = \frac{2}{Z} \langle E_1 E_2 \sin s\delta \rangle \tag{8.12}$$

where S is the Poynting vector; S_x and S_y are the x and y components of the Poynting vector; Z is the intrinsic impedance of the medium; E_1, E_2, and δ are as defined previously; and the angle brackets indicate the time average.

Normalized Stokes' parameters are often used and are obtained by dividing (8.9) through (8.12) by S, where $S_0 = I/S$, $S_1 = Q/S$, $S_2 = U/S$, and $S_3 = V/S$. If any of the parameters, Q, U, or V (or S_1, S_2, or S_3), have a nonzero value, it indicates the presence of a polarized component in the wave. The degree of polarization, d, is defined as the ratio of completely polarized power to the total power, or

$$d = S_1^2 + S_2^2 + S_3^2 \tag{8.13}$$

If $S_1^2 + S_2^2 + S_3^2 = 1$, the wave is completely polarized.

For a completely polarized wave, Stokes' parameters do not require time averages and the brackets denoting time average can be removed from (8.9) through (8.12).

Substituting into (8.7) and (8.8), the axial ratio and tilt angle in terms of Stokes' parameters are

$$ar = \frac{\sqrt{S_0 + S_3} + \sqrt{S_0 - S_3}}{\sqrt{S_0 + S_3} - \sqrt{S_0 - S_3}} \tag{8.14}$$

and

$$\tau = 0.5 \tan^{-1} \frac{S_2}{S_1} \tag{8.15}$$

Table 8.1 gives the values of the normalized Stokes' parameters for seven polarized wave states. Stokes' parameters are measurable using an antenna system and some amplitude signal processing.

Jamming signals may require Stokes' parameters to be measured to describe their polarization.

8.4 Measuring Polarization

There are a number of techniques to measure the polarization of a wave, as can be seen by the listing shown in Table 8.2. IEEE STD 149-1979 lists three broad categories and four measurement techniques. Kraus [1] listed three techniques, but two are essentially the same phase-amplitude techniques using different antennas. Basically there are three techniques that could be used at a collection site:

1. Polarization pattern method;
2. Phase-amplitude method;
3. Multiple antenna (amplitude-only measurement) method.

For wideband operation, the polarization pattern method is not as easily implemented as the other two methods. Modern receiver systems make phase-amplitude

Table 8.1 Normalized Stokes' Parameters for Seven Polarization Wave States

Normalized Stokes' Parameter	Completely Unpolarized Wave	Completely Polarized Waves					
		Linearly Polarized				Circularly Polarized	
		$\tau = 0°$	$\tau = 90°$	$\tau = 45°$	$\tau = 135°$	Left Hand	Right Hand
S_0	1	1	1	1	1	1	1
S_1	0	1	−1	0	0	0	0
S_2	0	0	0	1	−1	0	0
S_3	0	0	0	0	0	1	−1

Source: [2].

Table 8.2 Polarization Measurement Techniques

Kraus [1]	IEEE STD 149-1979	Kanareikin [4]	Hollis et al. [5]
1. Polarization pattern method using rotating linear antenna and includes a measurement of sense 2. Linear component method: measure amplitude of two orthogonal linear polarized components and phase angle between them 3. Circular component method: same as #2, only using circular polarized (with opposite senses) components.	Broad Categories 1. Techniques that give partial information about antenna's polarization 2. Techniques that give complete polarization information but require comparison to a polarization standard (comparison techniques) 3. Techniques that require no comparisons and provide complete polarization information on an antenna (absolute methods) Measurement Techniques 1. Polarization pattern method 2. Rotating source method 3. Multiple-amplitude component method 4. Phase-amplitude method	1. Polarization diagram method 2. Compensation method 3. Orthogonally polarized components phase and amplitude 4. Method of several or one antenna with variable polarization 5. Modulation method	1. Phase-amplitude of linear polarized components 2. Phase-amplitude of circular polarized components 3. Multiple antenna amplitude measurements 4. Polarization pattern to get tilt angle and axial ratio; use separate antenna for sense

measurements for fine-grain analysis. The multiple antenna, or amplitude-only, method makes use of Stokes' parameters and amplitude-only measurements from four antennas simultaneously or from one antenna capable of variable polarization.

8.4.1 Polarization Pattern Method

The polarization pattern method makes use of a rotating linearly polarized antenna. The received signal level is recorded at each position of the rotating antenna and is a measure of the component of the E field in the direction of the rotating antenna at any instant. If the radar is also scanning, then the rotation of the receiving antenna's polarization must be fast enough to complete several revolutions within the time the main beam is directed toward the intercept site.

These measurements provide a polarization pattern from which the polarization ellipse can be determined. From the polarization pattern, the axial ratio and tilt angle can be determined (see Figure 8.3).

To determine the polarization sense, an additional measurement is required. Generally, the outputs of two circularly polarized antennas with opposite senses (one right handed and one left handed) are compared in order to determine the polarization sense. Thus, by specifying the axial ratio, tilt angle, and polarization

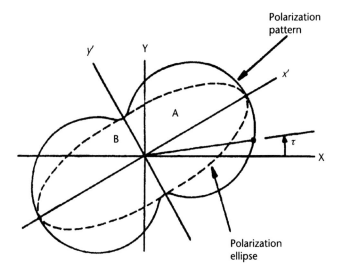

Figure 8.3   Polarization patterns and polarization ellipse.

sense, the polarization characteristics of a wave are completely described. Figure 8.4 shows photographs of the polarization pattern for linear, elliptical, and circularly polarized waves [6].

8.4.2   Phase-Amplitude Method

The phase-amplitude method requires that either two orthogonal, linearly polarized antennas or two circularly polarized antennas of an opposite polarization sense be connected to a receiver that can measure the amplitude of each component and the phase angle between them.

From a knowledge of E_1 and E_2, along with the phase angle, the axial ratio can be determined from (8.7), the tilt angle from (8.8), and the direction of rotation from (8.3) and (8.4). Kraus [1] has created a chart of polarization ellipses as a function of the ratio E_2/E_1 and the phase angle δ. In this chart, reproduced above

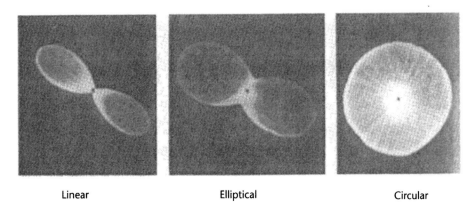

Linear   Elliptical   Circular

Figure 8.4   Photographs of polarization patterns. (*From:* [6]. © 1951 Horizon House, Inc. Reprinted with permission.)

as Figure 8.2, the wave is approaching the viewer; therefore, a clockwise wave approaching the viewer is a left-hand circular polarization and a counterclockwise wave is a right-hand circular polarization.

To make polarization measurements over a broad frequency range, one could use a broadband horn antenna as the feed for a parabolic dish with one probe for horizontal polarization and another for vertical. Crossed log periodic antennas could also be used as the feed to a dish. A receiver that measures the phase difference between two pulsed signals permits evaluation of the polarization in subsequent processing. Errors result if there is more than one signal in the receiver's pass band.

One method to display polarization is to mix the pulsed RF signals to a frequency within the bandwidth of an x-y oscilloscope. Monitoring the Lissajous figure gives an indication of the polarization. As an antenna sweeps (as in a circular scan) through the beam of a receiving antenna, the indicated polarization can be expected to vary, especially if multipath is present. When the peaks of the two beams are "looking" at each other, the polarization can best be measured (beam-on beam data).

8.4.3 Multiple Antenna Method

It is possible to measure polarization by amplitude measurements from multiple antennas without making phase measurement [7, 8]. The procedure outlined here based on Kraus [7] makes use of Stokes' parameters. As shown by (8.7) and (8.8) and (8.14) and (8.15), the parameters of a polarization ellipse can be derived from either the Stokes' parameters or from phase-amplitude measurements.

Let the antenna system consist of four linearly polarized antennas and two circularly polarized antennas with an opposite polarization sense, as shown in Figure 8.5. The received wave is incident on each antenna in a normal direction. The output power (signal level) from each antenna is measured either by separate receivers or one receiver sampling each antenna. The power received from each antenna may then be designated by:

P_x = horizontal antenna

P_y = vertical antenna

P'_x = slant 45° antenna

P'_y = slant 135° antenna

P_L = left-hand circular antenna

P_R = right-hand circular antenna

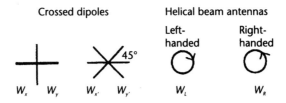

Figure 8.5 Six antennas used for polarization measurements of an unknown wave. Note that for ELINT uses, the antennas should have the same phase center. (*From:* [7]. © 1982 John Kraus. Reproduced by permission.)

All six antennas should provide an equal signal level to a completely unpolarized wave (i.e., have the same effective aperture). From (8.9) through (8.12),

$$S_0 = \frac{I}{I} = \frac{P_x + P_y}{P_x + P_y} \tag{8.16}$$

$$S_1 = \frac{I}{I} = \frac{P_x + P_y}{P_x + P_y} \tag{8.17}$$

$$S_2 = \frac{U}{I} = \frac{P'_x - P'_y}{P_x + P_y} \tag{8.18}$$

$$S_3 = \frac{V}{I} = \frac{P_L + P_y}{P_x + P_y} \tag{8.19}$$

Assuming the sums of each of the power pairs $(P_x + P_y)$, $(P'_x + P'_y)$, and $(P_L + P_R)$ are unity, the normalized Stokes' parameters S_1, S_2, S_3 are then related as shown in Table 8.3.

From Table 8.3, it is clear that the outputs of pairs of antennas are either summed (S_0) or differenced $(S_1, S_2,$ and $S_3)$ to obtain the normalized Stokes' parameters. Although the same data (normalized Stokes' parameters) could be obtained from four antennas versus the six used here, the use of four antennas complicates the calculations beyond the simple sum and differences of this method [7]. Six antennas are recommended and have proven useful in data collection systems [8]. Kanareikin [5] presents, in considerable detail, polarization measurement techniques and equipment implementations.

8.5 Cross-Polarization

Practical antennas radiate some energy with polarization orthogonal to that intended. This is referred to as cross-polarized radiation. The cross-polarized signal from a reflector antenna increases as the curvature increases. This is shown in Figure 8.6, which gives the cross-polarization level in terms of the power loss in the main pattern. Note that although the loss seems small, 0.1 dB represents 2.3% of the power, which may be observed by ELINT. In this figure, the half angle refers to the angle between a line representing the boresight direction of the reflector and a line drawn from the focus to the edge of the reflector.

Table 8.3 Relation of Normalized Stokes' Parameters to Antenna Pairs

Stokes' Parameters	Vertical and Horizontal Antennas	Slant Antennas	Circular Antennas
S_0	Sum	Sum	Sum
S_1	Difference		
S_2		Difference	
S_3			Difference

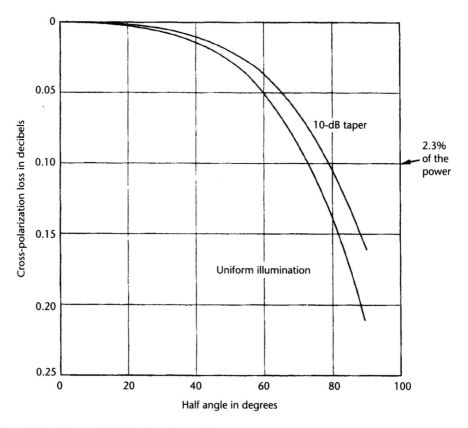

Figure 8.6 Cross-polarization loss for a reflector antenna.

The cross-polarized signal level may actually exceed that of the proper or co-polarized signal when sidelobes are received. If the ELINT site is unknowingly using a cross-polarized antenna, and then finds the peak signal level, this will not represent the true ERP of the radar. Because it is common to not be in the main beam of the radar, the effect of cross-polarized energy can be more significant for the ELINT observer than it is for the radar designer. Consider the measurements shown in Figures 8.7 and 8.8 (furnished by Advent Systems, Inc.). Figure 8.7 is the power received from an ASR-3 radar when using a vertically polarized antenna to match the vertical polarization of the ASR-3. The peak of the radar beam in elevation is 3° above the ELINT site; therefore, this pattern is taken 7.4 dB below the peak of the beam. Figure 8.8 shows the cross-polarized pattern taken with a horizontally polarized receiving antenna. Although the peak of the beam is about 10 dB lower on the cross-polarized pattern, the cross-polarized pattern is actually above the copolarized pattern at many azimuths. Note the dip in the center of the main beam of the cross-polarized pattern. Clearly, the ELINT site must examine the received signal using several polarizations on receive to learn about the transmitted polarization.

The polarization tilt angle is shown in Figure 8.9 for 2.6° on each side of the peak of the beam in azimuth. Three elevation cuts are shown: at the peak of the beam in elevation and 3° above and below the elevation boresight. Inside the 3-dB points of the azimuth beam, the axial ratio basically indicates vertical polarization.

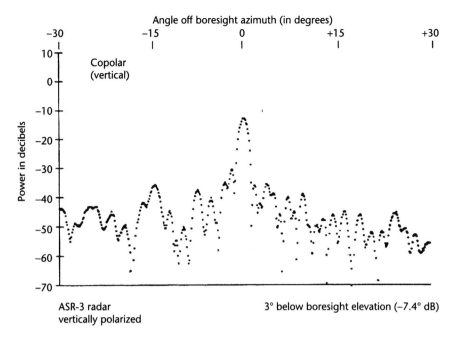

Figure 8.7 Copolarized pattern of an ASR-3 radar antenna.

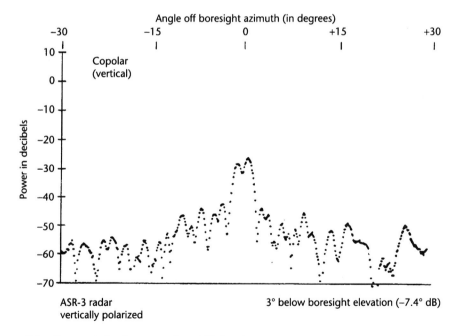

Figure 8.8 Cross-polarized pattern of an ASR-3 radar antenna.

However, the axial ratio indicates polarization varying from about 45° to 150° over just 5.2° of azimuth. This shows the importance of measuring the polarization in the main beam. Complete reversals of the polarization occur in the sidelobes of the pattern.

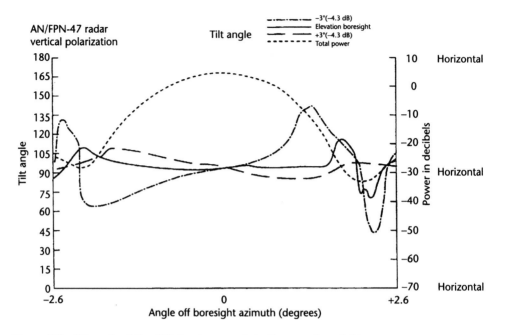

Figure 8.9   Tilt angle of the ASR-3 antenna polarization near boresight.

The consequences of unknowingly using a cross-polarized antenna on receive can be severe. Not only is the ERP greatly underestimated, but the bearing to the emitter, as determined by the peak signal when the receive antenna is rotated, can be in error. This happens because the peak of the response of the receive antenna in the cross-polarized mode is generally offset from the boresight of the copolarized beam. Likewise, the pulse shape of the weaker cross-polarized signal may be corrupted by multipath and noise.

8.6   Propagation Effects

As a wave propagates through the atmosphere, the polarization is rotated. This is due to the Faraday effect and is caused by free electrons. Figure 8.10 shows the amount of rotation for a two-way trip through the entire atmosphere at several elevation angles during the day and during the night. There is less rotation during the night because the Sun's energy is largely responsible for creating the free electrons. There is less rotation at the higher elevation angles because the path of the wave traverses less atmosphere vertically (90°) than at the horizon (0°). At higher frequencies, the rotation is quite small—less than 0.1 radian above 5 GHz in the worst case: of two-way propagation through the entire atmosphere at the horizon in the daytime. At 1 GHz, under the same conditions, there may be about 5 radians of rotation. At short range, and for one-way ELINT propagation, this effect is often negligible. At lower frequencies and long range, however, Faraday rotation must be considered and taken into account.

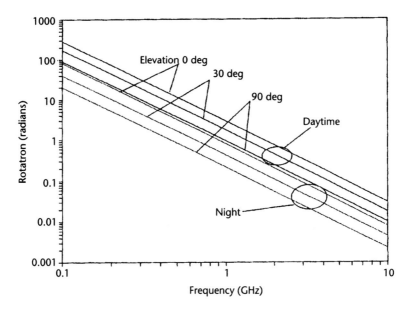

Figure 8.10   Two-way polarization rotation for a target at a height of 1,000 km.

8.7   System Aspects of Polarization

The choice of polarization for a radar design may be made on the basis of the function of the radar. Ground or sea reflections are much greater for horizontal than for vertical polarization. A long-range search radar can take advantage of this by obtaining twice the detection range at certain elevation angles and no coverage at other elevation angles [9]. For an airborne radar (or ground radar) to detect targets in or behind heavy rain cells, circular polarization is generally selected. There is a polarization effect on a radar cross-section that depends on the target area (e.g., a long thin wire produces little reflection when the incident field is linearly polarized at right angles to the wire axis and produces maximum reflection when the wave polarization is parallel to the axis of the wire). For any given target and aspect angle, there is a polarization of the incident field (wave) that gives a maximum echo (received signal). Conversely, there is a polarization that also gives a minimum signal. Therefore, use of polarization diversity by the radar could maximize the signal return. Some immunity from jamming is also obtained (unless a noise jammer uses two orthogonal linear antennas with noise generators to produce an unpolarized wave).

The selection of a particular polarization for the same function may be different from one location (country) to another. In the United States, for instance, television band antennas are horizontally polarized. This choice was made to maximize the SNR because most man-made noise sources are vertically polarized. However, in some other places, the same television band uses vertical polarization because it provides the maximum signal strength from a wave-propagation point of view. Propagation at various frequencies can sometimes dictate a particular polarization, as shown in Table 8.4 [10].

Table 8.4 Polarization Choice Versus Frequency

Frequency	Polarization	Comment
VLF	Vertical	Horizontal polarization will be cancelled by image antenna
HF	Mostly vertical	
UHF/VHF	Vertical or horizontal	Some horizontal may be used for specific application
Microwave (above 1,000 MHz)	Choice may be made based on function	Vertical provides maximum signal. Horizontal discriminates against man-made noise: 1. Circular for detection of targets in rain or for transmission through the ionosphere at any frequency due to Faraday rotation 2. Horizontal for clutter discrimination 3. Vertical to minimize lobing due to the ground 4. Systems design (antenna and radome design may affect polarization choice)

Polarization measurements can aid in understanding the radar function and in analysis of the ELINT data. Detection of a rotating linear polarization may lead to the conclusion that the radar has a conical scan mode.

Table 8.5 gives the polarization loss (take $10 \log P_R/P_{max}$ for the loss in decibels) for various transmit and receive antenna configurations. For two linearly polarized antennas, the polarization power loss is inversely proportional to the cosine of the angle between their polarizations.

Knowledge of basic information regarding wave polarization is essential. Sometimes polarization measurement is overlooked due to the need for special equipment. Advances in antennas, microwave components, and signal processing gear may mean that polarization measurements will receive more attention in the future. The interested reader is directed to [1] for basic information and to [3] for an excellent bibliography on all subjects related to antenna measurements.

8.8 Antenna Beam Shape

The usefulness of radar depends heavily on the ability to determine the direction to targets of interest. Indeed, even in the absence of target range, the direction

Table 8.5 Polarization Loss for Various Transmit/Receive Antenna Combinations

Transmit Antenna Polarization	Receive Antenna Polarization	Ratio of Power Received to Maximum Power
Vertical	Vertical	1
Vertical	Slant (45° or 135°)	1/2
Vertical	Horizontal	0
Vertical	Circular (right hand or left hand)	1/2
Horizontal	Horizontal	1
Horizontal	Slant (45° or 135°)	1/2
Horizontal	Circular (right hand or left hand)	1/2
Circular (right hand)	Circular (right hand)	1
Circular (right hand)	Circular (left hand)	0
Circular (right hand or left hand)	Slant (45° or 135°)	1/2

to the target may be sufficient information to engage it. Therefore, the angular measurement capability of the radar is a key aspect of its design.

This being the case, the ELINT analyst is charged with determining (as much as possible) all of the radar's directional properties. Many radar systems make use of a highly directional antenna that is used for both transmitting and receiving. The antenna pattern of this type of radar usually can be determined rather completely from ELINT. Systems that use separate transmit and receive antennas are less amenable to ELINT analysis because little can be learned about the receive antenna. The assessment of radar antenna capabilities and characteristics is much more successfully completed if the radar antenna can be seen or photographed in addition to the reception of the radar signal by an ELINT station. Photographs are often available from sources such as trade publications, export sales versions, parades, or photoreconnaissance. The fundamental difficulty is that there is often no way to associate a photograph of a particular antenna configuration to the radar signal received at an ELINT station. Analysis approaches that require knowledge of the antenna's physical configuration as well as the ELINT information are often used.

Except in the case of radar tracking a target, the radar may search for targets by moving its directional beam in azimuth and elevation—either mechanically or electronically. This motion of the beam past an ELINT station provides an indication of the antenna's radiation pattern. Determining the properties of the directional beam pattern is called beam analysis. Determining the types, rates, and angular extent of the beam motion is called scan analysis. Scan analysis is the subject of Chapter 10. However, insight into the beam pattern generally requires knowledge of the scanning action.

8.9   Basic Antenna Pattern Relationships

The antenna pattern can be characterized by parameters such as its 3-dB azimuth and elevation beamwidths, gain, and sidelobe levels. It can also be described by plots of the field strength (at a fixed distance from the antenna) as a function of angle. Historically, the antenna theory is based on the theory of the diffraction of light. As the antenna size increases with respect to the wavelength, the approximations improve.

Basically, the far field of the antenna pattern can be thought of as the sum of the contributions from various parts of the radiating[4] area. These contributions add vectorially, and the various phases create peaks and valleys in the radiation pattern. The radar antenna is usually designed to concentrate the radiation in a single direction,[5] both to locate the target directionally and to concentrate the available power for increased detection range. The availability of a sharply defined beam alone does not provide the azimuth and elevation directions to the target, because a means of accurately pointing the beam is also needed.

---

4.   While "radiation" is the term used, antenna structures for our purposes are reciprocal and can be used for reception as well as transmission.
5.   An important exception is the difference pattern of monopulse radar receivers.

The far field of the antenna pattern refers to the region beyond [11]

$$\frac{2D^2}{\lambda}$$

where D is the maximum antenna dimension and λ is the wavelength. Also, the distance must exceed eight wavelengths. Obviously, ELINT observations are in the far field.

Radar antennas are characterized by an *aperture* that is *illuminated* with electromagnetic energy. For example, a parabolic reflector combined with a feed horn produces a constant phase in a front of the reflector. The illuminated aperture, therefore, is thought of as a plane area directly in front of the reflector and feed assembly and normal to the conic axis or line drawn through the center of the reflector and its focal point. The illumination function describes the variation of the excitation across the aperture. The far field for many aperture illumination functions can be described as the Fourier transform of the illumination function. This means that narrow beams in the far field require large apertures. Likewise, illumination functions that are constant across the aperture and drop suddenly at the aperture edge produce high sidelobes, while those illumination functions that taper off gradually toward the edge of the aperture produce lower sidelobes but less sharply defined beams. Antenna theory shows that the sharpest beam (and highest gain) is obtained when the illumination is constant over the aperture. The beamwidth is defined as the angular width of the main beam between the points of the far-field pattern that are 3 dB below the peak. The directive gain of the antenna is the ratio of the peak of the far-field pattern to the total pattern averaged over the entire spherical angle of 4π. Thus, losses in the antenna are ignored in determining the directive gain. While the ELINT analyst may be able to determine the directive gain, the actual power gain (directive gain reduced by antenna losses) cannot be determined from ELINT alone.

For a given aperture, uniform illumination provides the maximum directive gain. The actual directive gain of an actual aperture therefore is less than this value:

$$G_D \leq \frac{4\pi A}{\lambda^2} = \frac{4\pi L W}{\lambda^2} \tag{8.20}$$

where:

G_D = directive gain of aperture

A = area of aperture (m^2)

L = length of rectangular aperture

W = width of rectangular aperture

λ = wavelength (m)

For uniform aperture illumination, the directive gain can be related to the 3-dB beamwidths of the pattern by the approximation:

$$G_D \approx \frac{4\pi}{\theta_{AZ}\,\theta_{EL}} \tag{8.21}$$

where:

G_D = directive gain

θ_{AZ} = azimuth bandwidth (rad)

θ_{EL} = elevation beamwidth (rad)

Equation (8.21) is the ratio of the area of a unit sphere to that part of the area covered by a narrow beam of angular extent $\theta_{AZ} \times \theta_{EL}$. Converting this to beamwidths measured in degrees requires multiplication by $(180/\pi)^2$. Combining (8.20) and (8.21) yields

$$LW \approx \frac{\lambda^2}{\theta_{AZ}\,\theta_{EL}} \tag{8.22}$$

Thus, estimating the beamwidths provides insight into the aperture's dimensions and vice versa. This is a valuable aid in deciding whether a photograph of an antenna is associated with a particular signal. Likewise, the sidelobe levels can be helpful in determining the character of the illumination function and the extent to which the aperture size exceeds the value given by (8.22).

8.10   Beam Patterns from ELINT

As a simple example, consider a search radar with an elevation fan beam scanning circularly in azimuth. The ELINT signal level at an ELINT station varies with the angular position and reaches a peak when the radar antenna is directed toward the ELINT station. This variation is repeated, and one main beam peak is separated in time from the next by the rotation time of the circular scan. If the signal level is accurately recorded over a wide dynamic range, the ELINT analyst obtains the azimuth antenna pattern. From this, the angular position from the main beam peak can be related to the time base of the plot because the separation of the main beam peaks corresponds to 360°.

It is difficult to maintain an accurate amplitude scale from the receiver antenna over a wide dynamic range. The sidelobe structure of interest may be obscured by receiver noise or multipath effects. Saturation of the receiver may make the 3-dB beamwidth difficult to determine directly; therefore, the time locations of the peaks and nulls can often be discerned more precisely than the −3-dB points. For most practical antennas, the locations of peaks and/or nulls can be related, at least approximately, to the 3-dB beamwidth.

For a rectangular aperture, the antenna gain as a function of angle was given in (7.5). From this the 3-dB azimuth beamwidth is given approximately as

$$\theta_{3\,dB} = \frac{2.78\lambda}{\pi L} \tag{8.23a}$$

Assuming the elevation angle is zero, the first null in azimuth is located at an angle such that

$$\frac{\pi L}{\lambda} \sin \theta_1 = \pi \qquad (8.23b)$$

or

$$\sin \theta_1 = \frac{\lambda}{L} \qquad (8.23c)$$

For small values of θ_1, the ratio of the 3-dB beamwidth to the width between the first nulls is

$$\frac{\theta_{3\,dB}}{2\theta_1} \approx \frac{2.78}{2\pi} = 0.44 \qquad (8.24)$$

For a circular aperture, a similar calculation yields $\theta_{3\,dB}/2\theta_1 \cong 0.42$, as indicated by the graph in Figure 8.11. Similar ratios can be computed for comparing the 3-dB beam width to the separation of the sidelobe peaks.

The 3-dB beamwidth is broadened by tapering of the illumination function to reduce the sidelobes. In fact, a radar having the −13-dB first sidelobes that character-

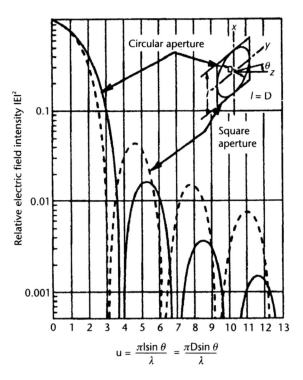

Figure 8.11 Far-field antenna patterns for circular and rectangular apertures (uniform illumination). (*From:* [11]. © 1970 McGraw-Hill, Inc. Used with permission of McGraw-Hill.)

ize a uniformly illuminated rectangular aperture would not be very useful if present on both the transmit and receive antenna patterns. Radar cross-sections of common targets can differ by many decibels, so that a large target return in such sidelobes could obscure a small target return in the main lobe. Likewise, strong clutter returns could arise due to the sidelobe energy striking the ground. Generally speaking, the peak sidelobe level for the two-way antenna pattern must be more than 40 dB below the main lobe for a practical radar system. In radar systems that use active transmit arrays, it is currently not practical to taper the illumination on transmit to reduce the sidelobes. Uniformly illuminated antennas can be termed "ultra high sidelobe antennas." In such systems, sidelobe reduction is achieved on receive by tapering the received signals. Note that in airborne radar systems, it is necessary to keep the aperture as small as possible. The need to avoid returns from the ground through the sidelobes means that the two-way sidelobes are maintained ~40 dB below the main lobe by tapering the aperture illumination function. In ground and shipboard applications, aperture size is not quite as limited. However, the ground or sea is much closer in range. These radars usually have two-way sidelobes more than 40 dB below the main beam, resulting in even greater widening of the beam relative to that of a uniformly illuminated aperture. Very few radars have antennas with one-way sidelobes more than 40 dB below the main beam. Those that do are termed "ultra low sidelobe antennas." As the technology of microwave transmit amplifiers improves and amplitude programming is available, tapering the transmit illumination function may occur on active arrays, which, at present, have high sidelobes on transmit. This is a favorable situation for ELINT work because it allows observation of the transmitted signal's sidelobes at higher levels; however, the sidelobe level on receive is not observable from ELINT. Nevertheless, knowledge of radar system requirements for good performance combined with the observation of the transmit sidelobe level gives insight into the likely range of sidelobe levels on receive.

The ELINT analyst can have a reasonable idea of the probable relationship between the aperture size, 3-dB beamwidth, and sidelobe level on transmit for various classes of radar systems. Uniform illumination produces the highest gain and the highest sidelobe levels. The natural question to ask next is what aperture distribution gives the highest gain and narrowest beamwidth for any given level of peak sidelobes? The answer is the Taylor distribution. To specify this distribution, it is necessary to choose the maximum level of the sidelobes and the parameter $\bar{n}$, which specifies the number of sidelobes having the maximum level. Figure 8.12 illustrates several such distributions, and Figure 8.13 illustrates the beam pattern for sidelobes 25 dB down and $\bar{n}$, equal to 5. From the viewpoint of radar design, this illumination function is difficult to realize for reflector antennas fed by conventional feeds. In fact, the Taylor distribution is generally associated with array antennas. Nevertheless, from the ELINT viewpoint, it provides an interesting limiting case. For an observed sidelobe level, the Taylor distribution gives the maximum possible gain and narrowest beamwidth for a specified antenna dimension.

For rectangular apertures, the Taylor pattern can be thought of as the superposition of sin $(x)/x$ beams. For circular apertures, the patterns are very similar and can be thought of as a superposition of the $J^1(x)/x$ Bessel function beams. Figure 8.14 shows the beam broadening and gain reduction for the Taylor distribution

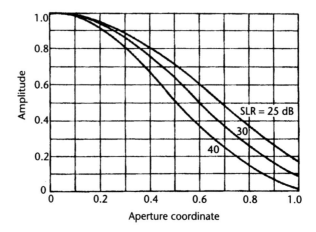

Figure 8.12   Taylor $\bar{n}$ aperture distributions.

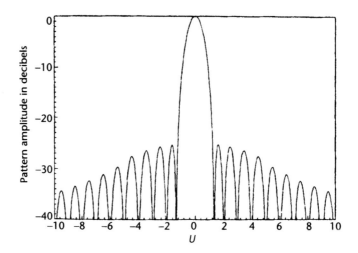

Figure 8.13   Taylor line source pattern for sidelobe = 25 dB, $\bar{n}$ = 5.

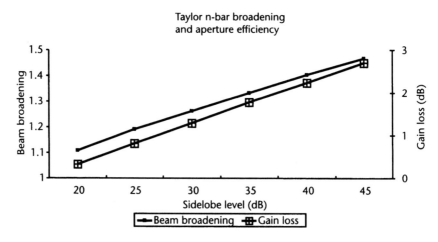

Figure 8.14   Taylor broadening and aperture efficiency.

in two dimensions. Here $\bar{n}$ has been selected to be as large as possible while still maintaining a monotonically decreasing aperture distribution. The gain reduction is computed as if the same aperture distribution and sidelobe levels are specified for both the azimuth and elevation directions. Note that the gain reduction is related to the aperture efficiency: gain reduction (dB) = −10 log (aperture efficiency). Aperture efficiency is a measure of the extent to which the aperture is fully illuminated—uniform illumination represents 100% aperture efficiency.

From the ELINT viewpoint, if the sidelobe level is measured, Figure 8.14 can be used to estimate the beam broadening factor (BBF) and gain reduction relative to a uniformly illuminated aperture of the same size. Then, if the azimuth and elevation beamwidths are also measured, one can estimate the antenna dimension using (8.23a) plus the knowledge of the BBF and the wavelength:

Antenna dimension = (0.88 BBF)(wavelength)/3-dB beamwidth)

The relationship of the width between the first nulls and the 3-dB beamwidth for the Taylor distribution is illustrated in Figure 8.15. For a uniformly illuminated antenna, this ratio is 0.44. If the sidelobe level is estimated and the width between the nulls is measured, Figure 8.15 provides a way to estimate the 3-dB beamwidth from the null-to-null beamwidth.

Although the theoretical basis for a low-sidelobe antenna design is available, practical considerations limit the extent to which the sidelobes can be suppressed. For reflector antennas, the feed at the focus of the reflector blocks part of the aperture and this raises the level of the sidelobes. Figure 8.16 illustrates this effect for a circular reflector illuminated with a Hansen one-parameter distribution [13] for a center-fed reflector. Suppose the design calls for a −40-dB sidelobe level, assuming no aperture blockage. (This is the point at the left of the lowest curve in Figure 8.16.) If the diameter of the feed is 0.1 times the diameter of the reflector,

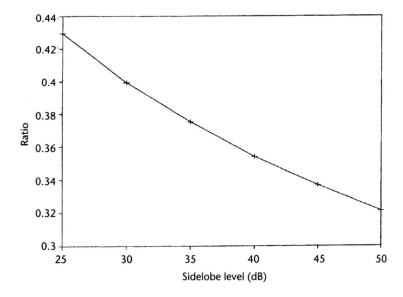

Figure 8.15 Taylor illumination: ratio of 3-dB beamwidth to null-to-null beamwidth.

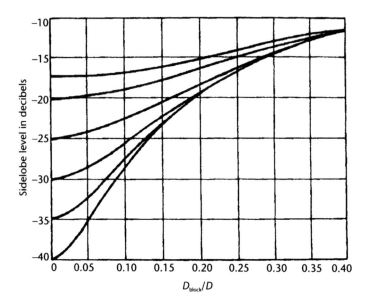

Figure 8.16   Sidelobe level versus the blockage diameter ratio for Hansen one-parameter circular distribution.

the sidelobe level will rise to about −28 dB. Of course, this blockage also reduces antenna efficiency. Many search systems use reflector antennas with offset feed. This greatly reduces aperture blockage compared to that for a center-fed reflector.

The sidelobe level rises and the gain is reduced by inaccuracies in the reflector surface. These effects are shown in Figures 8.17 and 8.18. Figure 8.17 shows how the sidelobe level is affected, and Figure 8.18 shows how the gain is affected. In

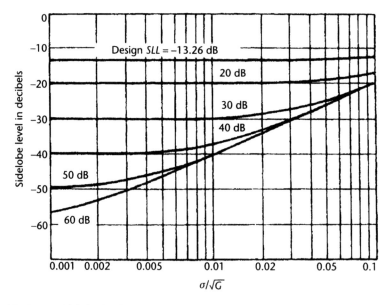

Figure 8.17   Mean sidelobe level versus universal factor.

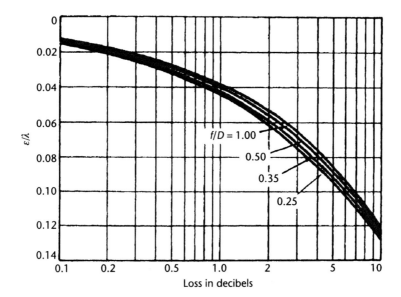

Figure 8.18   Reflector directivity loss versus rms tolerance.

these figures, the reflector imperfections can be related by noting that the rms tolerance (ϵ) in Figure 8.18 is 0.5 times the tolerance (σ) in Figure 8.17.

8.11   Beam Patterns of Array Antennas

The basic idea of any array is to create an illuminated aperture, just as any other antenna. These arrays use discrete elementary antennas spread over an area and control the signal to each one in order to form and steer the beam. The major difference to the radar designer (and to an ELINT observer) is the speed with which the beam can be moved—the problem of the inertia of a large structure is overcome. Such a beam motion (scanning) is considered in Chapter 10. The beam pattern of array antennas is affected by the discrete spacing of the elementary antennas (and their individual beam patterns) as well as by the methods for controlling the signal to each element. The discrete elements typically are equally spaced across an aperture with a spacing of one-half the wavelength. The directive gain is related to the aperture dimension [see (8.20)]; therefore, if the elements are spaced $\lambda/2$ apart, the gain is related to the number of elements by

$$G_D \leq \pi N_L N_W \tag{8.25}$$

where:

$N_W = 2W/\lambda$ = number of elements along the width of the aperture

$N_L = 2L/\lambda$ = number of elements along the length of the aperture

Also, because the 3-dB beamwidths are related to the aperture dimensions (approximately) by

$$\theta_{AZ} = \frac{\lambda}{L} \tag{8.26a}$$

and

$$\theta_{EL} = \frac{\lambda}{W} \tag{8.26b}$$

it can be said that the approximate number of elements in an array is

$$N_L \approx \frac{\lambda}{\theta_{AZ}} \tag{8.27a}$$

and

$$N_W \approx \frac{\lambda}{\theta_{EL}} \tag{8.27b}$$

Of course, these estimates should be adjusted for the nonuniform aperture illumination functions used for sidelobe control. This can be done using Figures 8.14 and 8.15.

The foregoing applies for beams directed normal to the aperture or *broadside*. As the beam is steered away from broadside, the beamwidth (and gain) changes because the effective aperture is smaller when viewed from a direction other than normal. The gain is reduced by a factor of cos θ, where θ is the beam pointing angle as measured from normal to the aperture.

The spacing of the elements need not be $\lambda/2$; however, it is economical to use the fewest number of elements possible. The maximum spacing is related to the appearance of multiple lobes (called grating lobes) in the antenna pattern. The element spacing required to prevent the appearance of grating lobes depends on how far from broadside the beam is to be pointed. The spacing limit is [11]

$$S \le \frac{\lambda}{(1 + \sin \theta)} \tag{8.28}$$

where:

S = element spacing

λ = wavelength

θ = angle away from broadside

Because $|\sin \theta| \le 1$, a spacing of $\lambda/2$ or less guarantees that only one main lobe appears. The maximum element spacing as a function of the maximum scan angle is shown in Figure 8.19.

The number of independent beam positions in a sphere is approximately equal to the antenna gain if the gain is constant. Away from broadside, the gain is reduced (and the beam broadened) so that the number of independent beam positions is reduced to approximately $(\pi/2)(N_L \cdot N_W)$ [1].

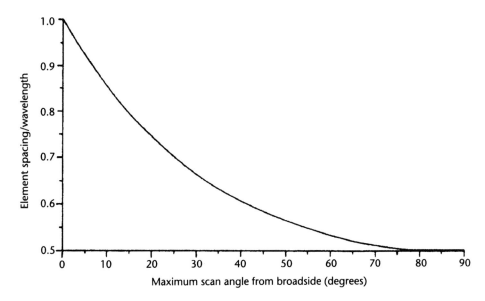

Figure 8.19 Maximum element spacing as a function of the maximum scan angle.

The control of the beam pointing angle is critical if electronic scanning is to be used. However, some radar systems use array antennas with fixed illumination functions that are pointed mechanically. The advantages are that good control of the aperture illumination is possible and the peak power required of each element is reduced by the number of elements used.

If the antenna illumination is controlled by digital techniques (e.g., digital phase shifters), the quantization errors act to reduce the gain and increase the sidelobes. The gain reduction [11] is shown in Table 8.6. The rms sidelobe level due to quantizing is approximately at an isotropic level (i.e., below the main beam by an amount equal to the main beam gain), increased by the gain loss shown in Table 8.6. Clearly, four to eight states in the phase shifter are sufficient to reduce quantization sidelobes below the sidelobe levels due to the illumination of the finite aperture [12]. This situation is to be expected by the ELINT analyst.

The energy lost from the main beam appears in the sidelobes, and the rms sidelobe level due to the quantization in the phase shifters is approximately given by the ratio of the loss in gain to the average main beam gain.

An illustration of the degradation that can occur is shown in Figure 8.20 [14]. The antenna pattern as designed is illustrated on the left. On the right is the pattern with ±2-dB amplitude errors, ±10° random phase errors, and 5% random element failures.

Table 8.6 Number of Phase Shift States Versus the Loss in Gain Due to Quantization

Number of Phase Shift States	Loss in Gain Due to Quantization
4	1.0 dB
8	0.23 dB
16	0.06 dB

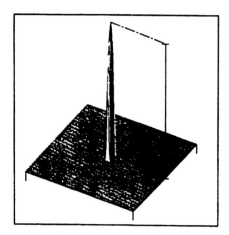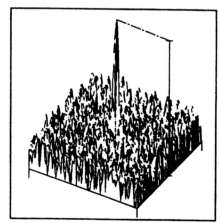

Figure 8.20   Effects of phase and amplitude errors. (*From:* [14]. © 1987 Horizon House, Inc. Reprinted with permission.)

8.12   Antenna Beam Summary

ELINT measurements of beam patterns are often made possible by the motion of the radar antenna beam due to its scanning (i.e., angular searching); therefore, the principles and techniques described in Chapter 10 are often used in conjunction with those described here. The importance of beam pattern information is clear. The directive gain is one of the three most important factors that determine the capabilities of a radar (the other two being the power transmitted and integration time). The 3-dB beamwidth largely determines the ability of the radar to resolve targets in the angular coordinates in space. This, in turn, heavily influences the type of search pattern the radar must employ. (A radar with a wide elevation beam and a narrow azimuth beam will search by moving the beam in azimuth, and vice versa.) Likewise, the sidelobe levels and angular positions can be important in developing effective jammers to work against the radar. The sidelobe levels also determine the extent to which the radar is affected by ambiguous angular responses. For example, a strong point target can appear at many angles as long as the sidelobe levels exceed the minimum detectable signal level for that radar's receiver.

An example helps illustrate the principles of beam analysis discussed here. Suppose an ELINT measurement shows a circular scan that repeats every 10 seconds. This means that the antenna rotates 360° in 10 seconds. The measurement also shows that the first nulls are separated by 0.1 second, which represents 3.6°. If uniform illumination were used, the 3-dB beamwidth would be 0.44 times the width between the nulls, or 1.58°. Because acceptable radar performance requires sidelobes that are lower than that of a uniformly illuminated aperture, Figure 8.14 can be used to find the ratio between the 3-dB beamwidth and the null-to-null beamwidth for any measured or estimated sidelobe level for the Taylor illumination function. If the peak sidelobe level is 30 dB down, a ratio of about 0.40 would be expected. This means the 3-dB beamwidth would be at least 1.44° based on the ELINT data and an assumed 30-dB sidelobe level.

Next, suppose that it is necessary to decide whether or not an observed ELINT signal could be radiated by a radar sold internationally, but whose specifications are not fully known. The dimensions of the reflector antenna are known from advertising to be 4.5m wide by 1m high. Suppose the frequency of operation from ELINT is 3 GHz. This means the wavelength is 0.1m. Using (8.23a), this means that the azimuth beamwidth (radians) is 0.88 times the ratio of the wavelength (0.1m) to the horizontal dimension of the antenna (4.5m)—assuming uniform aperture illumination. This works out to 0.0196 rad or 1.12°. Because the observed beam width from ELINT is 1.44°, this means that the beam broadening ratio is 1.44/1.12 = 1.29. Figure 8.14 shows that this beam broadening ratio is consistent with a −30-dB sidelobe level. From this evidence, it may be concluded that the observed beam is consistent with a reflector of this size and that the gain reduction expected due to a nonuniform aperture illumination in both azimuth and elevation is about 1.3 dB. This implies that −30-dB sidelobes in elevation are assumed. The corresponding elevation beamwidth based on the 1m height of the reflector and the 0.1 wavelength from ELINT is (0.88)(1.29)(0.1/1) = 0.114 rad or 6.5°.

To estimate the gain, either the dimensions or the beamwidths can be used. Using (8.21), the gain from the dimensions is

$$4\pi(LW)/(\text{wavelength})^2 = 5{,}564.8 \text{ or } 37.5 \text{ dB}$$

This assumes uniform illumination; therefore, the gain should be reduced by 1.3 dB due to the beam broadening caused by the nonuniform aperture illumination, resulting in 36.2 dB for the estimated antenna gain.

References

[1]   Kraus, J. D., *Antennas*, New York: McGraw-Hill, 1950, pp. 464–484.

[2]   Kraus, J. D., and K. R. Carver, *Electromagnetics*, New York: McGraw-Hill, 1973, Ch. 11 and 14.

[3]   "Test Procedures for Antennas," Section II, IEEE STD 149-1979, December 19, 1979.

[4]   Kanaraeikin, D., et al., *Polyarizatsiya Radiolokatsionnykh Signalev [Polarization of Radar Signals]*, Moscow: Sovetskoye Radio Press, 1966, English Translation of Chapters 4 and 11 are available from NTIS as AD-783 761, May 1974.

[5]   Hollis, J. S., T. J. Lyon, and L. Clayton, Jr., *Microwave Antenna Measurements*, Atlanta, GA: Scientific Atlanta, 1970.

[6]   Rumsey, V. H., et al., "Techniques for Handling Elliptically Polarized Waves with Special Reference to Antennas," *Proc. IRE*, May 1951, p. 533.

[7]   Kraus, J. D., *Radio Astronomy*, New York: McGraw-Hill, 1966.

[8]   Clayton, L., and S. Hollis, "Antenna Polarization Analysis by Amplitude Measurements of Multiple Components," *Microwave Journal*, Vol. 8, January 1965, pp. 35–41.

[9]   Skolnik, M. I., *Radar Handbook*, New York: McGraw-Hill, 1970, pp. 236–238.

[10] Blake, L. V., *Antennas*, New York: John Wiley & Sons, 1966.

[11] Skolnik, M. I., *Radar Handbook*, New York: McGraw-Hill, 1970, Ch. 9–13.

[12] Stark, L., "Microwave Theory of Phased-Array Antennas—A Review," *Proc. of IEEE*, Vol. 82, No. 12, December 1974.

[13] Jordan, E. C., (ed.), *Reference Data for Engineers: Radio, Electronics, Computer, and Communications*, 7th ed., Indianapolis, IN: Howard W. Sams & Co., 1985, pp. 32-44–32-55.

[14] Geideman, W. A., et al., "An All GaAs Signal Processing Architecture," *Microwave Journal*, Vol. 30, No. 9, September 1987, p. 132.

CHAPTER 9

LPI Radar and the Future of ELINT

9.1 What Is LPI Radar?

The meaning of "low probability of intercept" is not precise. The general idea is to make a radar system whose signal is below the level of threshold of detection of opposing ELINT receivers while still being able to detect targets at useful ranges. Proponents of LPI radar have said that the purpose is "to see without being seen." Consider the particular case of a warning receiver carried by the radar's target. Then the range from the radar to the intercept receiver is equal to the range from the radar to the target, and the receiver and the target are in the same part of the radar's antenna beam. The goal of the LPI radar is to detect the target at a range greater than the range at which the receiver can detect the radar's signal. Can this be achieved?

This is analogous to the situation where a person with a flashlight (the radar) is trying to find another person hiding in the dark. In this analogy, both the target and interceptor use their eyes as their receiver. If both have the same eyesight (or ability to detect light), the interceptor usually has the advantage and can see the flashlight beam (and even its sidelobes) from great distance—much greater than the distance at which the person using the flashlight can see the light reflected from the target. In this situation, the person with the flashlight can achieve LPI operation if the interceptor (or target) has very poor eyesight—equivalent to wearing dark glasses. This is analogous to the approach taken to designing LPI radar—the radar receiver can be nearly matched to the transmitted signal but the intercept receiver generally is not. Many details concerning LPI radar design as well as approaches to intercepting such signals are summarized in [1]. Earlier work on LPI radar and detecting its signals is contained in [2].

The signal strength available to the radar receiver varies as the negative fourth power of the range, and that available to the ELINT receiver varies as the negative second power of the range. Therefore, there is always a range short enough to cause the power available to the radar receiver to exceed the power available to the ELINT receiver. This is illustrated by the example in Figure 9.1, which shows the power available from a 1W transmission to the radar receiver (as reflected from a 1 m^2 target) and to an ELINT receiver in the radar's main beam. As can be seen, the range at which the two received powers are equal is at approximately 200m. The power received at a range of 10 km is about −89 dBm for the ELINT receiver and about −123 dBm for the radar receiver. To hide its main beam from detection at 10 km requires that the radar receiver respond to a signal level of

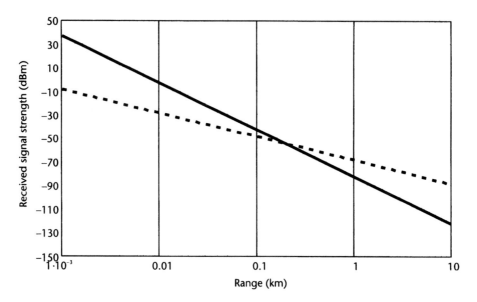

Figure 9.1 Comparison of received signal strength at radar received (solid line) and ELINT receiver (dotted line) in the radar's main beam. Transmit power of 1W; radar antenna gain of 30 dBi on both transmit and receive; target cross-section of 1 m^2; wavelength of 10 cm; ELINT receiver antenna gain of 0 dBi.

−123 dBm but also that the ELINT receiver fails to respond to a signal level of −89 dBm—in other words, a difference in sensitivity of 34 dB is required.

9.2 Radar and ELINT Detection of Signals

Radar detection range depends on the total energy returned from the target during the observation time. If the power returning from the target is constant over the observation time, the energy is the product of the average power and the observation time. If it is not constant, then the energy is the product of the average power times the observation time. For pulsed radar, the average power is the product of the peak power times the duty factor. The radar duty factor is the ratio of the pulse duration to the pulse repetition interval (PRI). However, unless pulse compression is used, radar range resolution is the pulse duration times the speed of light divided by 2. This means that narrow pulses provide fine range resolution *whether you want it or not*.

In addition, narrow pulses reduce the duty factor when the PRI is determined according to the unambiguous range and velocity requirements. The use of short duration pulses makes it difficult to get sufficient energy back from the target. Pulse compression is used to increase the average power while retaining the maximum range capability of long pulses and the range resolution of short pulses. As illustrated in Figure 9.2, a CW signal can have much lower peak power than a pulsed signal but have the same energy returned from the target.

Most of today's intercept receivers are designed to detect single pulses and then to form pulse trains and further analyze these pulses, whereas radar receivers are

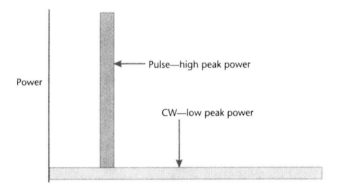

Figure 9.2   Illustration of pulsed and CW radar signals with the same average power.

designed to make use of all of the energy returned from the target prior to making a decision.

9.3   Matched Filter Theory

The maximum output SNR is obtained from a matched filter. The matched filter, when the noise is stationary white and Gaussian, is a filter whose impulse response is the same shape as the signal but reversed in time. For such a filter the maximum output SNR is equal to the input signal energy–to–noise spectral density ratio. This is the best performance the radar can achieve to perform its function. Of course, a truly matched filter may not be achieved in practice, and usually an "integration loss" is incorporated into the radar performance calculation. For a single pulse signal with a rectangular pulse shape processed by a receiver with a rectangular bandpass with the optimum bandwidth, there is a loss of about 0.85 dB relative to a matched filter [3]. The input SNR is

$$\text{SNR}_{\text{in}} = S/(kT_oNFB) \tag{9.1}$$

where S is the input signal power to the matched filter, k is Boltzmann's constant, T_o is 290° Kelvin, NF is the receiver noise figure, and B is the receiver noise bandwidth (assumed wide enough to pass the signal spectrum).

The output SNR from a matched filter is

$$\text{SNR}_{\text{mf}} = S\tau/(kT_oNF) \tag{9.2}$$

where τ is the integration time (equal to the duration of the signal).

The integration gain is the ratio of the output SNR to the input SNR, or

$$\text{Gain}_{\text{mf}} = B\tau \tag{9.3}$$

9.4   One Interception Strategy: Noncoherent Integration

For ELINT systems, the use of a matched filter is almost never practical. First, the waveform of interest is not known exactly, and second, the ELINT receiver usually attempts to preserve the signal's time, amplitude, and frequency characteristics so that it can be identified. In this case, the receiver of choice may consist of an envelope detector followed by a noncoherent integrator. (For example, a low pass filter.) This scheme is also often used in radar, for example, when a train of pulses is transmitted with random phase and returned from the target. Each pulse is processed by a matched filter, and then the individual pulse amplitudes are added without regard to phase in a noncoherent fashion. In the same way, multiple samples of the envelope of a radar signal at an ELINT system can be added noncoherently to enhance sensitivity. For example, suppose the signal-of-interest (SOI) is known to be present in some frequency band for some duration of time. The envelope of the signal plus noise in that band can be sampled, and these samples can be added noncoherently for an appropriate period of time and then compared to a threshold. The performance of such noncoherent integration is mathematically difficult to determine exactly; however, convenient approximations are available. For a large number of samples averaged, the distribution of the average approaches a normal distribution (consider the law of large numbers). In this case, the integration loss relative to coherent integration approaches the square root of the number of samples averaged. In the coherent (or matched filter case), the integration "gain" is proportional to the number (N) of independent samples integrated, or, in dB, $10 \log(N)$. This gain is reduced in the case of noncoherent integration by approximately $5 \log(N) - 5.5$ dB [2]; that is, for values of probability of detection and false alarm generally used in practical systems, noncoherent integration gain is approximately $5 \log(N) + 5.5$ dB for large N. For example, for $N = 100$, the coherent or matched filter gives an output SNR, which is $10 \log(100) = 20$ dB greater than the input SNR, whereas the noncoherent integrator gives an output SNR that is approximately $5 \log(100) + 5.5 = 15.5$ dB greater than the input SNR [4]. In this example, the loss relative to a matched filter is approximately 4.5 dB.

$$\text{Gain}_{nc} \cong 3.55 \sqrt{(B\tau)} \text{ or } 5.5 + 5 \log(B\tau) \text{ dB} \qquad (9.4)$$

In the absence of knowledge of waveform specifics, one strategy is for the ELINT equipment to integrate noncoherently for as long as the radar integrates. In this case, the loss relative to coherent radar is approximately

$$\text{Loss}_{ESM} \cong 5 \log(B\tau) - 5.5 \text{ dB} \qquad (9.5)$$

9.5   ESM and Radar Range Compared

From the one-way and two-way free space range equations (see Chapter 2), ranges at which the minimum signal strength is available at the ELINT receiver and at the radar receiver can be compared:

$$\frac{R_{\text{ELINT}}}{R_{\text{Radar}}} = R_{\text{Radar}} \left[\frac{S_{\text{Radar}}}{S_{\text{ELINT}}} \frac{G_{TE} G_{\text{ELINT}}}{G_T G_R} \frac{1}{\sigma} \right]^{0.5} \qquad (9.6)$$

where:

R_{ELINT} = signal detection range for the ESM receiver

R_{Radar} = target detection range of radar

S_{ELINT} = sensitivity of ELINT receiver

S_{Radar} = sensitivity of radar receiver

σ = radar target cross-section

G_{TE} = gain of radar transmit antenna in direction of ELINT receiver

G_T = gain of radar transmit antenna in direction of target

G_R = gain of radar receive antenna in direction of target

G_{ELINT} = gain of ELINT antenna in direction of radar

If the radar receiver uses matched filtering and if the ELINT receiver uses noncoherent integration over the same time interval, and if the receivers have the same noise figure, the ratio of ELINT receiver sensitivity to radar receiver sensitivity is given approximately by (9.5) for large values of $B\tau$. (Because radars do not achieve ideal matched filtering in practice, it is typical to reduce the performance of the radar from that predicted by matched filter theory by some amount; here, however, ideal matched filter performance is assumed.)

If the terms on the right of (9.6) are known, then the ratio of radar range to ELINT range can be calculated. Alternately, the minimum value of the ELINT antenna gain required could be calculated for given radar and ELINT ranges. If the main beam-to-sidelobe ratio of the radar antenna is known, as well as the radar antenna gain on receive, and also if the target's radar cross section is known, then the minimum antenna gain required for the ELINT equipment can be determined.

$$G_{\text{ELINT}} = G_R \left(\frac{G_T}{G_{TE}} \right) \frac{R_{\text{ELINT}}^2 \sigma}{R_{\text{Radar}}^4} \frac{S_{\text{ELINT}}}{S_{\text{Radar}}} \cong G_R \left(\frac{G_T}{G_{TE}} \right) \frac{R_{\text{ELINT}}^2 \sigma}{R_{\text{Radar}}^4} \frac{\sqrt{B\tau}}{3.55} \quad (9.7)$$

(Note that G_{TE}/G_T = SLR or ratio of sidelobe to main beam gain of the radar's transmit antenna.)

This framework can be used to investigate the feasibility of ELINT systems using noncoherent integration of signal energy to detect LPI radar signals—without knowledge of the details of the signal modulation. If the signal modulation is known, that added knowledge could improve the sensitivity of the ELINT receiver. Of course, there is a need to do more than just detect the presence of the radar signal. Information to determine the type of threat is generally required, and for this, some general waveform information is usually needed. Another key aspect is to obtain information about where the emitter is located using angle of arrival or time difference or frequency difference techniques.

If the transmit power of the radar is reduced, the radar target detection range decreases, but the ELINT detection range decreases even more. For example, if the transmit power is reduced by a factor of 16, the radar target detection range is reduced to one-half or its original value (the fourth root of 16), whereas the ELINT detection range is reduced to one-fourth of its original value (the square root of 16).

If the transmit power is reduced sufficiently, a point is reached where the radar range and the ELINT range are equal. If the ELINT intercept range is to be equal to the radar's target detection range, this condition requires setting (9.7) equal to 1 and solving for the range. The result is given in (9.8), assuming that the ELINT receiving station has an isotropic antenna ($G_E = 1$). This has been termed the "quiet" radar condition because its signal is detectable only at ranges shorter than the range at which it detects its target.

$$R_R = R_E = \left[\frac{S_{\text{ELINT}}}{S_{\text{Radar}}} \frac{\sigma(\text{SLR}) G_R}{4\pi} \right]^{0.5} \tag{9.8}$$

In this equation, R_R is the radar range in meters, R_E is the ELINT range in meters, σ is the radar target cross-section in square meters, SLR is the ratio of radar antenna main beam gain to sidelobe gain (assuming the ELINT observer is in the radar sidelobes), and G_R is the radar antenna gain on receive. The ratio of the ELINT receiver sensitivity to the radar receiver sensitivity is generally larger than 1 because the radar receiver is nearly matched to the transmitted signal and the ELINT receiver is not matched. The ELINT receiver is not matched partly because it must process a variety of signals and partly because it tries to preserve the characteristics of the transmitted signal. Preserving such characteristics as rise time, pulse width, and intra pulse modulation requires a larger bandwidth compared to the bandwidth of a matched receiver.

What values can be expected for the ratio of the sensitivity of the ELINT receiver to that of the radar receiver? For an energy detection wideband ELINT receiver, this is approximately the value given by (9.5) when the ELINT receiver's predetection bandwidth is matched to the radar's signal modulation bandwidth and when the ELINT receiver's noncoherent integration time is matched to the radar's integration time.

Assuming that the SNR required is the same at both the radar receiver output and the ELINT receiver output, the conclusion is that the approximate value of the receiver sensitivity ratio is the square root of the radar's time-bandwidth product divided by 3.55. The time bandwidth product is limited by target motion to $c/2V$ in order for the target to remain in one range cell for one coherent integration period. For target velocity of 300 m/s, this value is about 500,000. Therefore, the ratio of ELINT receiver sensitivity to radar receiver sensitivity is limited to about 23 dB for energy detection receivers adapted to the radar's waveform.

Consider a "quiet" radar capable of detecting a 10-m^2 target at 10 km. According to Figure 9.3 and (9.8), it must find a way to make the intercept receiver about 50 dB less sensitive than the radar receiver when the ELINT system's antenna gain is 1. Because of the limitation above of 26-dB sensitivity loss for a noncoherent receiver using the same integration time as a matched filter radar receiver, energy detection receivers should be sufficient to detect the main beams of radars designed

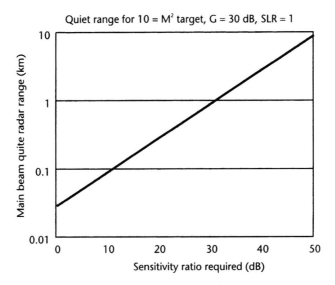

Figure 9.3   Quiet radar range as a function of receiver sensitivity ratio (δ). (ELINT receiver in the main beam of the radar uses an omni-directional antenna.)

to detect small targets at ranges in excess of 10 km. Designing a radar whose main beam is barely detectable at the same range at which the radar can detect its target is very difficult at militarily useful ranges and for the cross-sections typical of military targets. Of course, existing ELINT receivers may not be designed to cope with a signal from a "quiet" radar. In these circumstances, quiet radar operation may be achieved at greater ranges.

9.6   Some Pulse Compression Modulation Constraints

The use of energy detection receivers is necessary if the radar waveform is noise-like to the interceptor. While such waveforms can be used, their time sidelobes and Doppler properties may be far from ideal in any specific time interval; it is only over a long-term average that their ambiguity function approaches the ideal "thumbtack" shape. Generally, at present, radar designers prefer to control their ambiguity function by using specially designed, deterministic pulse compression waveforms such as linear FM (LFM) or binary phase shift keyed (BPSK). Polyphase shift keyed and nonlinear FM waveforms are also described in the literature but have not been deployed extensively. For example, the "best" BPSK waveforms are the Barker codes. These have time sidelobes at zero Doppler of maximum amplitude of 1. However, the longest known binary Barker code is only 13 bits long. It has been shown that odd length Barker codes with length greater than 13 do not exist; however, there are longer sequences of polyphase Barker codes. When pulse compression modulation other than random (or pseudo-random) noise is used, it is possible to process the intercepted signal in ways that outperform energy detection. This kind of processing permits the interception of known deterministic waveforms at greater range than wideband energy detection. Guosui et al. describe development of random signal radars [5]. Figure 9.4 shows the current concept of

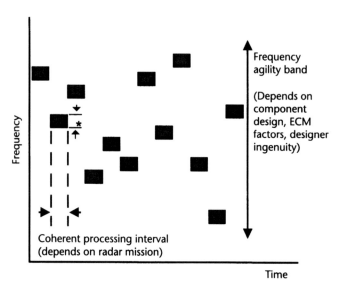

Figure 9.4 Challenging ELINT signal: frequency agile CW.

an LPI signal in terms of its frequency as a function of time. The individual "blocks" represent radar coherent processing intervals.

9.7   Interception Techniques Using the Envelope of the Received Signal

The use of noncoherent integration techniques in the interception process allows the receiver to operate without specific knowledge of the details of the transmitted waveform. For example, the particular pulse compression modulation in use may be unknown, but the approximate range resolution (and hence the signal's band-width) may be inferred from the function the radar performs. Some examples make clearer the processing required. An early example of using noncoherent integration to detect LPI signals is the use of rapid scanning of the RF band of interest using a swept narrowband receiver. Each time the receiver sweeps past the LPI signal, a sample of the envelope of the signal plus noise is obtained. Various types of noncoherent integration can be applied to these samples. Similarly, a channelized receiver can also be used to sample the energy in each of its frequency channels, and these samples can be combined noncoherently to achieve improved detection of the LPI signal. These two concepts are very similar. In the case of the swept receiver, the envelope samples are obtained sequentially at the different frequencies across the band covered, whereas in the case of the channelized receiver the envelope samples are obtained simultaneously from each channel. Hence the sample rate (for a given frequency channel) for the swept receiver is limited to the reciprocal of the sweep time but it is at the Nyquist rate for a channelized receiver. For a given signal duration, the channelized receiver provides more samples than a swept receiver.

One Rapid Sweep Superhet Receiver covered a 2-GHz band and swept that band in 256 steps of 8 MHz in 20 μs. Its sampling rate was 50,000 samples per channel per second. A similar channelized receiver with 256 channels each 8 MHz wide could provide 16 million samples per channel per second. The bandwidth of 8 MHz was selected to be wide enough to pass the pulse compression modulations then in use. (Many current threat radar systems use bandwidths of less than 10 MHz.) A common design is to test the sample against a threshold and declare the presence of a pulse if one sample exceeds the threshold. To provide 90% probability of detection and a 10^{-6} probability of false alarm requires a SNR of 13.2 dB. The computation of the average of the envelope over a number of samples and then comparing the average to a threshold allows the use of a much lower threshold to achieve the same probability of detection and false alarm. The approximate improvement in sensitivity is found from (9.4). If the waveform dwells at one carrier frequency for 2.5 ms and the bandwidth of the channels is 8 MHz, then $BT = 20,000$ and the sensitivity improvement is approximately 27 dB for the channelized receiver. The swept receiver provides a sample every 20 μs or 125 samples in 2.5 ms. Equation (9.4) gives the improvement in sensitivity as approximately 15.5 dB for the swept receiver. There is an added benefit of the swept receiver in eliminating interference from the pulses from ordinary radar signals. Sampling the signal for 80 ns every 20 μs means that most pulsed signals will not contribute much to the average over 2.5 ms. The probability of coincidence between the sampling gate of the sweeping receiver and the pulsing of the interfering radar is very low. Of course, this could also be seen as a drawback if one is interested in detecting both pulsed signals and modulated CW signals.

Another way of processing the samples of the envelope is to test each sample against a threshold and then require that a certain minimum number of samples cross the threshold out of a given number of samples tested—often called M of N detection or binary integration [2, 6]. Here M samples out of N must cross the threshold ($N \geq M$). This process is not quite as effective as computing the average of the samples. It has the effect of suppressing very strong signals of short duration—which is sometimes an advantage. The computation requires use of the binary probability distribution. The first step is to determine the probability of noise alone crossing the threshold at least M times out of N for a given probability that a single sample of noise crosses the threshold. The notation $P_{fa,1}$ denotes the probability that one sample of noise alone crosses the threshold, and $P_{fa,N}$ is the probability that at least M of N samples of noise alone cross the threshold. These are related by the binary probability distribution for N trials and for $P_{fa,11}$ the probability of "success" on one trial. Likewise, the probability that signal plus noise crossed the threshold on one trial is $P_{d,1}$ and the probability that signal plus noise crosses the threshold at least M times out of N tries is $P_{d,N}$. These are also related by the binary probability distribution. Some examples of the results are given in Table 9.1. There is an optimum value of M for any specific case. Values of M near $N/2$ or $1.5N^{0.5}$ have been suggested in the literature [6, 7]. The latter is used in Table 9.1. The required SNR has a broad minimum [6], and so choosing the exact optimum value of M is not critical.

Although there is no theoretical limit to the improvement as the value of N increases, there are important practical considerations; namely, as the threshold

Table 9.1 Examples of Integration Gain Using M of N Integration

$N = 8, M = 4$	$N = 16, M = 6$	$N = 32, M = 8$	$N = 64, M = 12$
SNR = 7.3 dB	SNR = 5.4 dB	SNR = 3.4 dB	SNR = 1.7 dB
Gain = 5.9 dB	Gain = 7.8 dB	Gain = 9.8 dB	Gain = 11.5 dB

M is selected as $1.5N^{0.5}$; gain compared to $P_{d,1} = 0.9$, $P_{fa,1} = 10^{-6}$ (single pulse SNR = 13.2 dB). N determined by signal duration during its coherent processing interval for $P_{fa,N} = 10^{-6}$, $P_{d,N} = 0.9$.

decreases, the value of $P_{fa,1}$ becomes larger and so does the value of $P_{d,1}$. Eventually the threshold is so low that there is not much difference between these two values. Then a slight change in the noise level could drastically affect the final values of $P_{d,N}$ and $P_{fa,N}$. This is illustrated in Figure 9.5, which shows the probability of both false alarm and detection at −1.5-dB SNR. If the threshold is selected at 2.65 normalized units, then $P_{d,1} = 0.12$ and $P_{fa,1} = 0.03$. With a relatively small difference between probability of detection and false alarm, a small change in the noise level could cause a drastic change in the performance of the system. After the M of N process (for $M = 24$, $N = 256$), the probability of detection and false alarm at SNR = −1.5 dB is shown in Figure 9.6. Now the probability of detection at the threshold of 2.65 is 90% and the probability of false alarm is 10^{-6}. Detection using a single sample would require the SNR to be 13.2 dB to give this same performance; therefore, the $M = 24$, $N = 256$ process provides $13.2 - (-1.5) = 14.7$ dB of processing gain. Coherent processing of 256 samples provides 24.1 dB of processing gain; hence the loss of the M of N process is 9.6 dB relative to coherent integration. The approximate gain expected from a noncoherent process as given by (9.2) is

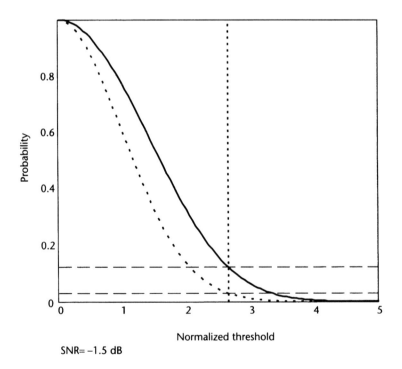

SNR= −1.5 dB

Figure 9.5 Probability of detection (solid) and false alarm (dotted). SNR = −1.5 dB. Threshold set at 2.65 yields $P_{fa,1} = 0.03$ and $P_{d,1} = 0.12$.

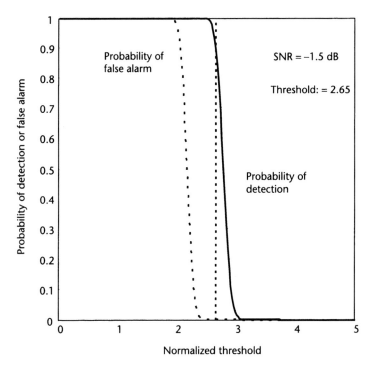

Figure 9.6   Probability of detection (solid) and false alarm (dotted) after M of N processing; $M = 24$, $N = 256$. Threshold of 2.65 yields $P_{fa,256} = 10^{-6}$ and $P_{d,256} = 0.90$ at SNR $= -1.5$ dB.

17.5 dB. Hence the loss of the $M = 24$, $N = 256$ process relative to ideal noncoherent integration is about 2.8 dB.

The case of $M = N$ is of interest for detecting long duration (or CW) signals. This means that the signal plus noise must be above the threshold for N consecutive samples. As shown in [2], the SNR required to provide a specified level of performance is

$$\text{SNR} = -\ln\left[\frac{1 - P_d^{1/N}}{1 - P_{fa}^{1/N}}\right]$$

For a large number of samples, this value approaches

$$-\ln\left[\frac{\ln(P_d)}{\ln(P_{fa})}\right]$$

as N approaches infinity.

For 0.9 probability of detection and 10^{-6} probability of false alarm, the required SNR approaches 6.9 dB as N approaches infinity. This illustrates how requiring consecutive samples to cross the threshold limits the processing gain. For this example, 20 consecutive samples provide a gain of about 6 dB—only 0.9 dB less than if the number of consecutive samples approaches infinity.

9.8 Narrowband Channels and Frequency Modulated Signals

Although linear FM is a common modulation used in communications and radar systems, any sort of FM can be thought of as linear over short time intervals. For FM signals, the question is what is the best intercept receiver bandwidth to use to intercept the signal and warn a user of its presence? Described below is the fundamental principal for designing a receiver for linear FM; and for nonlinear FM, the same approach can be used over portions of the signal when the frequency versus time variation is approximately linear. A technique for pulse compression of linear FM signals is described in [8]. The technique can accommodate any pulse compression ratio less than the square of the number of bandpass filters in a filter bank (generally implemented using an FFT). In radar applications a programmable pulse compressor allows the transmission of pulses with a variety of linear FM pulse compression characteristics. Of course, in radar, the frequency deviation and pulse duration are known and the receiver is programmed properly to receive the waveform transmitted. In intercepting unknown signals of the linear FM type, this information is not available to program the signal processor. Furthermore, the purpose of the intercept receiver is not pulse compression per se, but rather to detect the presence of the signal and measure its frequency deviation and time duration. An approach similar to the method of [8] can be applied to the design of an intercept receiver, although here the channel outputs are converted to amplitudes and no use is made of the phase information. The basic design of a digital channelized receiver is shown in Figure 9.7. Whether the channels are implemented in analog form or digital form is a matter of convenience.

For a signal whose frequency is linearly changing with time, the output of a bandpass filter whose bandwidth is considerably less than the frequency excursion is approximately a pulse whose duration is the channel bandwidth divided by the rate of change of the RF at its center frequency. For example, if the channel bandwidth is 1 MHz and the RF is changing linearly at the rate of 0.01 MHz/μs over a band of 100 MHz, the filter output will be a pulse of duration 1/0.01 = 100 μs, which occurs at the approximate time the RF signal passes through that channel's bandpass. Compare this to a wideband receiver covering the entire 100-MHz band in one channel. The noise level in one channel is increased from 1 to 100 MHz (20 dB), but the signal lasts 10,000 μs, not 100 μs. For a wideband receiver, the best sensitivity occurs when the bandpass filter is followed by an

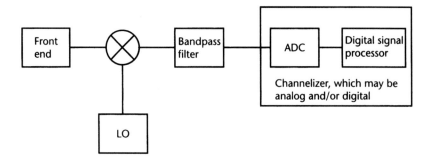

Figure 9.7 Channelized receiver configuration.

envelope detector and a video (lowpass) filter whose bandwidth is comparable to the reciprocal of the pulse length. The effective noise bandwidth of such a wideband receiver is approximately

$$\text{ENB} = \sqrt{2B_V B_{\text{RF}}}$$

where:

ENB = effective noise bandwidth

B_V = video (lowpass) bandwidth after the envelope detector

B_{RF} = RF bandwidth before the envelope detector

Other things being equal, the sensitivity of two receivers is given by the ratio of their effective noise bandwidths. The ratio of noise bandwidths is

$$\text{ENB1/ENB2} = \sqrt{B_{V1} B_{\text{RF1}}/(B_{V2} B_{\text{RF2}})}$$

For our example, $B_{\text{RF1}} = 100$ MHz, $B_{V1} \sim 1/1{,}000\ \mu$s. Also $B_{\text{RF2}} = 1$ MHz, $B_{V2} \sim 1/100\ \mu$s. We see that the ENB values for these two receivers are the same, and their ratio is 1. In general, the ENB for a receiver designed in this way for linear FM is given as

$$\text{ENB}lfm = \sqrt{\Delta f/\Delta t}$$

Here, Δf is the frequency excursion of the linear FM and Δt is the time over which the excursion occurs. Their ratio, R, is the rate of change of the frequency (slope) of the linear FM. Notice that if the channel bandwidth is the square root of the LFM slope, then the number of channels is

$$\text{Number of channels} = \Delta f/\sqrt{\Delta f/\Delta t} = \sqrt{\Delta f \Delta t}$$

Put another way, this says that the square of the number of channels is equal to the time-bandwidth product—which is precisely the basis of [8].

How narrow can the channel RF bandwidth be? If we reduce the RF bandwidth so that the signal is in the channel for a time less than the reciprocal of the channel bandwidth, we begin to lose signal amplitude. R is the rate of change of the RF and B is the channel bandwidth; we want B/R to be greater than $1/B$ or

$$B > \sqrt{R}$$

For $R = 0.01$ MHz/μs, B must be wider than 100 kHz. Consider the following example. A 100-MHz linear FM frequency excursion occurs over 10 ms ($BT = 10^6$). The receiver may have the same effective noise bandwidth for any choice of receiver RF bandwidth from 100 MHz down to 0.1 MHz; *provided that the video filter bandwidth is approximately R/B—the reciprocal of the "pulse duration" present at the channel outputs.* For 0.1-MHz channel bandwidth, there are such pulses at the outputs of 1,000 channels to test for the presence of signals. To keep the total false alarm rate the same as in a single wideband channel system, the threshold of each narrowband channel would necessarily be set a bit higher than that for a single channel.

A receiver designed for linear FM has an effective noise bandwidth of the square root of the FM slope over a wide range of choices for RF bandwidth, provided appropriate video filtering is used. The sensitivity of a single channel of this receiver depends on the FM slopes expected, not on the total frequency deviation or the time duration of the FM waveform. Figure 9.8 illustrates how this principle may be applied to a channelized receiver design.

There are several advantages of using channel bandwidths smaller than the frequency excursion of the linear FM. First, the character of the FM (namely the slope) can be determined by examining the sequences of pulses at the outputs of the narrowband channels. Second, by considering the outputs of many channels before making a decision as to whether a signal is present or not, sensitivity can be improved compared to the sensitivity of a single channel. In principle, this is the same concept as in pulse compression; however, for interception purposes, one may ignore phase and other waveform details but achieve less than optimum sensitivity. For example, one way to improve sensitivity is to declare a signal is present if some number of channels exceeds a threshold in a certain length of time, regardless of their order. Suppose we had a system with 256 channels and required that at least 60 of them exceed a threshold in any 10-ms window of time. The threshold on the individual channels can be lowered and the result is a sensitivity improvement of about 18 dB for 256 narrowband channels followed by M out of N detection logic. (In this example, $M = 60$ and $N = 256$.)

Logic can be used to combine the signals from all of the channels in other ways. For example, one could add their power spectra (ignoring their phase differences). Another way is to use frequency versus time templates. This is similar to the Hough (or Radon) transform, which sums the channel outputs along straight lines of all possible slopes and starting points. The slope is a key parameter of the signal. Accurate determination of the slope can help sort and identify several linear FM signals present in the same band at the same time; in other words, *postdetection processing of the outputs of a bank of narrowband channels can enhance sensitivity*

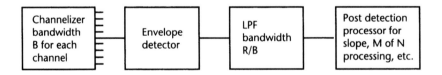

The bandwidth of each channel, B, is to be greater than the square root of the FM slope, R, and less than the linear FM excursion, delta f. In that case, the effective noise bandwidth *a* channel is the same for any choice of B and is equal to the square root of the FM slope, R, for which it was designed.

Figure 9.8 Processing after the channelizer.

and characterize the slope of linear FM signals. In a modern design, the channels could be realized by predetection digitizing followed by an FFT and magnitude computation, and then possibly followed by averaging the FFT amplitudes. Comparison to a threshold can also be used, and the threshold crossings counted in an M of N (or similar) detection process. Slope measurement is accomplished by examining the sequence of the narrowband channel threshold crossings in a manner similar to the Hough transform, or similar to the method of [8].

If the signal uses nonlinear FM, the instantaneous frequency could be approximated using linear segments. The largest slope determines the minimum channel bandwidth. The same techniques can then be applied for signal detection. To find the character of the FM, various templates of appropriate shape could be tried, as is done using the Hough transform to look for arbitrary shapes.

Many FMCW systems use multiple FM slopes (including zero slope) to make range measurements with differing resolution and ambiguity and to deal with Doppler shifts. The waveforms of interest may also switch slopes in different modes. To deal with these signals with a receiver having a single value of channel bandwidth, the channel bandwidth is selected as the minimum to accommodate the steepest slope. Then there would need to be different video bandwidths or postdetection processes (e.g., differing values of M and N) to accommodate the different pulse durations caused by the different slopes. In the situation where the steepest slope requires a channel bandwidth so wide that the character of the smallest slope cannot be determined, it would be necessary to use two or more channel bandwidths (i.e., multiple FFT lengths).

The use of channel bandwidth equal to the square root of the slope is optimum in the sense that it is the narrowest bandwidth for which the output of the channel is not reduced by the rapid sweeping of the signal through the channel. For linear FM, once the channel bandwidth becomes less than the square root of the FM slope, there is a reduction in the signal amplitude (as well as the noise level). Because the time spent sweeping through one channel is less than the impulse response of that channel, the signal is not present for the entire time that the filter requires to build up to its full output amplitude. The signal energy input to the filter is the signal power multiplied by the time the signal is present in the band of that filter, which is (channel bandwidth/slope). The noise energy at the input is the noise power $kT_o(NF)$(channel bandwidth) multiplied by the time the noise is present, which is 1/(channel bandwidth). Therefore, the signal-to-noise energy ratio is S(channel bandwidth/slope)/$[kT_o(NF)]$. This is the maximum SNR present at the channel output when the channel bandwidth is less than the square root of the slope. When the channel bandwidth is greater than the square root of the slope, the output SNR becomes equal to the input SNR or $S/kT_o(NF)$(channel bandwidth).

When the channel bandwidth is less than the square root of the slope, the signal frequency sweeps through several channels during the impulse response time of the channel filters, and there will be several adjacent channels excited during that time. For example, if the slope is 1 MHz/μs, in 32 μs the signal will sweep though 32 MHz. The square root of the slope is 1 MHz. Suppose the channel bandwidth is set at 31 kHz instead of 1 MHz. This reduces the energy into each filter by 1/32 or about 15 dB. But notice that 1,000 of the 31-kHz channels will be excited in 1 μs instead of 32 1-MHz channels. If it were possible to use M of

N or other detection logic on the 1,000 31-kHz frequency channels excited by the sweeping signal, the overall detection performance would be improved by about 18 dB, so the net gain is about 3 dB relative to 1-MHz channels. (But note that the slope is obscured.) During the next 32 μs, a different set of 1,000 channels (centered 32 MHz higher in frequency) will be excited, and so on, until the end of the sweep occurs. If the sweep lasts longer than 32 μs, there will be additional channel outputs available to combine. Suppose the sweep lasts 256 μs. Then there will be 8 sample periods of 32 μs available during each LFM sweep, and these could be combined (for example, by using 4 of 8 detection logic) to gain another 6 dB—but the proper way to combine these is not known unless the slope is known.

Carrying this to the limit, suppose the channel bandwidth is so narrow that the impulse response time of the channels is as long as the sweep time; for example, if the sweep time is 256 μs, the channel bandwidth could be reduced to about 4 KHz. Now there is only one sample available from each of the 64,000 4-kHz channels during one sweep, and the detection logic must combine the outputs from 64,000 frequency channels. Now there is no information about the slope contained in the timing of the filter outputs, and knowledge of the slope is not needed to combine the channel outputs for detection of the presence of the signal. Giving up the knowledge of the slope and using very narrowband channels provides about the same detection performance. These three cases are summarized in Table 9.2.

9.9 Predetection Processing Methods to Detect Linear FM and Other LPI Signals

Many of the LPI radars advertised to date make use of linear FM as the pulse compression waveform—not the random signals mentioned above. Linear FM is also commonly used in altimeters. One method of ranging is to mix the received signal with the transmitted signal. Then, for a stationary target, the beat frequency is equal to the product of the FM slope and the time delay due to the range. For this type of signal, the matched filter in the radar becomes a mixer (to derive the beat frequency) followed by a filter bank to determine the value of the beat frequency and hence the range. The bandwidth of the filters can be as narrow as the reciprocal of the sweep time of the FM. For moving targets, the effects of the Doppler shift must be taken into account [9]. For example, the Wigner-Hough Transform (WHT) or Radon-Wigner Transform (RWT) and the Radon Ambiguity Transform (RAT)

Table 9.2 Comparison of Three Bandwidth Channels

| | | | | Detection Performance for a Linear FM Segment | | |
| | | | | Integration | Straddle | |
Channel BW	Noise	Signal	M/N	Gain	Loss	Net Gain
1.0 MHz	Ref	Ref*	60/256	15 dB	1.5 dB	0
0.031 MHz	–15 dB	–15 dB	1,024 & 4/8	21 dB	2.1 dB	3.9 dB**
0.004 MHz	–24 dB	–24 dB	64,000	24 dB	1.5 dB	7.5 dB**

*The 1-MHz channels and the 0.031-MHz channels require knowledge of the correct slope to achieve the M of N processing gain.
**Requires raising the threshold to achieve the same false alarm rate. This reduces the net gain by ~2 dB.

can achieve nearly coherent detection performance of linear FM waveforms [10, 11].

Pace [1, Ch. 9] explores the application of the Wigner-Ville Distribution (WVD) to detect the presence of linear FM signals and other signals in noise. The WVD has the greatest concentration of energy of any two-dimensional time-frequency distribution for linear FM. The WVD of a signal $x(t)$ is given by

$$W_x(t, f) = \int x(t + \tau/2)x^*(t - \tau/2)e^{-j2\pi f \tau}\, d\tau \tag{9.9}$$

The ambiguity function is given by

$$A_x(\nu, \tau) = \int x(t + \tau/2)x^*(t - \tau/2)e^{j2\pi\nu t}\, dt \tag{9.10}$$

or

$$A_x(\nu, \tau) = \iint W_x(t, f)e^{j2\pi(\nu t + \tau f)}\, dt\, df \tag{9.11}$$

Equation (9.11) shows that the ambiguity function in (9.10) is the two-dimensional Fourier transform of the WVD in (9.9). The radon transform of the WVD is found by computing the integral of the time-frequency distribution along straight lines at different angles. Hence, for linear FM, when one uses the correct angle (or FM sweep rate) and starting time, a large value will be obtained. At other angles and/or starting times the contribution to the integral will be largely due to noise. Figure 9.9 shows a contour plot of the WVD of a linear FM signal 1 μs long with an FM deviation of 20 MHz, starting at 30 MHz and extending to 50 MHz. Figure 9.10 shows the radon (or Hough) transform of the WVD in Figure 9.8. The peak in Figure 9.10 occurs at the slope and starting frequency of the linear FM. Figure 9.11 shows the same thing as Figure 9.10 but with a 0-dB SNR. It shows the enhanced detectability of the signal using the WVD and radon transforms. Detection would be achieved by establishing a threshold value for the amplitude of the radon-WVD process. Note that the original data window must include the starting time and ending times of the signal and the bandwidth must include the entire FM sweep. The minimum number of input data points required is determined by the time-bandwidth product of the signals of interest. The radon transform of the WVD results in a two-dimensional space of FM slope versus starting frequency (or starting time).

Consider the ambiguity function (AF) instead of the WVD. Because the AF is a correlation function, its peak is always at zero delay and zero frequency shift. For linear FM, there is a ridge in the AF extending from the origin at an angle corresponding to the slope of the linear FM. If the radon transform of the AF is computed, the result is a one-dimensional plot of amplitude versus slope. This replaces the two-dimensional plot of amplitude versus slope and starting frequency (or time) obtained from the radon transform of the WVD.

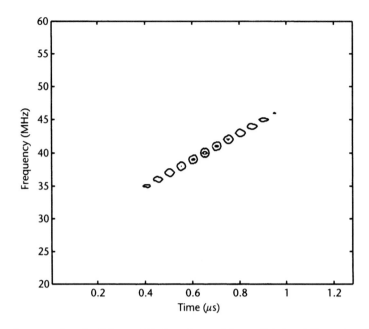

Figure 9.9 Contour plot of a Wigner-Ville Transform of linear FM signal (1-μs duration; 20-MHz linear FM deviation; $BT = 20$).

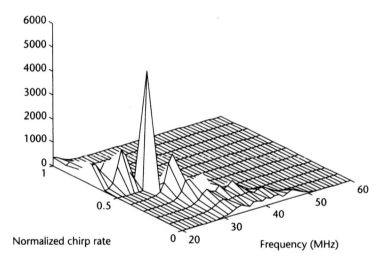

Figure 9.10 Radon (or Hough) transform of WVD in Figure 9.9. Linear FM sweep starts at 30 MHz and extends to 50 MHz.

Figure 9.12 shows a contour plot of the AF of an FMCW signal with up and down slopes equal to 1 MHz/μs. The original signal consisted of two up 32-MHz and down 32-MHz cycles of linear FM lasting a total of 128 μs. Figure 9.13 shows the RAT, which is the radon transform of the AF shown in Figure 9.12. The peaks due to the linear FM up-slope and down-slope are clearly visible at an input SNR of −10 dB. The one-dimensional nature of the plot of the magnitude versus slope should be compared to the two-dimensional plot of magnitude versus slope and starting frequency of Figure 9.10. There is more information in the plot of Figure

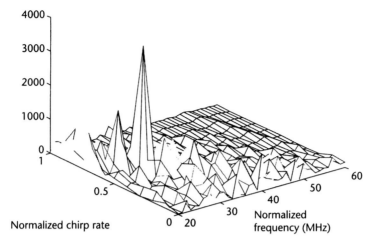

Figure 9.11   Same as Figure 9.10 but with 0-dB SNR. (Signal peak is clearly visible.)

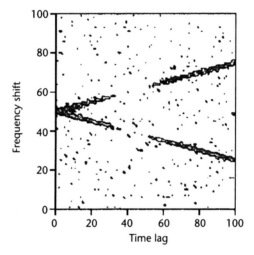

Figure 9.12   Contour plot of ambiguity function of FMCW signal (up-slope = down-slope = 1 MHz/μs). Horizontal scale: max. time lag = 800 ns. Vertical scale: zero frequency shift = 50; max frequency shift = ±1.6 MHz.

9.10; however, if the task is to detect the presence of linear FM and find its slope, then the RAT of Figure 9.13 is sufficient and requires less processing.

The performance of the RAT relative to a matched filter is shown in Figure 9.14 [11]. If the input signal-to-noise energy ratio is above 0 dB, the loss approaches 3 dB. Achieving signal energy greater than the noise spectral density determines the length of time the interceptor must integrate. If this time is less that the duration of the LFM sweep, then Figure 9.14 can be used to determine the loss relative to a matched filter when using the RAT. If the time required exceeds the duration of the FM sweep, then detection may still be possible; however, Figure 9.14 does not apply directly. Instead, a way must be found to combine the energy of two or more sweeps. This may cause additional losses, as indicated in Figure 9.15. Here the integration time was increased sufficiently to allow reliable detection of the presence

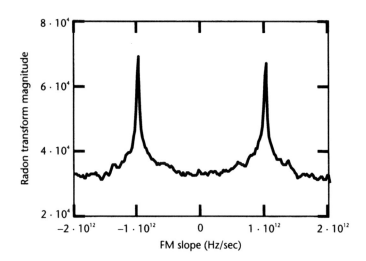

Figure 9.13 Radon transform of ambiguity function shown in Figure 9.12. SNR = –10 dB. (Duration of waveform: 128 μs; BT = 4,096.)

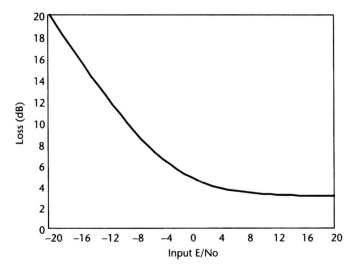

Figure 9.14 Loss of RAT relative to a matched filter (maximum lag assumed to be less than the LFM sweep time).

of the weak signal—that is so that the output SNR was above about 13 dB. As can be seen, worst-case performance is about 10 dB below that predicted for a sweep time longer than the duration of the integration.

Another type of signal processing often suggested for detecting LPI signals is to compute the cyclostationary spectral density (CSD). This is defined as the Fourier transform of the cyclic autocorrelation function [1, Eq. (11.4)]:

$$R_x^\alpha(\tau) = \int x(t + \tau/2)x^*(t - \tau/2)e^{-j2\pi\alpha t}\,dt \qquad (9.12)$$

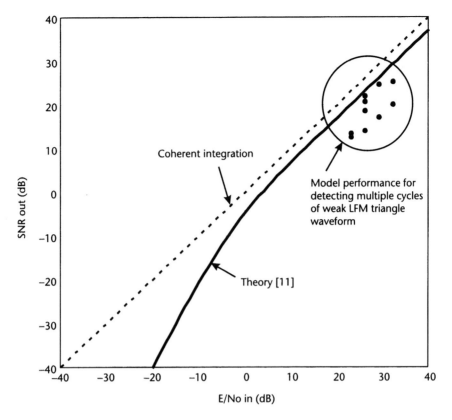

Figure 9.15   Modeled performance of RAT relative to a matched filter. (Low input SNR requires multiple cycles of the LFM waveform, resulting in more loss than predicted.)

In (9.12), the variable α is called the cyclic frequency. Now notice that (9.12) is actually the same as (9.10) if the symbol α is changed to ν—meaning that the CSD is the same as the ambiguity function. The spectral correlation density is the one-dimensional Fourier transform of the cyclic autocorrelation function:

$$S_x^\alpha(f) = \int R_x^\alpha(\tau)e^{-j2\pi f\tau}\, d\tau \qquad (9.13)$$

The CSD is a two-dimensional density function of cyclic frequency and Fourier frequency. For example, if the signal had a triangular frequency versus time variation, and if the variable τ extended over a large number of cycles of the triangular variation, then there would be peaks when α is a multiple of the repetition rate of the triangular modulation and f is at the center or carrier frequency of the signal. The emphasis of the CSD is to find periodicities in the cyclic correlation function (or AF)—for example, the chip rate of a BPSK waveform or the repetition rate of a periodic FM waveform. For example, a noise-like binary sequence using a fixed chip rate would be detectable by this type of processing.

However, consider again the concept of the programmable pulse compressor [8]. In this method of pulse compression, the channel (or FFT) outputs are combined using a parabolic phase weighting to provide pulse compression for any pulse compression ratio equal to or less than the square of the number of channels. While in radar the pulse compression ratio is known and selected in advance, the same technique can be used to create simultaneously pulse compression processing for a variety of time-bandwidth products. In a manner similar to the WHT or RAT, the pulse compression ratio that provides the highest output would also provide the means to detect the presence of a LFM signal.

9.10   ELINT Receiver Requirements for Interception of Low Peak Power Signals

The implications of using the various interception techniques described above for ELINT are that those wideband signals must be digitized and then processed in parallel in a variety of ways. For example, there could be one path for high peak power pulsed signals, another for energy detection, another for LFM detection, another for BPSK detection, and so on. Note that LPI signals are below the noise in a wideband receiver. While they can be detected using a variety of processing algorithms, it will be necessary to set the LSB of the A/D converter far below the receiver noise to preserve these signals. This implies a reduction in dynamic range or the need for more resolution (bits) from the A/D converter. A major requirement for LPI radar detection by a general purpose ELINT receiver is more bits and sampling rates in excess of 1 GHz. The high sampling rates are required to search wide bands, such as the radar's frequency agility band, and/or to provide good probability of interception in a short time, not so much because the radar's coherent (or instantaneous) bandwidth is wide but because the center frequency is unknown and/or the signal uses frequency agility. Wideband communications signals using very short pulses may be used to transmit missile guidance data or otherwise be a part of the ELINT signal environment. These could be very difficult to detect if random pulse spacing known only to the receiver but not to the interceptor is superimposed on the data. Wideband radar signals using short pulses are less of a problem to the ELINT community. In addition to the problem of excessively fine range resolution as the pulse duration becomes short, there is the need to get the same amount of energy back from the target as when a longer duration pulse is used. This means that except for very short range radars, the transmitted pulses are of higher amplitude and can be more easily detected. LPI radar trends are to reduce peak power (and increase the duty factor) to avoid detection by current ELINT receivers designed for single pulse detection (i.e., those which use peak power for single pulse detection).

Because LPI radar is a qualitative term, *quiet radar* is suggested here as a quantifiable alternative. Low sidelobes are important to avoid detection of the radar signal except in the main beam—especially as a defense against anti-radiation missiles. Power management and atmospheric attenuation shielding are often considered as LPI techniques but are of very limited benefit in search radar. Power management is useful in altimeters and some tracking systems. When used, the transmit power is reduced to the level needed to maintain track. Atmospheric shielding is of limited use due to the two-way path length of the radar. In a homogeneous atmosphere, the radar two-way path has more attenuation than the one-way path to the intercept receiver unless the interceptor is located more than twice the range from the radar's target. Of course, if high altitude ELINT collectors are used, the atmospheric effects are largely eliminated as the density of the atmosphere decreases.

The ELINT challenge is frequency agile, noise-like signals of modest bandwidth transmitted with high duty factor (as shown in Figure 9.4). The instantaneous bandwidth is not likely to exceed that of conventional radars having the same function and in every application will probably be less than 500 MHz. Randomness in the radar's pulse compression waveform makes it necessary for the ELINT receiver to use energy detection techniques. Today's radars typically transmit the same pulse compression code or waveform in every coherent integration interval during one mode. Tomorrow's radars may choose to transmit different waveforms from one coherent processing interval to the next. In this case, the ratio of the ELINT receiver's sensitivity compared to that of the radar's receiver need not be greater than the square root of the radar's time-bandwidth product

Wideband and ultra-wideband radar are loosely defined in terms of the fractional bandwidth used by the radar. F_{high} and F_{low} are the upper and lower 3-dB points of the radar's bandwidth. The fractional bandwidth, FB, is given by

$$FB = 2\frac{F_{high} - F_{low}}{F_{high} + F_{low}} \tag{9.14}$$

Wideband radar has a fractional bandwidth greater than 10%, and ultra-wideband greater than 25%. Recently, a new definition of UWB radar has been suggested, as follows [12]:

> Ultrawideband radar is a radar set having range resolution (Δr) much smaller than the target length, L, along the radiated direction.

Ultra-wideband applications include underground probing (for example, mine detection), very short range (a few meters) such as proximity sensors and automatic braking devices, and short range (< 100m) such as intrusion alarms and through-the-wall detection. Due to limited energy on target, it is not likely that long range UWB radar will be deployed. The use of ELINT receivers to detect such signals must be weighed against the cost of doing so relative to the threat posed by these signals. Due to their current short range, monitoring these signals is also best done at short range. In this case, the need to sort them from among a myriad of other signals is reduced and their high peak power may make detection relatively easy at short range.

References

[1] Pace, P. E., *Low Probability of Intercept Radar*, Norwood, MA: Artech House, 2004.

[2] Wiley, R. G., *Electronic Intelligence: The Interception of Radar Signals*, Dedham, MA: Artech House, 1985, Chapter 2.

[3] DiFranco, J. V., and W. L. Rubin, *Radar Detection*, Englewood Cliffs, NJ: Prentice Hall, 1968, p. 184.

[4]   DiFranco, J. V., and W. L. Rubin, *Radar Detection*, Englewood Cliffs, NJ: Prentice Hall, 1968, p. 389.

[5]   Guosui, L. et al., "Development of Random Signal Radars," *IEEE Trans. on Aerospace and Electronic Systems*, Vol. 35, No. 3, July 1999, p. 770.

[6]   Tsui, J. B. Y., *Digital Techniques for Wideband Receivers*, Norwood, MA: Artech House, 1995 (Section 9.12).

[7]   Skolnik, M., *Radar Handbook*, New York: McGraw Hill, 1970.

[8]   Thor, R., "Programmable LFM Signal Processor," Patent Number 4,591,857, May 27, 1987.

[9]   Stimson, G. W., *Introduction to Airborne Radar*, 2nd ed., Raleigh, NC: Scitech Publishing, 1998, Chapter 13.

[10] Ozdemir, A. K., and O. Arikan, "Fast Computation of the Ambiguity Function and the Wigner Distribution on Arbitrary Line Segments," *IEEE Trans. on Signal Processing*, Vol. 49, No. 2, February 2001, p. 381.

[11] Jennison, B. K., "Detection of Polyphase Pulse Compression Waveforms Using the Radon-Ambiguity Transform (RAT)," *IEEE Trans. on Aerospace and Electronic Systems*, January 2003, p. 335.

[12] Immoreev, I. Y., and J. D. Taylor, "Ultrawideband Radar Special Features and Terminology," *IEEE A&E Systems Magazine*, May 2005.

CHAPTER 10

Antenna Scan Analysis

10.1 Introduction

The radar antenna beam pattern typically covers only a small portion of the angular region of interest. The antenna must couple the energy between the transmitter and space. To observe targets over a portion of space larger than covered by the main beam, the antenna beam is moved to observe different angles at different times. Often the spatial coverage is thought of as requiring a certain number of beam positions. For hemispheric coverage, there are approximately N_{bp} independent beam positions required:

$$N_{bp} = \frac{2\pi}{\theta_{AZ}\,\theta_{EL}} \tag{10.1}$$

where:

θ_{AZ} = 3-dB azimuth beamwidth (rad)

θ_{EL} = 3-dB elevation beamwidth (rad)

The motion of the beam (angular position versus time) is called scanning. The scanning pattern is designed to cover the entire volume of interest to the radar system. The size of this volume can vary depending on the purpose of the radar at the time. If searching a hemisphere is necessary to acquire a target, a smaller search volume can subsequently be used to refine its location. If the radar can track the target, scanning may cease altogether. Such constant illumination of the target by the radar can be a warning to the target that tracking is occurring and an attack may soon occur. In the case of an electronically steered array (ESA), tracking may be indicated by repeated bursts occurring at the same beam position (or having the same amplitude when observed by an ELINT receiver).

In mechanically scanned radar systems, the scanning pattern is typically periodic. In electronically scanned systems, the search pattern may be periodic; however, it can change rapidly in response to events in the environment. In addition, multiple targets can be tracked by assigning a certain fraction of the available pulses to track them. This lengthens the time required to complete one search pattern and also makes the beam motion appear to be partially erratic or random. Mechanically scanned systems make up the bulk of all existing radar systems. However, the deployment of ESAs is proceeding rapidly due to:

1. Rapidly decreasing costs of digital memory and computational power;
2. Development of precision microwave integrated circuits;
3. Development of solid-state power sources at microwave frequencies.

Some radar systems may use manual (operator) control to position the antenna. Perhaps, when the radar is acquiring a target to be tracked, the nonperiodic beam motion may confuse an ELINT observer. The usual case, however, is that the radar performs a periodic beam motion pattern that is amenable to ELINT analysis almost immediately. Of course, thousands of different ELINT observations may be required to catalog all of the combinations of angular coverage, scan speeds, and patterns available to the radar.

The scanning function's overall character is summarized by the following parameters:

1. Plane of scanning;
2. Slew rates of the beam motion (angular change per unit of time);
3. Dwell time at one beam position;
4. Period of time required to complete the scan (if periodic).

Although these general characteristics are important, it is usually necessary to describe the scanning pattern in more detail. The common types have been given names (such as circular, sector, and raster) and different parameters are important in describing each of these types.

10.2   Some Principles of Searching

Theoretically, it makes no difference whether the transmitted radar signal is formed into a narrow beam or not. The key to the search process is the energy received from a target. A single, broad-transmitted beam covering the whole search sector combined with multiple narrow receive beams can achieve the same result. The received echo power is then integrated over the same total time as would have been taken to point a narrow transmit beam sequentially to cover the sector. This approach is rarely followed because:

1. It is costly to implement many parallel receive channels.
2. The long integration times required are difficult to implement, especially for moving targets with significant Doppler variations.
3. Radially moving targets may fly through a particular range cell before the integration is completed.

Many radars, therefore, use the same antenna and beam shape to both receive and transmit.

Beam motion is typically in the same direction as the longest dimension of the antenna aperture because that is the direction of the narrowest antenna beam. For example, a cut paraboloid antenna wider than it is high has a narrow azimuth beamwidth and a broad elevation beamwidth. This antenna would typically be rotated (scanned) in azimuth. This search radar could provide target azimuth and range, but target elevation information is missing unless the radar includes multiple beams in elevation or some type of elevation scanning along with azimuth scanning. In a traditional radar installation, such a two-dimensional search radar would be used together with a height finder having a tall but narrow antenna aperture that searches in elevation at a particular azimuth (to which it is directed by the search radar). The height finder then provides elevation and range information. Combining the data from the two radars locates the target in three dimensions. This illustrates a fundamental principle of search: it is quicker to search a large volume coarsely and reserve high-resolution searching for those regions of the volume where a target has been coarsely located. In this example, the search proceeded sequentially by dimension—first the azimuth dimension was searched, then the elevation dimension.

Suppose a given resolution cell requires illumination for a certain time, T_{cell}, to determine if a target is in a cell. If the number of such cells in a given volume is $(N_{AZ} N_{EL})$, then examining all of the cells requires searching each cell and this requires a time equal to $(N_{AZ} N_{EL}) T_{cell}$. If the dimensions are searched sequentially, the time required is $(N_{AZ} + N_{EL}) T_{cell}$—a dramatic reduction. If the number of cells is the same in each dimension, the reduction is a factor of $N_{AZ}/2$. This fundamental idea is the driving force behind most of the search schemes now in use.

Usually, the radar scans a given angular region so that the same amount of time is devoted to each part of the region and the search is begun at some convenient portion of the region. In some cases, the time per cell is reduced if there is less need to detect at long range in that region. For example, an air search radar may need to detect targets at long range (say, 300 km) near the horizon (at low elevation angles) but only need to detect targets at short range (say, 30 km) at high elevation angles because aircraft are limited to flying at some known maximum altitude. Clearly, better results can be obtained if the search takes into account any a priori knowledge of where the target may be located. For example, it would be useful to search along the predicted trajectory of a projectile when the launch time and velocity vector are known. If the probability of detection is related to the time spent searching a particular cell, and if the a priori probabilities of the target being in each cell can be assigned, an optimum search strategy can be formulated [1]. While present radar designs generally do not use such a strategy, electronically scanned, computer-controlled arrays have the potential for such optimal search.

10.3   Relationships Among Scan Rate, Maximum Unambiguous Range, and Energy on Target

As described in Chapter 2, the radar designer must carefully choose the waveform parameters when designing the radar. The ELINT analyst can make use of these same design criteria to understand the signal and determine the function of the radar system. The total time to cover the search region must be selected to meet operational criteria. Obviously, if the region is searched only once a day, this is not acceptable for an air-defense radar. For repetitive scans, one can immediately know that the radar needs to obtain target information at least once per scan cycle to do its job. The maximum unambiguous range of the radar is determined by the PRI (see Chapter 2). It is given by

$$R_u = (c) \, \text{PRI}/2$$

The time spent illuminating one target is important because the ability to detect a target depends on the amount of energy that hits it. The energy is proportional to the number of pulses sent in the direction of the target. Clearly, the faster the beam moves in angle, the fewer the number of pulses that strike a particular target. Suppose the number of pulses received as the beam moves past the bearing of the ELINT site is determined. This is referred to as the number of *pulses per beamwidth* (PPBW). Then the time spent by the radar "looking" in the direction of one target is

$$T_{\text{cell}} = (\text{PRI})(\text{PPBW})$$

The number of cells to be searched in some manner is determined by the antenna beamwidths and the size of the angular region to be searched. For an antenna having certain beamwidths, the number of cells is

$$N_{\text{AZ}} N_{\text{EL}} = \frac{\text{AZ sector}}{\text{AZ beamwidth}} \cdot \frac{\text{EL sector}}{\text{EL beamwidth}}$$

Now it is clear that the total scan time is

$$T_{\text{scan}} = T_{\text{cell}}(N_{\text{AZ}} N_{\text{EL}})$$

Consider the following example. Suppose a $100° \times 50°$ (AZ)(EL) region is to be searched with a $1°$ beamwidth in both azimuth and elevation. There are then 5,000 beam positions. Suppose the radar designer wished to have 16 pulses strike the target to take advantage of integration to reduce the peak power of the transmitter. Suppose the maximum range to be searched is 300 km. This requires a 2-ms PRI. The total scan time is then

$$T_{\text{scan}} = (5,000)(16)(0.002) = 160 \text{ sec}$$

Clearly, this design is not acceptable for an aircraft detection system because the targets can move an appreciable distance during that time. At 200 m/s, an aircraft would move 32 km between radar scans. Because the scan time, beamwidth, PPBW, and PRI are observable and because the extent of the scan can often be determined, these relationships can be used to check for consistent values among these ELINT parameters. If a fan beam is used in elevation so that all elevations of interest are illuminated in one azimuth beam position, the number of beam positions to be searched is reduced to 100 and the scan time required is reduced to 3.2 seconds—a much more practical value obtained by giving up knowledge of the target's elevation.

10.4   Fan Beam Scanning: Circular and Sector

An antenna beam that is wide in one direction and narrow in the perpendicular direction is called a fan beam. Such a beam is used for scanning in one direction only. As noted, the classic scan pattern for a long-range search radar is circular scanning in azimuth. This type of radar is expected to provide azimuth bearing and range information, but not elevation. To achieve coverage of all elevations, a wide elevation beam is needed. To provide good azimuth resolution, a narrow azimuth beam is needed, as shown in Figure 10.1(a). The antenna aperture (as noted in Chapter 8) is therefore wide in the horizontal direction and narrow in the vertical direction. The elevation beam should extend to the maximum altitude to be searched at the minimum range of interest.

The direction of rotation could be either clockwise or counterclockwise without affecting the radar performance. If the ELINT analyst wished to determine the direction of rotation, two intercept antennas and receivers could be used to determine which antenna first intercepts the main beam of the radar, provided that it can be determined on which side of a line connecting the interceptors the radar is located.

The speed of the azimuth scanning motion depends on the performance required of the radar. The detectability of targets is basically dependent on the energy that strikes the target, which for a pulsed radar is proportional to the number of pulses

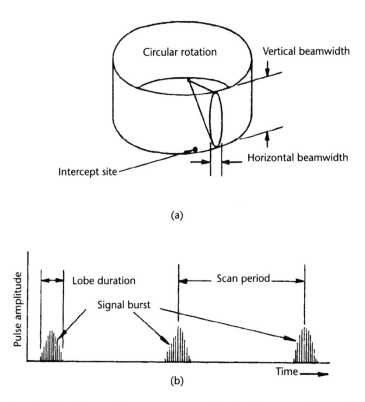

(a)

(b)

Figure 10.1   Typical circular scan intercept pattern: (a) circular scan region; and (b) amplitude versus time pattern.

used to illuminate the target as the beam sweeps past. An important parameter of the radar is, therefore, the number of PPBW. For a circular scan this is given by

$$\text{PPBW} = \frac{(\text{Scan Time})\theta_{AZ}}{(360)(\text{PRI})} \qquad (10.2)$$

where:

      PPBW = pulses per beamwidth

    Scan Time = time required to scan 360° (one revolution)

      θ_{AZ} = 3-dB azimuth beamwidth (degrees)

      PRI = radar pulse repetition interval (average)

The expected pulse amplitude versus time for such a radar appears as shown in Figure 10.1(b).

For long-range search radars, more energy is needed on target; therefore, more PPBW are to be expected. However, a longer pulse interval is also needed so that a large unambiguous range can be achieved. Both of these require a longer scan time. Typical long-range search radars (with a range capability of hundreds of kilometers) have scan times in the 10- to 30-second range. The long rotation time does not hamper the performance of long-range search because, at such ranges, targets need not be immediately tracked or engaged. Shipboard surface search or ground-controlled intercept radars might have somewhat less detection range capability, but they could make use of the more frequent scan capability. Circular scan times up to 10 seconds might be typical, with correspondingly shorter pulse intervals. It would be rare to find a 360° circular scan completed in as little as 1 second.

A circular scan is only one of a number of commonly used periodic scans. If only a portion of the full azimuth circle is of interest, the radar may scan only a sector. A sector scan can occur in two forms: bidirectional and unidirectional. It can also be performed in azimuth or elevation (or any other orientation). A bidirectional mechanical scan is simply a reversal of the direction of the scan when an angular limit is reached. This is often used in height-finding radars that have a tall, narrow aperture to produce a narrow elevation beam. Such "nodding" height finders are commonly used along with circular scanning search radars to obtain elevation angle measurements on targets at azimuths specified by the search radar. Therefore, such elevation sector scanning is often accompanied by nonperiodic slewing in azimuth. The height finder should have range capabilities similar to that of the search radar with which it is associated. (The display of a height finder is typically an elevation versus range map.) Targets are associated with those detected by the search radar by noting the range as measured by both the search radar and the height finder. Often, the pulsing of the two radars is synchronized.

The intercepted scan pattern for a bidirectional sector scan depends on the position of the intercept receiver, as shown in Figure 10.2. A horizontal bidirectional sector scanning in azimuth is most often observed in aircraft search radar modes

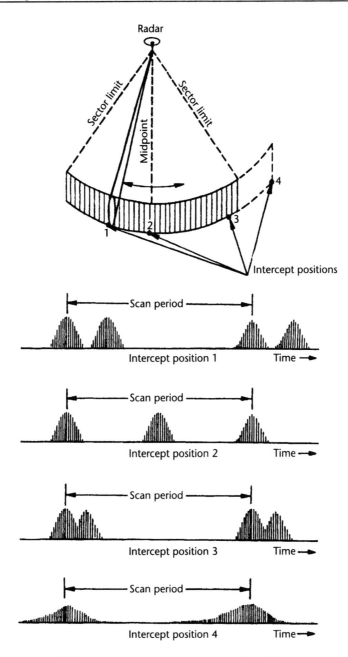

Figure 10.2   Intercepted bidirectional sector scan patterns at four different receiver locations.

or shipboard radars. This is the situation when the scan patterns indicated in Figure 10.2 are encountered at ground or shipboard intercept sites located at different azimuth positions with respect to the radar. With height finders, the situation for ground-level ELINT sites is generally that of position 3 or 4 in Figure 10.2. The other situations in Figure 10.2 can be encountered from an airborne ELINT platform if a height-finding radar set is being intercepted. Unfortunately, positions 3 and 4 present a difficult situation from which to determine the time between the 3-dB points (or even between the nulls). Position 3 can cause an estimate of up to twice the actual time between the 3-dB points. Position 4 reveals only sidelobe levels. (The scan period can be estimated from position 4, however.)

The sector width cannot be directly determined from the amplitude-versus-time data alone; only the sector-to-beamwidth ratio can be determined. This ratio is given by the ratio of the sector period to the time between the 3-dB points. Determining the angular width of both the sector and the beam is usually done by estimating the beamwidth from the antenna dimensions. Then the sector width is determined from the sector-to-beamwidth ratio. For height finders, the sector width can be estimated from the function. If the maximum range is known (it could be estimated from the PRI) and the maximum altitude of the targets of interest is known, then the maximum elevation limit is known at the maximum range. Usually, the height finder must cover high elevations only to ranges much closer than the maximum range. Typical sector limits are a few degrees below the horizon to 30° or more above the horizon, as indicated in Figure 10.3. If the range from the interceptor to the height finder is known and the elevation of the ELINT receiver is changed, the scan pattern changes as shown in Figure 10.2. The sector width could then be estimated from the altitude change needed to produce a given change in the scan pattern.

Unidirectional sector scanning is used when the radar function requires rapid scanning (such as rapidly changing, short-range situations). Reversing the motion of the aperture takes time and results in spending too much valuable search time at the sector edges. Unidirectional scanning can be implemented by moving the antenna feed, not the whole reflector. The idea is shown in Figure 10.4. RF energy enters the rotating feed at the center. Several feed horns are spaced around the rotating mechanism, but only one at a time couples to the center waveguide and faces the reflector. No matter where the intercept site is located within the sector, the same scan pattern is observed. The effect is that of a circular scan at the feed, made to cover a sector by the geometry of the feed and reflector mechanism.

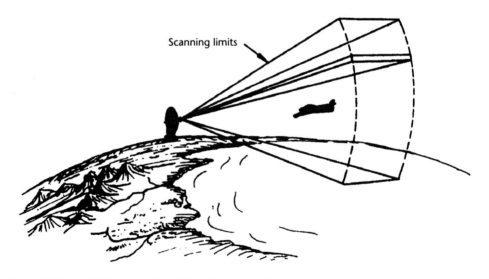

Scanning limits

Figure 10.3   Height finder with a bidirectional elevation scan.

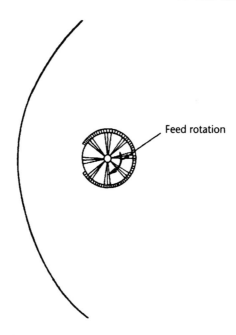

Figure 10.4 Unidirectional sector scan mechanism.

Usually, the high speed is an indication that circular scanning is not being observed. Sometimes, slight variations in the effectiveness of the various feed horns cause an amplitude variation from main beam to main beam that allows the ELINT analyst to infer the number of horns in use.

Circular scanning radars also can include multiple beam systems. The beams are pointed at different elevation angles and thus provide a rough indication of the target altitude. Each beam is often transmitted at a separate RF, and multiple transmitters and receivers are used for each beam. Typically, four to six separate beams are used with narrower beam patterns near the horizon. The ELINT station must be equipped with multiple receivers to demonstrate conclusively the simultaneous use of more than one beam. Of course, the signal power of the beams whose sidelobes are directed toward the ELINT site will be lower than that of one whose main beam is received.

Another type of circular scan system designed to measure elevation is the V-beam radar. This is an obsolescent approach, but it is still in use. The V-beam uses two reflectors with one mounted at an angle to the other (typically slanted 45°). The two beams are transmitted at different frequencies and form a V pattern in space. The time interval between the detection in one beam and the other depends on both the target range and altitude. The V-beam can also include multiple beams in each leg of the V, as shown in Figure 10.5. An ELINT intercept of the V-beam radar using a single narrowband receiver channel appears exactly the same as an ordinary circular scan radar. Two receiver channels are needed to demonstrate the V-beam property. The antenna scan patterns from the two receiver channels will appear as shown in Figure 10.6. The beam offset shown in Figure 10.6 depends on the beam-to-beam angular separation at the horizon plus the angular separation due to the V-beam pattern if the ELINT receiver is at a higher elevation.

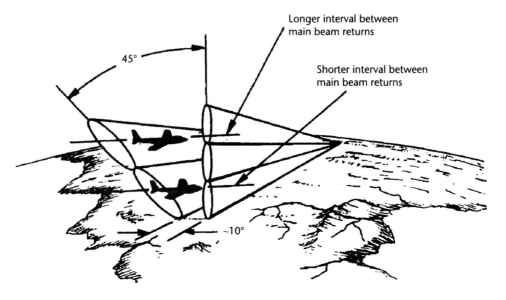

Figure 10.5 V-beam and multiple-beam spatial coverage.

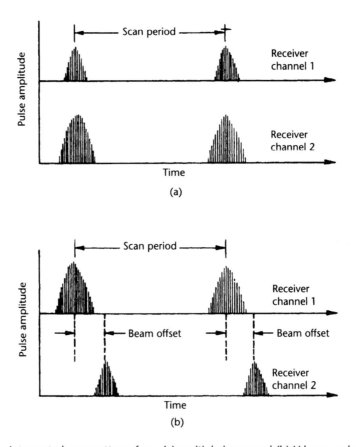

Figure 10.6 Intercepted scan patterns from (a) multiple-beam and (b) V-beam radars.

10.5   Pencil Beam Scanning: Raster, Helical, and Spiral

A radar beam with nearly equal azimuth and elevation beamwidths is called a *pencil beam*. To search effectively with such a beam requires motion in both azimuth and elevation. One common way to do this is through a raster scan, where a line is scanned in one direction (e.g., azimuth) over a certain sector, and then a second line is scanned in azimuth but at a different elevation angle. Figure 10.7 shows a four-line raster pattern with the lines horizontal (they could also be vertical or at any other angle).

The important parameters needed to describe a raster are its overall angular dimensions, the number and orientation of the lines, and the time required to complete one raster pattern. As illustrated by the intercepted scan patterns in Figure 10.7, the number of lines and the time period of one raster are usually determined rather easily. The time per line can then be estimated. The angular length of one line can also be estimated by multiplying the beamwidth by the ratio of the time per line to the time for one beamwidth. Because a raster is usually designed so that there are no gaps between the lines, the angular extent of the raster perpendicular to the line direction is approximately the beamwidth in that direction multiplied by the number of lines.

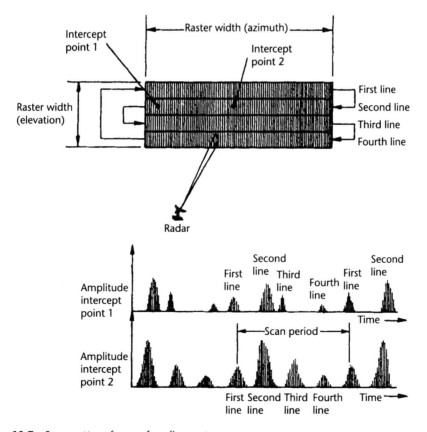

Figure 10.7   Scan pattern from a four-line raster.

A helical scan is a search pattern raster with horizontal lines but where each line covers 360°. Actually, the helical scan consists of a pencil beam circularly scanning with the elevation of the beam increasing by about one beamwidth during each azimuth revolution, as shown in Figure 10.8.

A spiral scan is used for searching a restricted region of space, like a raster scan; but the beam starts at a central point and spirals out from there, as shown in Figure 10.9. The signal amplitude versus time can be quite complex and varies with the intercept point. The spiral shown rapidly returns to the center and is generally outward. However, radars are also designed to spiral out and then spiral back in. The intercepted scan pattern reveals the basic period required to complete the scan. Some airborne intercept radars use a spiral scan for searching. A ground-based ELINT site, therefore, may be in the antenna sidelobes and must deal with a moving emitter.

10.6   Tracking Scans and Monopulse

A radar system is said to be performing tracking if it has located a target and is following its movement in range, elevation, and azimuth. One technique for doing this is conical scanning. When conical scan is used for tracking a target, the peak of the main beam never quite strikes the target. Instead, the peak of the main beam rotates circularly about the target, with the target being the center of the circle. The line from the radar through the peak of the main beam is called the beam axis. The motion of the beam axis forms a cone in space, as shown in Figure 10.10 [2]. The motion of the target causes the center of the cone to move as the radar follows the motion. The ELINT antenna will not be in the main beam of the radar unless it is on the target being tracked. For all other intercept points, the received

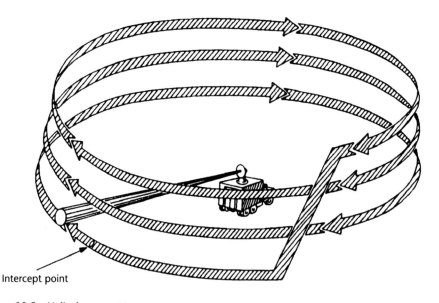

Intercept point

Figure 10.8 Helical scan pattern.

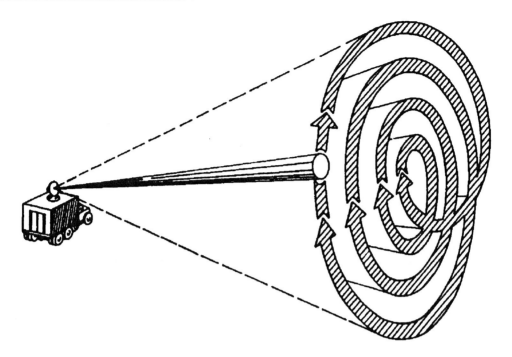

Figure 10.9 Spiral scan pattern.

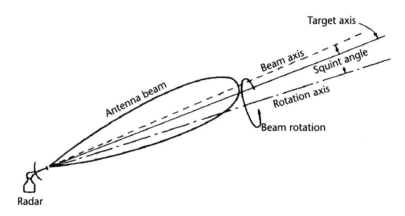

Figure 10.10 Conical scan. (*From:* [2]. © 1970 by McGraw-Hill, Inc. Used with permission of McGraw-Hill.)

pulse amplitude will vary approximately sinusoidally in step with the conical scan. This variation will be much less if the ELINT antenna is on the target, because the conical scan system tracks by feedback loops that attempt to maintain the target position at the center of the cone, where equal amplitude from the main beam exists.

Usually, the ELINT analyst can easily determine the period of the conical scan rotation (or its reciprocal, the conical scan rate). Typically, the conical scan rate is in the range of 25 to 80 Hz (revolutions per second). The conical scan rate is selected so that the target movement during one rotation period is limited to a small fraction of the beamwidth. The other parameters of interest for such a tracking radar are the beam axis slewing rates that can be achieved and the squint angle (the angle between the beam axis and the target axis of the cone formed by the conical scanning of the beam axis).

The slewing rate can be determined only from observing the radar performance in situations where the geometry is known. For example, if the ELINT antenna is on the target that the radar is tracking, abrupt aircraft maneuvers can test the limitations of the tracking system.

The squint angle is difficult to determine through ELINT. Clearly, the squint angle must be on the order of the 3-dB beamwidth because the radar loses target-detection capability if the cone angle is too large, that is, the tracking point is too many decibels down on the antenna pattern.

The key conical scan parameter is the rotation rate. If a jammer transmits a pulsed signal with amplitude variations at the same rate as the conical scan rate, the tracking circuitry can be disrupted.

An ELINT operator can easily identify a conical scan by the sound alone. The 25- to 80-Hz amplitude modulation of the basic pulse repetition frequency creates a warbling sound. Naturally, the tracking condition also causes a continuous signal with no noticeable amplitude modulation when the ELINT receiver is on the target being tracked.

A conical scan can also be implemented by providing scanning of only the receive beam. This is called *conical scan on receive only* (COSRO). One common implementation is to transmit through the sum port of a monopulse-type horn feed cluster. On reception, the azimuth and elevation difference ports are sampled with weights varying sinusoidally at the scan frequency ω_s to form a composite Δ channel:

$$\Delta = \delta_{az} \cos \omega_s t + \delta_{el} \sin \omega_s t$$

This is combined in a hybrid to create $\Sigma + \Delta$, which represents a receiving beam offset from the Σ-channel axis by a fraction of the beamwidth and rotating about the axis at ω_s. This implementation avoids mechanical vibration and rotation of the polarization vector, and permits rapid and variable scan rates.

It is difficult for an ELINT analyst to determine the conical scan rate (or even the existence) of COSRO. Although the transmitted signal is not scanned, there is usually enough mismatch and reflection in the microwave system to impose small AM on the transmitted signal, disclosing the scan rate to ELINT. A tracking system that does not exhibit any other type of tracking scan technique may be presumed to use some type of receiver technique for tracking purposes. In this case, a high SNR, possibly combined with a high-resolution spectrum analysis, may reveal the COSRO rate.

After locating a target, the conical scan tracking system must make the transition from the search scan to the tracking scan. To hasten this process, the conical scan may be left on during the search process. The search scan may be of the raster, helical, or spiral variety. A helical plus conical scan is shown in Figure 10.11. (Such scans are sometimes called Palmer helical or Palmer raster.) The main difference observed at the ELINT site is the relatively fast variation in pulse amplitude at the conical scan rate superimposed on the normal received scan patterns.

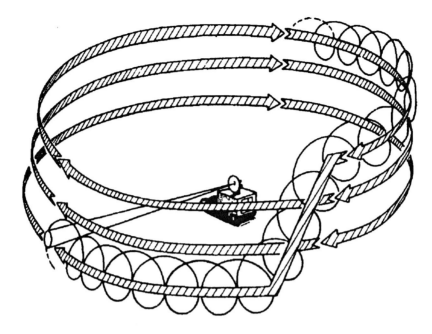

Figure 10.11   Helical plus conical scan.

Historically, the conical scan systems were preceded by lobe switching systems where the beam position could be rapidly switched slightly from right to left (and up and down). At first this switching was done manually and the operator adjusted the beam axis to make the returns from the various beam positions equal. Such *lobing* was done either with the combined receive/transmit beam or on receive only [dubbed *lobe on receive only* (LORO)]. ELINT analysis can reveal lobing on transmit by the discrete steps in the received signal strength. The resultant signal amplitude would have a square-wave-like modulation. LORO is more difficult to detect, but high SNR intercepts may reveal slight variations in the transmitted beam caused by the switching of the LORO antenna system components.

The disadvantages of all of the conical scan and lobing techniques include the susceptibility to tracking errors due to fluctuations in the echo amplitude and problems of tracking at long range. The conical scan or lobing rate should be relatively fast for good tracking, but at long range, the pulse interval becomes so long that only a few pulses occur in time to complete one conical scan or lobing cycle. Furthermore, when the echo from a pulse transmitted in one direction returns from a target far away, the beam will have moved to another direction.

These disadvantages were overcome by the development of the monopulse. (This term does not mean that only one pulse is used, but rather, in essence, lobing is performed on receive only using a single pulse.) Because monopulse is essentially a receiver technique, its use cannot be directly revealed through ELINT analysis. A tracking radar that does not exhibit any other scanning technique may be presumed to use monopulse, especially if it is of recent design. The relative performance of monopulse, compared to the other techniques, is summarized in Table 10.1.

Another important type of tracking radar is track-while-scan (TWS). One TWS method uses two fan beams and orthogonal sector scans to cover two overlapping

Table 10.1 Relative Performance of Some Tracking Systems

Parameter	Monopulse	Conical Scan	Sequential Lobing
Effective beamwidth Normalized to monopulse	1.0	1.4	1.3
Beam offset loss (two-way)	0	−3 dB	−6 dB
Maximum update rate	PRF	PRF/4	PRF/4
Typical update rate	PRF	PRF/10	PRF/4

Source: [2].

sectors using the orthogonal scans, as shown in Figure 10.12. The target being tracked is kept in the center of the overlapping scanned areas by feedback systems to the antenna drive motors. TWS is used in virtually all two-dimensional surveillance radars to provide azimuth and range data. It can also be used for three-dimensional data [as in most ground controlled approach (GCA) precision approach radars] by time-sharing a single transmitter between azimuth and elevation sector-scanning antennas. Other TWS systems can be implemented in receive-only configurations by using the sector scanning antennas to feed the receivers and a third nonscanning antenna for transmit. In this case, the transmit beam must be broad enough to cover the entire sector scanned by the receive antennas. In the receive-only system, a single transmit frequency is used; whereas in the TWS system using two transmitters, two frequencies must be used.

Like the conical scan and lobing systems, TWS systems are susceptible to jamming based on amplitude modulation at the scan rates. TWS on receive-only is a remedy for this. TWS radars (and monopulse radars) are used in missile systems for target and missile tracking. As such, they are of high interest to ELINT analysts. The parameters of interest for the TWS system include the beamwidths, sectorwidths, and sector scan directions. (For example, both of the sector scan directions could be tipped 45° to create a diamond-shaped overlap area. The important factor is that the two scans are orthogonal.) The receive-only versions make determination of such parameters very difficult using ELINT techniques alone. If two transmit beams are scanned, the problems are the same as described for sector scanning in Section 10.3. In the scan-on-receive-only systems, the transmit beam is relatively broad and yet the nonscanning character of a tracking system is evident.

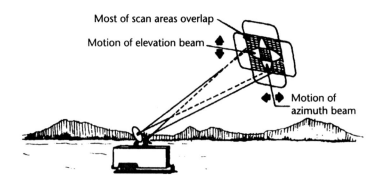

Most of scan areas overlap

Motion of elevation beam

Motion of azimuth beam

Figure 10.12 Track-while-scan mechanisms.

10.7   Electronic Scanning

Because electronic scan systems are often computer controlled, analysis of their scanning patterns really amounts to determining what computer programs exist in the software library of such a multiple-function array radar. The functions of the various programs as well as the beam-steering technique and its limitations are of interest. In addition, ELINT observations can define many signal parameters whether or not their functions are apparent. These parameters include pulses per beam position, the period for a repeated scan pattern, and possibly the number of discrete beam positions. A commonly used search scan is the raster type, which has many of the same properties as mechanically scanned systems.

Note that electronic scanning in one dimension (e.g., elevation) is often combined with mechanical scanning in another (e.g., azimuth). The ELINT observer can easily distinguish electronic scanning and multiple functions by the large variety of periods and patterns observed.

Periodic programs are often used for searching. They may be interrupted when targets are detected for verification routines—typically, a pause at a particular beam position to allow more pulses to hit the target. Verification may be done more than once (e.g., within several successive search patterns). After verification, target tracking is often performed by directing the beam toward each target on some schedule. These dwells also interrupt the search function so that the more targets there are, the longer it takes to search the volume of space for more targets. Ultimately, a limit is reached so that either the searching stops or a maximum number of tracks is reached. The number of pulses used during each track beam dwell, as well as the period of time between the track dwells, should be determined by the ELINT analyst. This can sometimes be done because the ELINT analyst can associate a given pulse amplitude with fixed radar beam angles (for a constant intercept geometry). Thus, while the actual beam angle remains unknown, the return to the same beam position can be determined by analysis of the pulse amplitude pattern. The ELINT analyst may also discover that the RF, pulse interval, and pulse duration may be programmed differently. Also, the analyst may find that the radar performs different functions. These variations may correlate with beam position. A good example is a frequency-steered array with the beam elevation a function of the RF. The pulse interval may be longer at low elevation angles near the horizon, where a longer unambiguous range is required. As the elevation increases, the pulse interval can be shortened so that the maximum altitude searched remains fixed. This reduces the time required to complete the search pattern. To maintain the same average power, the pulse duration may be reduced in proportion to the pulse interval. Finally, because the maximum detection range is reduced at high elevation angles, less energy is required for target detection. Therefore, the number of pulses per beam position may also be reduced. The energy reduction is partially accomplished by reducing the pulse duration. Because the PRI reduction is in direct proportion to the range reduction, but the energy required varies as the fourth power of the range, a significant reduction in the number of pulses per beam position is possible.

The key parameter characterizing the multifunction ESA is the beam dwell time. This is often the same as the CPIs of the radar, or the beam dwell time may be as long as several CPIs. The CPI reveals what the coherent integration time is and this helps the ELINT analyst estimate the detection capabilities of the radar.

10.8   Scan Measurement and Analysis Techniques

Historically, scan analysis was first performed by using headphones and a stopwatch. The received signal pulse repetition frequency is usually in the audio range (especially on older radars, although not on pulse Doppler systems). When the main beam was directed toward the ELINT site, the strong tone burst was easily identified and the stopwatch was started. On the next burst (or some number of bursts later) the watch was stopped, and hence, the scan period was determined. Sector, conical, and other scans can also be determined in this way: a sector scan by the syncopation due to an intercept point away from the sector midpoint; a conical scan by the warbling amplitude modulation; and a spiral scan by the fading and then strengthening of the main beam tone bursts. The time resolution was in the 0.1- to 0.01-sec range. As wider bandwidth chart recorders became available, this became the most prevalent method for scan analysis. A problem common to both audio analysis and chart recorder analysis is that the duty cycle of most radar signals is quite low—on the order of 10^{-3}. The amplitude of such narrow pulses is greatly reduced by the lowpass filtering action of the audio system or chart recorder. It is therefore necessary to incorporate pulse stretch circuitry, such as a sample-and-hold circuit that is discharged after a fixed time (e.g., 100 μs), ahead of the recorder or audio system.

In systems that digitize and store pulse arrival time, amplitude, and duration, many options for analysis are available. Time-domain analysis techniques are the most useful because the scan action of a radar is usually designed from the view of accomplishing sequential search or track functions, which are best conceived (at present) in the time domain. Scan analysis techniques tend to be intuitive and heuristic rather than analytical. This is partly a reflection of the fact that radar scanning schemes themselves are not strongly based in any theory, but historically have been developed in response to the need to search in ways that could be accommodated by the available hardware.

Bidirectional sector scanning can be distinguished by the sidelobe structure. Real antenna patterns are not symmetrical, and because sector scanning involves reversing the direction of the beam motion, these reversals can be discovered easily by the mirror-image appearance of the scans. This allows distinguishing bidirectional sector scans from unidirectional sector scans. Conical scanning can be implemented by moving the feed in a circular pattern while maintaining a fixed orientation of the feed, as shown in Figure 10.13(a). Then the polarization remains fixed. This same effect can be obtained by keeping the feed fixed and rotating the reflector dish by one-half of the conical scan cone angle. The rotating feed of Figure 10.13(b) produces rotating polarization.

Multipath can cause problems in determining scan patterns. It may cause large signal amplitudes to occur that may be interpreted erroneously as additional main beam illuminations. For fixed ground ELINT sites, these are caused by reflections from local objects (e.g., water tanks and mountains), and they can be anticipated.

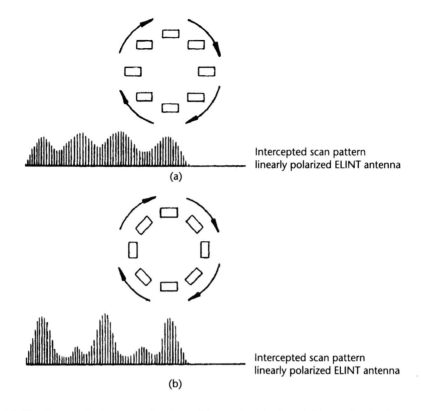

Figure 10.13   Two conical scan mechanisms: (a) nutating feed; and (b) rotating feed.

For intercepts from airborne radars, or especially airborne ELINT stations, reflections from aircraft propellers can be confusing because the propeller can cause amplitude modulations similar in appearance to conical scanning at rates of 40 to 60 Hz.

These hints for scan analysis indicate that experience is very important in properly interpreting the pulse amplitude versus time variations: ELINT scan analysis is more an art than a science.

10.9   A Three-Dimensional Search Example

A unique radar that electronically searches in elevation while circularly searching in azimuth illustrates a number of points about scanning in particular and search in general. (If you are unfamiliar with pulse compression, you may wish to read Chapter 11 before continuing with this example.) The combination of electronic scanning in elevation with mechanical scan in azimuth is a design frequently used by three-dimensional search radars. The electronic scan in elevation can be achieved using a constant delay in the feed to each row of an array antenna. Then changing the carrier frequency changes the phase of the signal to each row, which, in turn, changes the elevation pointing angle of the main beam. However, the changing carrier frequency, if done rapidly within the pulse, can also be used for pulse compression [3].

From the ELINT point of view, the S-band signal exhibits a circular scan with a period of, say, 10 seconds. Conventional analysis provides the information that the azimuth beamwidth is 1.5°. The pulse duration is 30 μs, and the pulse amplitude varies many decibels about a peak, whose position within the pulse at a ground site is near the leading edge of the pulse. When intercepted from an airborne receiver, the peak is later in the pulse. There is a large amount of frequency modulation within the pulse—with a frequency deviation of 120 MHz. The frequency deviation within the region of the peak pulse amplitude is always about 8 MHz regardless of the location of the peak within the pulse. The PRI is long, like that of a search radar. Suppose a photograph were available of a square array antenna, which is thought to be the source of this signal.

The task at hand is to explain the capabilities of this radar in light of what is known about the signal. The frequency deviation of 120 MHz would provide range resolution of about $c/(2$ bandwidth) or 1.25m, where c is the speed of light. This is too fine for general search applications, which are indicated by the 10-second scan interval. If a photograph were available of the antenna showing a square array, that would indicate that the elevation beam width is the same as the azimuth beamwidth, which is 1.5°. Such a beam would necessarily be scanned in elevation to make an effective 3-D search radar. Furthermore, the array means that electronic scanning is possible.

A simple type of electronic scan is elevation scanning by frequency, which in this case must occur within a single pulse. If that were the case, the peculiar amplitude peak would correspond to that portion of the 30-μs pulse that was most directly aimed toward the interceptor's elevation. In this case, the ratio of the 120-MHz frequency deviation to the 8-MHz deviation in the amplitude peak of the pulse indicates that there are about 15 beamwidths of 1.5° scanned in elevation—possibly 0° to 22.5°. Furthermore, a pulse-compression technique could be used to provide 1/8 MHz or 0.125 μs of range resolution or 18.75m. The elevation angle of the target would be indicated by the frequency at which the peak echo amplitude occurred. The pulse-compression ratio would be the time bandwidth product represented by the duration of the peak amplitude portion of the pulse times its bandwidth of 8 MHz. The 8-MHz portion represents 1/15 of the total pulse duration of 30 μs or 2 μs. Hence, the pulse-compression ratio would be on the order of $2 \times 8 = 16$. (This example is based on a three-dimensional search radar originally made by Plessey.)

References

[1]   Stone, L. D., *Theory of Optimal Search*, New York: Academic Press, 1975.

[2]   Skolnik, M. I., *Radar Handbook*, New York: McGraw-Hill, 1970, Chapter 21.

[3]   Radford, M. F., and R. Greenwood, "A Within-Pulse Scanning Height Finder," *IEE Int. Conf. Radar-73*, London, October 23–25, 1973, pp. 50–55.

CHAPTER 11

Intrapulse Analysis

11.1 Introduction

Many of the ELINT signals of interest consist of pulses—particularly radar signals. Determining the character of the intrapulse modulation gives valuable insight into the radar's function and design. The term "intrapulse" refers to the shape of the pulse envelope (or amplitude modulation function) and also to the frequency or phase variations within the pulse. If the carrier is not frequency modulated, the pulse is sometimes referred to as a CW pulse. (This term is sometimes used in radar literature and simply means a pulse with no intentional frequency or phase modulation.) If there is no intentional FM or PM within the pulse, the envelope of the pulse determines the range resolution of the radar. If there is intentional FM or PM, demodulating it (or determining its bandwidth) gives insight into the range resolution or other functions of the radar. Examining the unintentional AM and FM or PM can sometimes provide insight into the transmitter type and may help to identify the signal.

The radar waveform designer, of course, does not consider the individual pulses as an isolated parameter but must consider the entire modulating function, including the pulse intervals and phase or frequency modulation. The radar waveform determines its ambiguity function,[1] which is essentially the autocorrelation of the waveform with itself as it would appear if reflected from a point target having range delay τ and Doppler shift f_d. This function is plotted in three dimensions as magnitude versus range delay and radial velocity (Doppler shift).

If the signal at the ELINT site is carefully analyzed in terms of the amplitude and frequency modulation (including pulse-to-pulse phase variations), an ambiguity function for the radar can be estimated. The actual radar ambiguity diagram also depends on the processing performed by the radar's receiver (which cannot be determined through ELINT) and the antenna scan, which may determine the length of time the signal is directed toward a target.

It is customary in ELINT analysis to examine the pulse envelope in some detail by itself because so many radars use waveforms that consist of a repeated transmission of identical (or nearly identical) pulses. This helps to identify the signal type or at least its likely function. The pulse envelope shape must be observed at the output of a relatively wideband detector. In the past, pulse shape could be preserved using oscilloscope photographs or wideband analog tape recorders (e.g., 5- to 10-MHz bandwidth). At present, high-speed A/D converters are used. Some

---

1. See [1], Chapter 3, for a discussion of the ambiguity function.

typical A/D specifications are listed in Figure 11.1. At this time, A/D converters having 8 bits per sample operate up to 4 gigasamples per second (GSPS). Sample rates of about 400 megasamples per second (MSPS) at 12 bits per sample are also available.

The results of using a high-speed A/D converter to capture a single pulse are shown in Figure 11.2. A 200-MHz digitizing rate has been used to capture the predetection IF pulse (centered at about 50 MHz), along with the output of an AM detector (labeled "SQR DET") and the FM within the pulse from the output of a frequency discriminator (labeled "DISCRIM"). The predetection (IF) pulse

Intro Date	Res. (#Bits)	#Chn.	Max. Rate (MSPS)	SFDR (dBc)	ENOB (bits)	SINAD (dB)	SNR (dB)
Oct-99	8	1	1500	54	7.51	46	46.8
Jun-99	8	1	1000	52.3	7.55	46.2	47
May-01	6	2	800	41.5	5.65	35.2	36.5
Aug-99	8	1	600	57.5	7.65	46.8	47.1
May-99	8	1	500	59	7.1	—	44.5
May-01	6	2	400	45.5	5.75	36	36.6
May-83	10	1	250	68.3@fin=180MHz	—	56@fin=180MHz	56.3@fin=180MHz
May-99	8	1	250	59	7.1	—	44.5
May-83	8	1	250	69.1@180MHz	—	48.7@fin=180MHz	48.8@fin=180MHz
May-83	10	1	210	66.3@fin=180MHz	—	56.5@fin=180MHz	57@fin=180MHz
May-83	12	1	170	78@fin=65MHz	—	64.9@fin=65MHz	65.2@fin=65MHz
May-83	12	1	170	73@fin=100MHz	—	63.6@fin=100MHz	64.3@fin=100MHz
May-83	10	1	170	67@fin=180MHz	—	56.7@fin=180MHz	57.1@fin=180MHz
May-83	12	1	125	77@fin=100MHz	—	64.4@fin=100MHz	64.8@fin=100MHz
Aug-02	10	2	120	67	—	57.5	58.5
Aug-01	10	2	105	72	9.36	58.1	58.5
Jan-01	10	1	105	72	9.3	58	58.5
May-83	15	1	100	90@fin=15MHz	—	74.9	75.1
May-02	8	2	100	66	—	48.1	48.3

Figure 11.1   Some high-speed A/D converters.

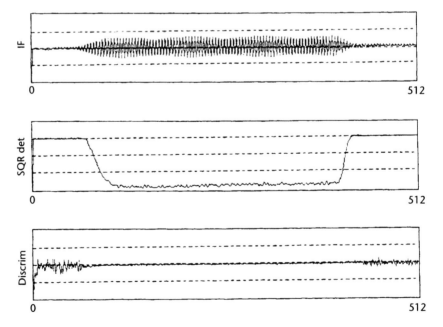

Figure 11.2   IF, AM, and FM digitized data (512 samples at 5 ns = 2.56 μs).

looks as if there is some modulation present; however, the AM and FM outputs show an ordinary unmodulated pulse. The apparent modulation on the IF signal is a beat between the 200-MHz digitizing rate and the fourth harmonic of the 50-MHz IF signal. (This is a consequence of having only four samples per IF cycle.) To capture the intrapulse AM and FM simultaneously requires multiple high-speed digitizers and the associated memory. It is common to have two channels available. Three or more channels is less common but sometimes useful. While the AM and FM waveforms may be obtained through digital demodulation of the predetection IF samples, analog detection methods may offer the advantage of wider dynamic range (especially in the case of an FM limiter-discriminator combination).

The actual pulsing of the radar is performed in a variety of ways [1, Ch. 7]. The details of the envelope shape concern the radar designer mainly for controlling the frequency spectrum outside the radar's band of interest. Interference with other radar or communications systems can result if poorly shaped pulses are transmitted. The ELINT analyst can gain insight into both the radar's capabilities and the modulation design by examining the details of the pulse envelope.

11.2 Pulse Envelope Parameters

In describing the pulse envelope it is convenient to use time-domain descriptors that are easy to measure and can be related to radar performance. The usual choices are the pulse duration, rise and fall times, and intentional or unintentional variations in pulse duration (or even pulse amplitude). The pulse duration is usually defined as the time between the half-power points of the pulse. For a linear detector, these are the points at 0.707 of the peak pulse voltage. For a square law detector, the voltage point of interest is at 0.5 of the peak pulse voltage. For any particular receiver detector, a calibration curve should be made to determine the half-power point voltage as a function of the peak pulse voltage.

Likewise, the rise time and fall time are the time between the 10% and 90% power points on the rise of the pulse and between the 90% and 10% power points on the fall of the pulse. In all cases, the percentages apply to the average amplitude of the pulse. A photograph of three successive radar pulses is shown in Figure 11.3. The photograph, taken from a moving platform, shows pulses from a circular scan search radar. The pulse amplitude, therefore, changes slightly. Because the signal is very powerful, receiver noise is not significant. However, multipath can cause variations in the fine structure along the top of the pulse.

Such a photograph of individual pulses is quite different from that of numerous pulses overlaid, which results when an ordinary oscilloscope setup is used. Single trace photographs (or at most a few traces) are important for an analysis of the pulse envelope details. If the amplitude is changing drastically due to scanning, the successive pulse envelopes will appear to have different pulse durations and rise and fall times, as indicated in Figure 11.4.

In the past, the difficulty in recording the pulse envelope over the wide dynamic ranges encountered meant an emphasis on oscilloscope and photographic techniques. Photographing successive pulses (as shown in Figure 11.3) requires a simple device such as a counter and a digital-to-analog (D/A) converter to offset the trace

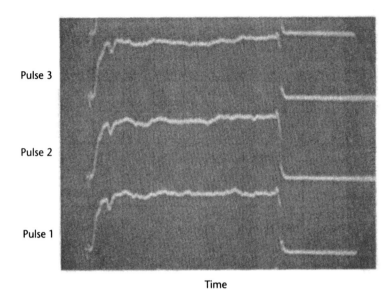

Pulse duration = ≈ 6 μs
Horizontal scan = 1 μs/division

Figure 11.3 Envelopes of successive radar pulses.

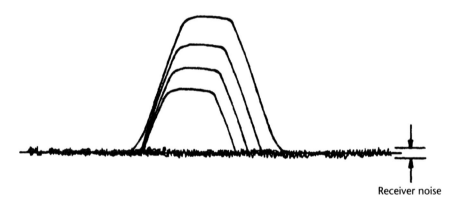

Figure 11.4 Multiple trace photograph effect on pulse envelope.

before (or after) each pulse. The scheme is shown in Figure 11.5. Each pulse triggers the sweep. After a delay expires, the counter advances, increasing the voltage from the D/A converter. This offsets the traces vertically on the screen in preparation for the next pulse. The camera shutter remains open during the entire process.

11.3 Envelope Parameter Measurements

High-speed A/D converters with sufficient resolution and sampling rate allow the digital capture of pulse envelopes. Precise measurements often require interpolation between the samples.

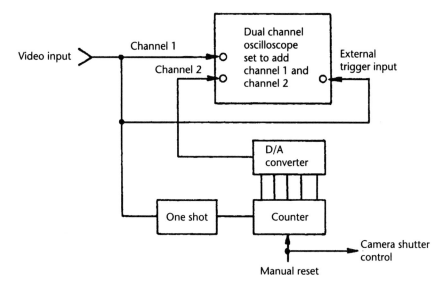

Figure 11.5   Raster technique for photographing successive pulses.

Automatic digital measurements of the pulse duration and rise and fall times must be interpreted carefully because changing irregularities in the pulse shape due to multipath and noise can cause wide variations in the readings. This is particularly true of time interval measurements. The problem becomes one of accurate threshold detection at, say, the 50% points, which must be properly located in spite of amplitude variation.

One type of threshold can be constructed using a delay and a sample and hold, as shown in Figure 11.6.

Rise and fall time measurements are difficult to automate due to the short times involved (often less than 50 ns). The digital approach can be used if the times are not too short relative to the time between the samples of the A/D converter.

The bandwidth of the ELINT receiver and A/D converter is very important in making the rise and fall time measurements. For good pulse fidelity, oscilloscopes incorporate amplifiers having an amplitude response that rolls off very slowly with

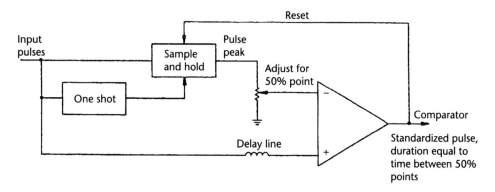

Figure 11.6   Half-power threshold scheme.

frequency. The pass band limit is often specified so that the 12th harmonic of the specified pass band limit is attenuated by no more than 75% [2].

The key point to be determined is the effect of the receiving system on the observed rise time. Figure 11.7 illustrates the effect of the various filters on the shape of the output pulse (private communication with D. K. Barton, February 2005).

11.3.1 Rise and Fall Times[2]

The rise time τ_r of a pulse is defined between the 10% and 90% voltage levels. When normalized to the bandwidth, $B\tau_r$ varies as shown in Figure 11.8, as a function of the noise bandwidth B_n and half-power bandwidth $B_{3\,dB}$ (which are identical for the rectangular filter). The curves for that filter contain ripples, resulting from the interaction of the $(\sin x)/x$ spectrum of the pulse with the edges of the rectangular bandpass. However, for $B\tau \gg 1$ the normalized rise time is essentially constant at $0.9/B$ for the rectangular filter, and $0.7/B$ to $1.1/B$ for the other filters. For all filters τ_r is reduced for $B\tau < 2$. This results from the fact that output peak amplitude is reduced more rapidly by the narrow bandwidth than is the slope of the leading edge. For $B\tau \ll 1$, $\tau_r < 0.7/B$, a result of the reduced peak outputs for the Gaussian and single-pole filters. For the more usual filters that are approximately matched or wider than matched, the well-known approximations given in [3–5], are applicable:

- From [3, p. 40]: $\tau_r = 0.7/B$, where B is the IF bandwidth;
- From [4, p. 413]: $\tau_r = (0.7 \text{ to } 0.9)/B3$, where $B3$ is the half-power IF bandwidth;
- From [5, p. 464]: $\tau_r \gg 1/B$, where the bandwidth definition is unspecified.

Note that Terman [4, p. 289] gives a value for video amplifiers as $\tau_r = (0.35 \text{ to } 0.45)/B_v$, where B_v is the (lowpass) video bandwidth measured from zero. This agrees with the factor $(0.7 \text{ to } 0.9)B_{3\,dB}$ for IF bandwidth that is measured on both sides of the center frequency.

A rule that has been used states that the rise time that results from cascading n stages (each of which is free of overshoot) is given by [6]

$$\tau = \left(\tau_1^2 + \tau_2^2 + \ldots \tau_n^2\right)^{0.5} \tag{11.1}$$

If the actual rise time of the transmitted radar pulse is τ_T and the rise time of the ELINT receiver and display system is taken as τ_R, the measured rise time, τ_M, is

$$\tau_M = \left(\tau_T^2 + \tau_R^2\right)^{0.5} \tag{11.2}$$

or

$$\tau_T = \left(\tau_M^2 - \tau_R^2\right)^{0.5} \tag{11.3}$$

2.   The author thanks Mr. David Barton for portions of the material in this section.

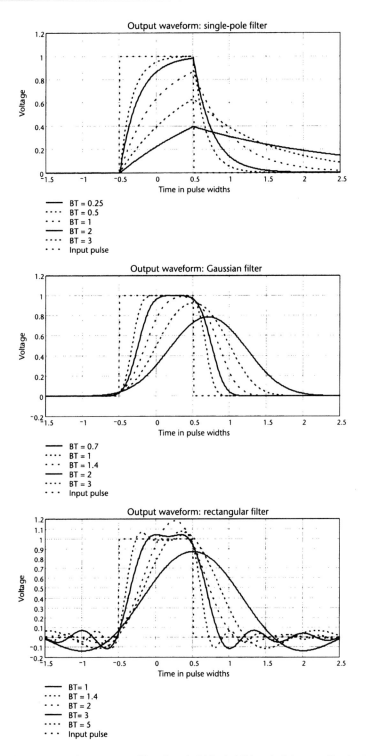

Figure 11.7   Output waveform versus filter bandwidth (additional delay applies in the cases of realizable Gaussian and rectangular filters).

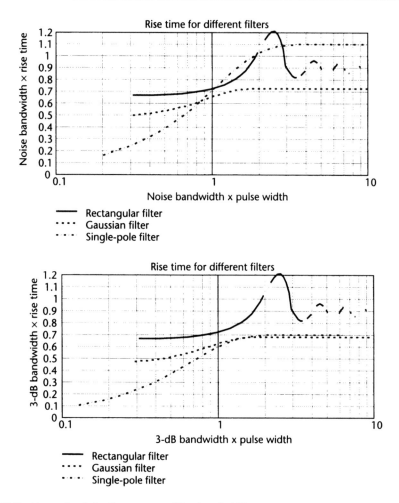

Figure 11.8 Normalized rise times versus filter bandwidth.

Equation (11.3) provides a means for estimating the transmitter rise time given the measured rise time and the inherent rise time of the receiving system. Figure 11.9 shows a plot of the ratio of the transmitter rise time to the measured rise time as a function of the ratio of the measured rise time to the receiver rise time. As can be seen, good estimates of the transmitter rise time require that the observed rise time be more than 1.5 times the receiver rise time.

Consider an example to illustrate the use of (11.3). The first step is to apply a laboratory pulse generator with a very short rise time to the input of the ELINT equipment and measure the rise time of the equipment. Suppose this is 50 ns. Next, suppose an ELINT signal is measured and found to have a rise time of 70 ns. Then, (11.3) gives the estimated rise time of the transmitted signal as about 49 ns—considerably less than was observed at the receiver output and slightly less than the rise time of the equipment itself. If this seems contradictory, remember that the laboratory measurement was made with essentially a zero rise time pulse applied and that the observed output was 50 ns. Increasing this to 70 ns requires an input rise time of about 49 ns.

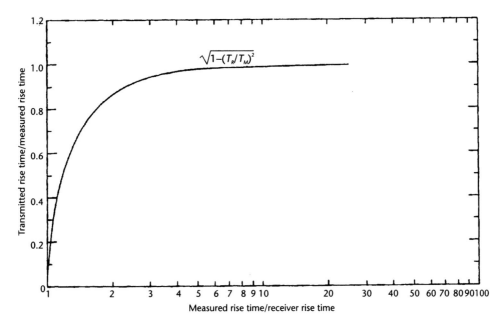

Figure 11.9 Rise time of transmitted signal related to measured rise time.

The receiver rise time should be measured prior to deployment. However, for planning purposes, the approximate relationship of this rise time to the receiver bandwidth may be used:

$$\tau_R \approx \frac{0.7}{B_R} \tag{11.4}$$

where:

B_R = receiver bandwidth (IF bandwidth or, for most systems, twice the video bandwidth)

τ_R = receiver 10% to 90% rise time

Equation (11.4) applies exactly for a simple resistor-capacitor (R-C) filter. It also closely approximates the response of a cascade of R-C filters or of any filter whose step response has less than a 5% overshoot [6].

It should be noted that the maximum slope of the leading edge of the pulse occurs at the half amplitude point and is approximately equal to the amplitude divided by the rise time. For the usual case of an ELINT receiver with bandwidth considerably larger than the reciprocal of the pulse duration, the slope/amplitude ratio ranges from about 0.9 to 1.1 times the reciprocal of the rise time [private communication with D. K. Barton, February 2005].

Historical Note. A pulse analysis technique designed to ease the burden of the ELINT analyst was the multiple gun oscilloscope [3]. By displaying several traces with different sweep speeds on a single screen, the pulse duration and pulse interval could be estimated at a glance without adjusting the controls. Full-screen sweep times of 5, 50, 1,000, 5,000, and 50,000 μs were used on one model. The first two covered most pulse durations of interest, while the last three covered the most PRIs of interest. The PRI scales were sometimes calibrated in pulses per second rather than units of time. Some of these pulse analyzers used traces with nonlinear sweeps (log or exponential) to cover wide signal parameter ranges.

The accurate estimation of pulse shape parameters is one of the more difficult tasks faced by the ELINT analyst. It is essential to have available calibration signals of known rise time, pulse duration, and fall time to calibrate the entire receiver, detector, digitizing, display, and measurement chain.

11.4   Some Radar Performance Limits Related to Pulse Envelope

The elementary radar equation that relates range resolution to pulse duration is

$$\text{range resolution} \sim (\text{speed of light})(\text{pulse duration})/2$$

In addition, for moving target indication (MTI) systems using delay line cancellers, the MTI improvement factor is limited by the presence of pulse duration jitter [1, p. 1748]:

$$I(\text{dB}) = 20 \log (\text{pulse duration/pulse duration jitter})$$

The envelope of the RF spectrum is related to the pulse shape. Figure 11.10 shows the approximate relationship between the RF spectrum envelope and the rise (or fall) time of a trapezoidal pulse as a function of the pulse duration. Figure 11.11 shows the same relationship for several other pulse shapes [1, p. 2920]. The electromagnetic interference to other signals that may be caused by a powerful radar

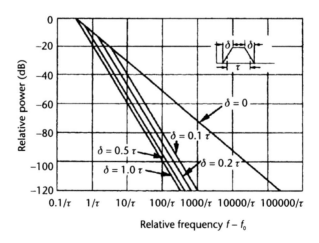

Figure 11.10   RF spectrum envelope for trapezoidal pulses (for rise time δ_1 = fall time $\delta_2 = \delta$). (*From:* [1]. © 1970 by McGraw-Hill, Inc. Used with permission of McGraw-Hill.)

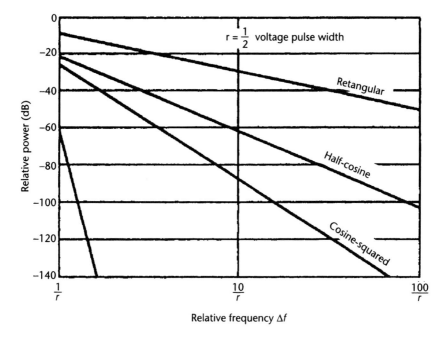

Figure 11.11   RF spectrum envelope for several pulse shapes. (*From:* [1]. © 1970 by McGraw-Hill, Inc. Used with permission of McGraw-Hill.)

makes it necessary for the transmitter designer to carefully control the spectrum roll-off. This often determines the pulse rise and fall times observed through ELINT.

To see how important this can be, consider the situation where two radars operate in close proximity (e.g., at the same airport). It is not unusual for the main beam power to be above +100 dBm and for the receiver to be sensitive to –100-dBm signals. Therefore, if the two systems are not to interfere with each other, there must be 200 dB of isolation between the transmitter of one and the receiver of the other. Suppose 100 dB is obtained by putting one on one side of the airport and one on the other side. The remaining 100 dB of isolation is to be obtained by operating the radars at different frequencies.

If a radar were to transmit rectangular pulses, the upper curve in Figure 11.10 gives the attenuation as a function of the frequency separation. Suppose the pulse duration is 1 μs. Then the horizontal scale of Figure 11.10 is expressed in megahertz. To obtain 100 dB of isolation when transmitting rectangular 1-μs pulses requires about 20,000 MHz of frequency separation. If radar 1 is at 2 GHz, radar 2 would be at 22 GHz! However, most radar systems have band-limiting elements in the transmission path, and hence, the actual pulse becomes more nearly trapezoidal, with rise and fall times set by this bandwidth. If a trapezoidal pulse with a rise time of 10% of the pulse duration is used, the spectrum falls 20 dB per decade instead of 10 dB per decade and the required frequency separation is reduced to about 300 MHz—a much more practical situation.

Because radar transmitters are generally high-power saturated amplifiers, the more rounded pulse shapes of Figure 11.11 are not generally used. (A Gaussian-shaped pulse has a very narrow spectrum, but it is infinitely long in time and impractical to generate.)

By carefully observing the pulse envelope, the ELINT analyst can infer something about the ability of the radars to operate in close proximity and also perhaps the minimum frequency spacing of RF channels if these are used.

Sometimes, the spectrum width of the pulse is of interest. For example, this might be useful for determining the spectrum width of a noise jammer to counter the emitter. For an ideal trapezoidal pulse with the rise time equal to the fall time, the 3-dB bandwidth of the spectrum in terms of the pulse duration, as measured between the 50% points, is shown in Figure 11.12. For a rectangular pulse, the 3-dB bandwidth is 0.88 divided by the pulse duration, while for a triangular pulse (the rise time and fall time equal to the pulse duration) the 3-dB bandwidth is about 0.64/pulse duration. The first spectrum nulls for a trapezoidal pulse are always separated by 2, divided by the pulse duration, regardless of the ratio of the rise and fall times to the pulse duration. Often the fall time is longer than the rise time. The 3-dB bandwidth of a trapezoidal pulse having a fall time three times the rise time is also indicated in Figure 11.12. In this case, the spectrum does not have a null near the reciprocal of the pulse duration; however, a deep minimum occurs there (see Appendix A).

11.5 Multipath Effects

Multipath may drastically alter the pulse envelope and also the apparent frequency variations within the pulse. Multipath can be modeled simply as a summation of

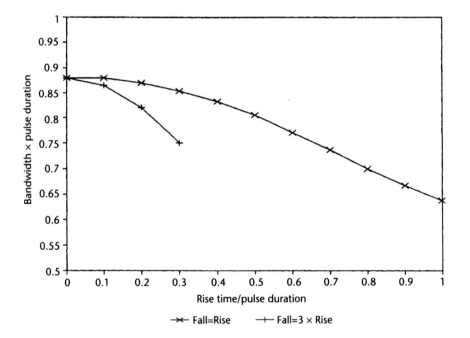

Figure 11.12 Bandwidth of trapezoidal pulses.

delayed replicas of the direct path signal with various amplitudes and phases. In any given situation the multipath effect can vary significantly from pulse to pulse. If the radar or the intercept platform is moving, movement of a significant fraction of a wavelength during the interpulse period is sufficient. For example, at 3 GHz the wavelength is 0.1m and a typical PRI is 1 ms. Moving a quarter-wavelength (0.025m) in 1 ms corresponds to 25 m/s or 90 km/hr. Even if both the radar and interceptor are not moving, the rotation of the antenna beam causes the amplitude of the signal reflected from different objects to change. This is especially true if the intercepted signal is from the radar's sidelobes.

Although the effects of multipath are most often associated with the observed pulse envelope (AM), the observed FM is also affected. Both effects can be demonstrated in the laboratory with the help of a microwave delay. Figure 11.13 shows the effect of adding equal amplitude signals at S-band—the original signal and the same signal delayed 300 ns. An adjustable line was used to vary the phase of the delayed signal, as shown in the Figure 11.13. On the AM pulse shape, the multipath may add or subtract, depending on the phasing of the delayed signal. To understand the intrapulse FM, consider that the onset of the multipath causes a discontinuity in the phase. The FM waveform is the derivative of the phase. Therefore, the discriminator output contains a transient at the beginning and end of the multipath. The transient may be positive or negative depending on the sign of the phase change.

The multipath shown in Figure 11.13 is quite extreme: the delayed signal and direct signal are of equal strength. Nevertheless, it serves as a warning to those

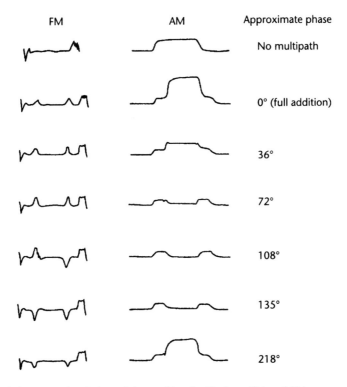

Figure 11.13 Laboratory simulation of the multipath effect on AM and FM.

who wish to make use of the pulse shape: both the envelope and the phase can be significantly affected by the multipath. Even the most careful analysis of the intercepted waveform may fail to reveal the transmitted pulse shape under these conditions.

The strongest multipath comes from reflecting objects near the transmitter and receiver and along the path between them. Normally, the radar site will be on a hill or otherwise away from large objects if at all possible. Likewise, the ELINT receiver will also be located away from nearby large objects if possible. The path between these sites may have hills or cities nearby that are unavoidable. The location of reflecting objects determines the delay of the reflected signal. If the range difference is 300m, the arrival time will be delayed by 1 μs. (The speed of light is about 300 m/μs.) If the delay is less than the pulse duration, there will be a multipath distortion of the pulse shape. If the delay is more than the pulse duration, two separate pulses will be observed and neither will be affected by the multipath. It is clear that the leading edge of the pulse is the best hope for observing at least part of the pulse free of the effects of multipath and that the trailing edge may consist entirely of multipath effects.

11.6   Intrapulse Frequency and Phase Modulation

Intentional intrapulse phase and/or frequency modulation is quite common in modern radar. Perhaps most radars of recent design use such modulation for pulse compression purposes. (This does not mean that half of the radars manufactured recently have pulse compression. Many radars are designed but never manufactured, and many more inexpensive than expensive radars are manufactured.) But it is a common technique used to increase radar range while maintaining good range resolution without increasing the peak power of the radar. The frequency or phase may also exhibit unintentional modulation and it may be combined with intentional modulation.

The radar's pulse compression ratio is the ratio of the transmitted pulse duration to that of the pulse duration at the receiver output after compression. The pulse compression ratio can be estimated from ELINT because it is always less than or equal to the time-bandwidth product of the radar pulse. By measuring the pulse duration (using the pulse envelope) and by measuring the bandwidth of a single pulse (by examining the output of a frequency discriminator or phase detector or by spectral analysis), the ELINT analyst can determine the time bandwidth product.

The time bandwidth product of conventional radar that does not use pulse compression is about unity (when the pulse duration is measured at the −3-dB level). The bandwidth is about equal to the reciprocal of the pulse duration. Radar that uses pulse compression is likely to have a time bandwidth product of at least 10. It is seldom worth adding the cost and complexity of pulse compression to a radar design unless at least this much compression is used. (As a counter example, the modified APY-2 AWACS radar uses a five-element Barker code to improve its ability to resolve range ambiguities caused by the high-PRF waveform.) Typical radars use time bandwidth products between 10 and 1,000. Some special-purpose very long-range and/or very fine-range resolution radars have time bandwidth products in excess of 100,000. There is an upper limit for the radar's time bandwidth product due to target motion. If the target moves radially with velocity, v, and if the range resolution of the radar (range cell extent) is distance, d, the amount of time the target will be in that range cell is $T = d/v$. Achieving this range resolution requires a bandwidth of $B = c/2d$, where B is the bandwidth and c is the speed of light. Thus, if the target is to stay in the range cell for the entire pulse duration, the time bandwidth product is limited to $BT < c/2v$. If the velocity is 500 m/s, the upper limit for BT is 300,000.

The most common types of modulation are phase reversals and linear FM. For a phase reversal signal, the range resolution is easily determined as being equivalent to that of conventional radar with the pulse duration equal to the minimum time between the phase reversals. The waveform for a 13-bit Barker code is shown in Figure 11.14. The predetection IF signal shows some drop in amplitude at the times of the phase reversals. This is usually due to a restricted bandwidth in the transmitting path, which causes the envelope to drop to zero as the phase is reversed. The waveforms marked Phase I and Phase Q are a result of mixing a limited version of the predetection signal with local oscillator signals that are 90° out of phase and nearly at the same frequency as the IF. It is easy to see the phase reversals in these waveforms. For linear FM (chirp), there would be a quadratic phase progression that would show up as an increasing frequency of the Phase I and Phase Q signals. But the more usual approach would be to use a frequency discriminator to observe the modulating signal directly. The predetection signal is often digitally demodulated.

For many methods of digital FM demodulation, the first step is to generate in-phase (i) and quadrature (q) samples from the predetection IF samples. This can be done either in the time domain by convolving the IF signal with the impulse

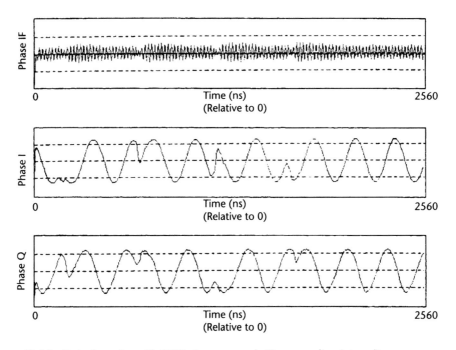

Figure 11.14   Part of a pulse with 180° phase reversals (5-ns sampling interval).

response of the Hilbert transform, or in the frequency domain by multiplying the fast Fourier transform (FFT) of the IF samples by $-(j)\,\text{sgn}\,(2\pi f)$ and then taking the inverse FFT. (This function is the Fourier transform of the impulse response of the Hilbert transform.) Once the i and q samples are available, the phase samples can be found by using the following equation:

$$\text{phase} = \arctan(q/i) \tag{11.5}$$

$$\Delta\phi = \arctan(q_2/i_2) - \arctan(q_1/i_1) \tag{11.6}$$

The frequency is the phase rate of change; the average frequency over the sample interval is the change in phase over the sampling interval divided by the sampling interval. Because the phase is determined over an interval of 360°, it is necessary to *unwrap* the phase (i.e., add 360° when the phase change from one sample to the next is too great). This can be avoided by using the expression for the difference of two arctan values:

$$f = \frac{1}{2\pi}\frac{d\phi}{dt} \approx \frac{\Delta\phi}{2\pi\Delta t} = \frac{q_1 i_2 - q_2 i_1}{2\pi(t_2 - t_1)(i_1 i_2 + q_1 q_2)} \tag{11.7}$$

The frequency value computed using (11.7) is actually the average frequency over the time interval between the two sets of (i, q) values rather than the true instantaneous frequency. To approximate the instantaneous frequency, the frequency should not change significantly during the time between the samples.

To see the precise relationships between the amplitude, phase and in-phase, and quadrature functions, it is useful to represent a phase-modulated signal as

$$s(t) = a(t)\cos[2\pi f_o t + \phi(t)] \tag{11.8}$$

where $a(t)$ is the amplitude or envelope, f_o is the carrier frequency, and $\phi(t)$ is the phase modulation; that is, the integral of the frequency modulation. Expanding the cosine function gives

$$s(t) = a(t)\cos\phi(t)\cos(2\pi f_o t) - a(t)\sin\phi(t)\sin(2\pi f_o t) \tag{11.9}$$

$$s(t) = i(t)\cos(2\pi f_o t) - q(t)\sin(2\pi f_o t) \tag{11.10}$$

from which:

$$i(t) = a(t)\cos\phi(t)$$

$$q(t) = a(t)\sin\phi(t)$$

$$a(t) = \sqrt{i^2(t) + q^2(t)}$$

$$\phi(t) = \arctan[q(t)/i(t)]$$

One of the main problems with digital FM demodulation is the limited dynamic range when A/D converters of only a few bits of resolution are used. Analog discriminators, when preceded by a limiter, provide a much wider dynamic range over which accurate FM demodulation is accomplished. One drawback is the nonideal discriminator curve (frequency-to-voltage transfer function), which varies from system to system. This means that the same signal, when processed by two different receiving systems, produces slightly different outputs. Of course, this may also occur with digital receiving systems, where equalization may be required to make two systems produce the same output.

11.6.1   Choosing the Receiver Bandwidth

Even though the modulating signal for a phase or frequency modulated signal is band-limited, the modulated signal is not band-limited.

The bandwidth required to preserve the fine structure of the modulation is about twice the deviation plus about 10 times the rate of modulation. This is shown in Appendix B, Section B.2. (This bandwidth requirement is sufficient to retain sidebands in the sinusoidal FM spectrum that are no more than 40 dB down.) If the pulse duration is on the order of 1 μs and there are five cycles of modulation across the pulse, a discriminator bandwidth of 50 MHz is needed. Of course, the wider the bandwidth, the more noise in the output (which may then obscure the FM).

Whether or not analog discrimination or digital demodulation is used, noise will be present in the FM output. If the signal is large compared to the noise, the phase angle variation due to noise is given (to close approximation) by the quadrature component of the noise divided by the signal amplitude. In the case of a limiter-discriminator combination, the weaker noise is reduced relative to the signal so that the output SNR increases by about 3 dB. Therefore, the performance of this type of discriminator is enhanced somewhat compared to linear demodulation methods. Figure 11.15 shows the amount of frequency noise at the output of a

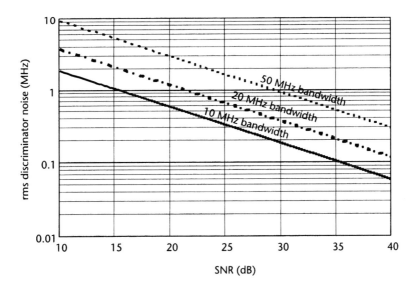

Figure 11.15   Discriminator rms noise output as a function of SNR input.

digital or analog discriminator with no limiter as a function of the input SNR. In this calculation, the bandwidth after the discriminator is matched to the bandwidth at the input. Because the output noise rises in proportion to the square of the frequency, lowpass filtering the discriminator output can significantly reduce the output noise at the cost of obscuring the FM waveform. For a given input SNR, there is actually more signal level present in a wide input bandwidth than in a narrow one. The video filter is able to eliminate so much of the noise that the output SNR is greater than the input SNR for wide input bandwidths. (This is apparent from (B.53) and (B.50) in Appendix B.)

As can be seen, the SNR required for rms discriminator output noise to be less than 100 kHz is quite large. This situation can be improved by post discrimination video filtering if the modulation rate allows.

11.7 Intentionally Modulated Pulses

It is customary to include the measurements of intentional phase or frequency modulation-on-pulse with other measurements of the RF, such as pulse-to-pulse frequency agility and the carrier frequency itself. For this reason, a discussion of some of the techniques used to analyze these characteristics is included in Chapter 14. Of course, many of the same techniques are used regardless of whether the modulation is intentional or not.

Intentional intrapulse modulation is nearly always for the purpose of pulse compression. The modulations most often used are linear FM (chirp) and binary phase reversals, sometimes in the form of Barker codes. An example of a linear FM waveform is shown in Figure 11.16, and an example of a 13-bit Barker code BPSK signal is shown in Figure 11.17. Other modulation types include stepped frequency modulation (or FSK), a discrete form of linear FM and phase shift keying (PSK) using phase steps of other than 180°. These are sometimes called polyphase codes to distinguish them from the more common binary codes. Both FM and PSK must deal with the problem of time sidelobes—these are "false" peaks in the output of the pulse compression processor caused by partial correlation of the received echoes with the transmitted waveform at time shifts (range delays) at other than the true range. For linear FM, it is common to use weighting networks at the output of the pulse compression filter to decrease these undesired outputs in a manner similar to the reduction of sidelobes in antenna patterns by tapering the aperture illumination. In the case of BPSK, the Barker codes have time sidelobes of maximum amplitude of 1, with the correlation peak amplitude equal to the length of the code. However the longest known Barker code is 13, which limits pulse compression using these codes to a maximum of 13:1. Longer BPSK sequences can be used at the price of higher time sidelobes. For example, random signals can be used with time sidelobes on average near zero but on any particular random code transmitted the time sidelobes are expected to be approximately the square root of the code length. Polyphase codes that use small phase steps often approximate the performance of linear FM. Note that linear FM has a quadratic phase progression. Nonlinear FM may be used, often for the purpose of reducing time sidelobes. The smallest frequency steps that result in orthogonal segments are such that there is

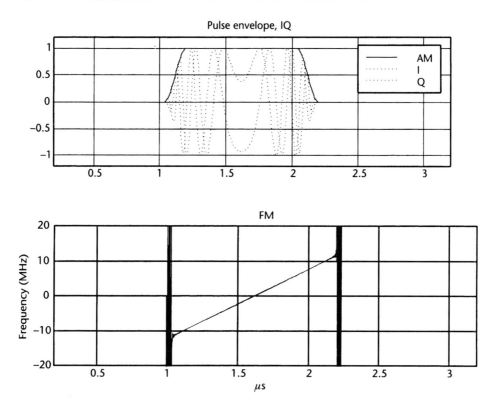

Figure 11.16   Modeled linear FM. Pulse width = 1 μs; Δf = 20 MHz (BT = 20).

a $\pm 90°$ phase accumulation over the duration of the constant frequency step. Thus, if the step duration is T, the minimum frequency shift is such that $FT = \pm 90°$. These frequencies are transmitted in a manner to provide minimum shift keying (MSK) waveforms, sometimes called Taylor codes. The N-length sequence Taylor coded waveform consists of phase modulated chips in the I and Q components of the signal at increments of $\pm 90°$. Each chip is comprised of a one-half cycle sine wave envelope. Listed in Table 11.1 are the known Taylor coded waveforms as given in [7]. Figure 11.18 illustrates the waveform for the 28A Taylor code.

From the ELINT viewpoint, it is important to determine the time bandwidth product of the signal and also to determine what kind of modulation is used. This information helps identify it and also helps determine its function.

Most radars have pulse compression ratios in the range of 10 to 1,000, with a few specialized radar systems for missile defense or for mapping having very large time bandwidth products up to 100,000 or more.

Once the intentional modulation has been analyzed, additional information can be obtained from the unintentional modulation. For example, it may be possible to tell one radar from another by the amount of nonlinearity found on its LFM waveform or by the amplitude ripple observed on the top of the pulse due to bandpass filtering a BPSK signal prior to transmission.

There is a need for a measure of distortion and/or how well two waveforms match. Another way of looking at this is to ask the question how much bandwidth is needed in an ELINT system to avoid distortion of the signals of interest. Several

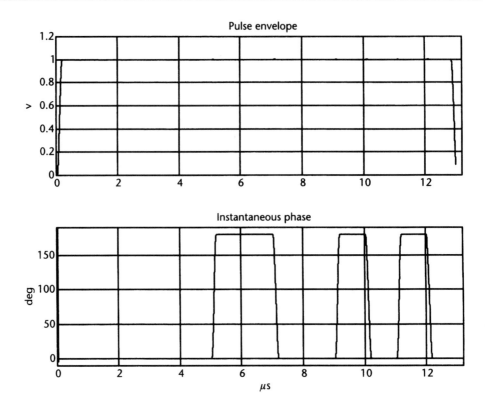

Figure 11.17 Modeled Pulse with 13-bit Barker code.

Table 11.1 The Known Taylor Codes

Code Length, N	Phase Sequence
13	1 j −1 −j 1 −j 1 −j 1 −j −1 j 1
25	1 j −1 j −1 j −1 −j −1 −j 1 j −1 −j −1 j 1 −j −1 j −1 j 1 j 1
27	1 −j −1 −j −1 j 1 j 1 −j 1 j 1 −j 1 j 1 −j 1 j 1 j −1 −j −1 −j 1
28A	1 j 1 −j 1 −j −1 j −1 j 1 j −1 j 1 j −1 j 1 j −1 j −1 −j 1 −j 1 j
28B	1 j −1 j −1 j −1 −j −1 −j 1 j −1 −j −1 j 1 −j −1 j −1 j 1 j 1 −j 1 −j

answers have been given already, based on the forms of Carson's rule developed in Appendix B and, for CFM, the distortion as measured in the peak and rms frequency deviation for the distorted and undistorted signal have been investigated (see Table B.2). A more general method of comparing signals for distortion measurement of other purposes is to use the normalized Euclidean distance (NED). For two signals $s_1(t)$ and $s_2(t)$, this is defined as

$$\text{NED} = \frac{\sqrt{\int\limits_{t_1}^{t_2} [s_1(t) - s_2(t)]^2 \, dt}}{\sqrt{\int\limits_{t_1}^{t_2} [s_1(t)]^2 \, dt} + \sqrt{\int\limits_{t_1}^{t_2} [s_2(t)]^2 \, dt}} \tag{11.11}$$

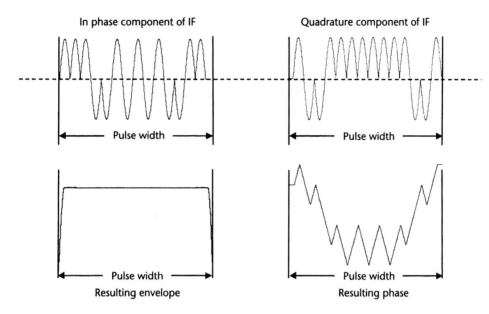

Figure 11.18   Modeled 28A Taylor code pulse.

For comparing waveforms, it is then convenient to adopt some maximum value of NED as the threshold as the criterion for deciding two signals are the same (or matched). This NED threshold value can also be used to determine what bandwidth is needed for the ELINT system in order that the output matches the input. For CFM, the relationship between modulation index and NED when only the first pair of sidebands is preserved is shown in Figure 11.19.

11.8   Incidental Intrapulse Shape—Uses and Causes

The traditional ELINT measurements include pulse duration; rise and fall times; and time bandwidth product, intrapulse phase, or frequency modulation for pulse

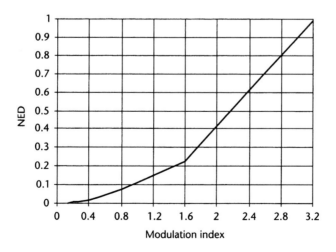

Figure 11.19   NED for CFM when the bandwidth is limited to the first sideband.

compression systems. Also included are the variations of these values (e.g., pulse duration jitter). There is additional information contained in the fine structure of the AM and PM wave shape, but the nature and use of this information depends on many factors. One of the most important of these is the SNR. To discover the fine structure of the intrapulse modulation through ELINT may require SNR values in excess of 40 dB. Once the "true" shapes are unmasked from the effects of noise and multipath, it may be possible to recognize which type of transmitter is in use (e.g., magnetron, klystron, and solid-state). Some have thought that finely resolved pulse shape data can be used at the front end of EW systems to deinterleave pulse trains even if they come from the same emitter type. Unfortunately, this would require high SNR values for single pulses. Bringing the pulse shapes out of the noise usually requires averaging together many pulses, which implies that they have already been deinterleaved.

Such averaging involves taking many individual wave shapes, aligning them in time and frequency, and then averaging them point by point across the waveform ensemble. The alignment process is critical. The simplest technique is to align the 3- or 6-dB points on the envelope and then average both the AM and FM waves based on that alignment. A more sophisticated method is to start with that alignment and then shift each waveform right or left a few samples to determine the alignment that gives the largest cross-correlation (or smallest NED). This alignment of pulse envelopes and averaging the envelopes is a noncoherent ensemble average of pulses having high individual SNR values (e.g., 20 dB). It is expected that the SNR improves at least as the square root of the number of pulses averaged—better than that if the SNR is high to begin with. If the individual pulses have a 20-dB SNR and a 40-dB SNR is desired, it may be necessary to average 1,000 pulses. Collecting that number of pulses may or may not be practical. A circularly scanning search radar may provide on the order of 25 to 50 main beam pulses. Hence, 20 to 40 scans could be required to collect 1,000 main beam pulses. If the SNR is high to begin with, the averaging may improve the SNR by a factor equal to the number of pulses averaged. Note that, for a scanning signal with varying amplitude, it is not very useful to include pulses that are too far below the peak. Consider the coherent ensemble average of two pulses of amplitudes a_1 and a_2 in the presence of noise of power N. Coherent here means that the peaks of the cycles of the predetection (or IF) sine waves for the two pulses are aligned in time. Adding these signals corrupted by noise directly gives a sine wave whose peak is now the sum of the peaks of the two signals and whose noise power is now $2N$. The SNR of the sum is then

$$\text{SNR}_{\text{ENS}} = \frac{(a_1 + a_2)^2}{4N} \tag{11.12}$$

With no weighting of the amplitudes, it is easy to see that if the amplitude of the weaker pulse is small, the SNR of the ensemble may be less than the SNR of the stronger pulse alone. That is, for SNR improvement, it is necessary that

$$\text{SNR}_{\text{ENS}} = \frac{(a_1 + a_2)^2}{4N} \geq \frac{a_1^2}{2N} \tag{11.13}$$

This implies that the weaker pulse amplitude must be at least 0.414 times the amplitude of the stronger pulse (or no more than 7.66 dB down) in order to improve the SNR of the ensemble. If a weighting factor is used, this limitation can be removed. Consider using a weighting factor of k times the amplitude of the weaker signal:

$$\text{SNR}_{\text{ENS}} = \frac{(a_1 + ka_2)^2}{2(1 + k^2)N}$$

Differentiating the SNR with respect to k and setting the derivative equal to zero shows that the value of k for maximum SNR of the ensemble average is

$$k = \frac{a_2}{a_1}$$

This means that the best ensemble average is formed by scaling the amplitudes of the weaker signals by the ratio of the amplitude of the weaker signals to the amplitude of the strongest signal. Clearly, signals that are (say) 20 dB below the amplitude of the strongest one will not contribute much to the overall SNR of the ensemble. Also, misalignment of the cycles of signals that are weak can easily result in SNR degradation instead of enhancement. Thus, practical considerations limit the ensemble average to the relatively strong pulses. Such ensemble averaging is often needed to bring out the character of the unintentional intrapulse frequency or phase modulation—or the unintentional variations of intentional modulation from the ideal—for example, incidental nonlinearity of a chirp signal.

The causes of unintentional intrapulse modulation are often inherent in the modulation sensitivity of high-power transmit tubes [6]. The major effects are: (1) pushing; (2) pulling; and (3) temperature, aging, and poor maintenance.

Pushing is caused primarily by variations in the output of the modulator and high-voltage power supply. Pulling is caused by variations in the load impedance seen by the tube primarily for pulsed oscillators (magnetrons). An example is a poor rotating joint in the waveguide to the antenna. The other effects are more slowly changing; for example, changes in the intrapulse wave shape as the tube warms up (changes over several hours) or as it ages over a period of weeks, months, or years. An example of poor maintenance, which may be seen in the wave shape, would be if the high-voltage power supply output decreases and no action is taken to adjust it.

The effects of different types of modulators on incidental intrapulse modulation are summarized in Table 11.2. There is not much incentive to reduce such effects if there is no significant improvement in radar performance. The greatest variations are caused by the modulator rate of rise voltage, ringing, and droop. Although the rate of rise is usually thought of in terms of the envelope, the frequency variations of a magnetron can be significant—it is even a strong determinant of the mode in which magnetron oscillations occur. In fact, many FM artifacts may also be seen in the envelope as well (i.e., there is a definite correlation between the FM and AM wave shapes). Both the amplitude and phase change as a function of the

Table 11.2 Modulator Interpulse Effects by Type

Type	Switch	Application	Pulse Shape	Intrapulse Expected*
Line	Thyratron silicon controlled rectifier (SCR)	Most common	Ripples	5
Hard tube	Capacitor coupled	Fairly common	Good	3
	Transformer coupled	Not often used	Fair	4
	Modulator anode (floating deck)	Usually high power	Excellent	2
	Grid	Widely used	Excellent	2
SCR	SCR and magnetic cores	Becoming more common	Ripples	5

*5 = significant; 1 = minimal.
Source: [8].

applied voltage. For example, a 1% ripple in the high-voltage modulation of a 10-GHz magnetron would cause a frequency deviation of 200 to 600 kHz at the same rate as the ripple [8]. For klystrons and traveling wave tube (TWT) amplifiers, the FM is more commonly specified in terms of phase variation per unit change in voltage. For a TWT, a 1% ripple in the cathode to anode voltage may cause a phase change of 5° or an amplitude change of 0.1 dB [9].

11.9   Comparing Wave Shapes

For FM and AM, the NED can be a useful measure of difference between two waveforms. These differences can be obtained to compare each pair of waveforms. Once the differences are available, a histogram of the differences can reveal the ability of the waveform to separate the data into classes. Figure 11.20 illustrates such a histogram of differences. In this example, to properly associate the pairs that belong to the same class (intraclass), it would be necessary to set a NED threshold of about 0.4 units. Unfortunately, this would also include about 24% of the pairs that do not belong together (interclass). If a threshold of 0.2 is used, all of the interclass pairs will be rejected, but only about half of the intraclass pairs will be accepted. This is typical of such classification schemes. There is a trade-off between correctly grouping those that belong together and falsely including those that do not belong together. Hence, two thresholds may be needed. For example, if the NED is less than 0.2, the pulses may be said to match. If the NED is between 0.2 and 0.4, the pulses may or may not match. If the NED is greater than 0.4, the waveforms are said to not match.

The receivers used to compare waveforms must be well matched in amplitude and group delay across their pass bands, and the signals must be tuned to near the center of the channel. Equalization is generally needed to match receivers and digitizers well enough to achieve small NED values when the same test signals are applied to each receiver. Figure 11.21 shows the phase error of the demodulated FM signal as a function of time alignment error between the upper and lower sidebands of a circular FM signal. The phase shifts allowed for NED values of 0.1 and 0.2 are indicated. Of course, as the modulation rate of the FM increases, the

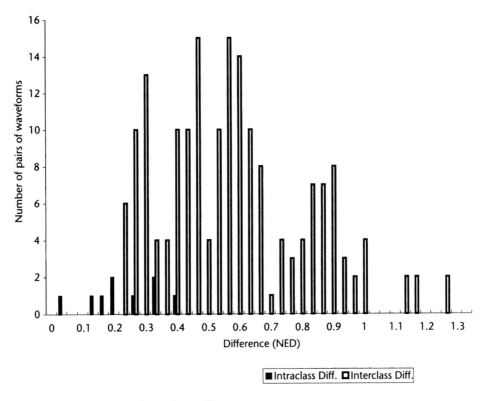

Figure 11.20 Histogram of waveform differences.

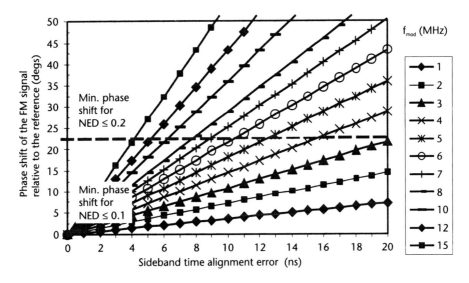

Figure 11.21 Phase shift of the demodulated FM related to sideband time alignment error.

more critical is differential delay for one sideband relative to the other. For example, if the receiver is to account for less than NED = 0.1 of mismatch for an FM signal with a 15-MHz modulation rate, the differential delay between the sidebands (30 MHz apart in frequency) must be less than 2 ns.

References

[1] Skolnik, M. I., *Radar Handbook*, New York: McGraw-Hill, 1970.

[2] Shackill, A., "Digital Storage Oscilloscopes," *IEEE Spectrum*, July 1980, pp. 22–25.

[3] Chance, B., et al., *Electronic Time Measurements,* Vol. 20 of MIT Radiation Laboratory Series, New York: McGraw-Hill, 1949. Also, "Radiation Lab Series: CD-ROM Edition," Norwood, MA: Artech House, 1999.

[4] Terman, F. E., *Electronic and Radio Engineering*, 4th ed., New York: McGraw-Hill, 1955.

[5] Skolnik, M. I., *Introduction to Radar Systems*, New York: McGraw-Hill, 1961.

[6] Valley, G. E., Jr., and H. Wallman, *Vacuum Tube Amplifiers*, Vol. 18 of MIT Radiation Laboratory Series, New York: McGraw-Hill, 1949, p. 77. Also, "Radiation Laboratory Series: CD-ROM Edition," Norwood, MA: Artech House, 1999.

[7] Blinchikoff, H. J., and J. W. Taylor, "Quadriphase Code—A Radar Pulse Compression Signal with Unique Characteristics," *IEEE Trans. on Aerospace and Electronic Systems*, Vol. 24, No. 2, March 1988.

[8] Boyd, J. A., et al., *Electronic Countermeasures*, Los Altos, CA: Peninsula Publishing, 1978, pp. 233–234.

[9] E. M. Galey, "Intrapulse and ELINT System Design," *Journal of Electronic Defense*, October 1989, pp. 87–106.

CHAPTER 12

Pulse Repetition Interval Analysis

12.1 Introduction

Pulse repetition interval timing is one of the major aspects of ELINT signal analysis. The radar's performance is affected significantly by the PRI timing used. In addition, many warning receivers and jammers make use of properties of the PRI. Finally, PRI data can generally be recorded easily for off-line analysis in the laboratory. Most PRI values are such that modest bandwidths can be used. Generally, the video or envelope bandwidth is of interest for recording or digitizing the PRI (and pulse duration) data. It is normally the case that the video bandwidth is approximately one-half of the two-sided predetection or IF bandwidth; however, as noted in Chapter 3, in the case of a wideband receiver, the video bandwidth may be small compared to the RF bandwidth of the receiving system.

The radar pulse is usually repeated; however, the interval between pulses may change. It is common to speak of the PRI used by radar even though the interval itself may not be repeated. In this discussion, the term PRI is used to mean the time from the leading edge of one radar pulse to the leading edge of the next radar pulse. For radar systems that use groups (or bursts) of several closely spaced pulses, the PRI refers to the time from the leading edge of the first pulse in one group to the first pulse in the next group. This may be referred to as the pulse group repetition interval (PGRI).

Just as close scrutiny of the pulse envelope shape gives insight into some aspects of the radar's performance and the mechanization of the modulator, close scrutiny of the pulse intervals gives insight into other aspects of the radar's performance and the mechanization of the pulse timing circuitry.

A subject that arises frequently in connection with PRI analysis is the use of countdown circuitry so that the PRI is a multiple of some shorter interval. To provide for accurate range calibration, the radar designer may find it convenient to synchronize the transmission of the radar pulse with the oscillator generating the timing marks. The PRI is therefore some multiple of the period of this oscillator. Early countdown circuitry involved using analog multivibrator circuits to establish a minimum pulse interval. The radar's transmission occurred on the first pulse from the timing oscillator after the expiration of the analog delay. Depending on the stability of the delay, one of a number of different multiples of the timing oscillator period might be selected as the PRI. Digital countdown circuitry is much less prone to counting errors.

Another facet of PRI analysis is dropped or weak pulses. This is more a problem of magnetron radars than of other types. A typical manufacturer's specification may be for 0.01% up to 1% missing pulses [1]. While this may make no noticeable difference in the performance of the radar, a few missing pulses in a burst subjected to ELINT analysis can cause confusion. Of course, weak pulses near the detection threshold may also be dropped with probability equal to (1 – probability of detection). If the threshold is set to achieve a probability of detection equal to 90% at some amplitude, then for that amplitude there is a missed pulse probability of 10%. A good rule of thumb is to set the threshold so that the probability of detection is greater than (1 – missing pulse fraction) to avoid missing significantly more pulses than those inherently missed by the transmitter.

12.2 Common PRI Categories

The variety of PRI schemes is seemingly infinite. There are, however, a number of schemes used sufficiently often that these categories have been given names. Generally, the ELINT analyst explores the PRI variations and attempts to place the radar emitter into one of the categories listed in Table 12.1. For each of these categories, the ELINT analyst emphasizes different aspects of the PRI variations. As indicated in Table 12.1, each PRI category is associated with different radar functions.

12.2.1 Constant PRI

The radar has a nearly constant PRI if the peak variations are less than about 1% of the mean PRI. Such variations are considered incidental in that they generally

Table 12.1 Common PRI Variations

Type	Typical Function	Remarks
Constant	Common search or track radars	Variations typically less than 1% of the average PRI value. Very stable constant PRIs are associated with MTI and pulse Doppler systems
Jittered	Reduces the effects of some types of jamming	Large variations—up to about 30% of the average PRI
Dwell and switch	Resolve velocity (or range) ambiguities especially in pulse Doppler radars	Bursts of pulses with several stable PRIs switched from one burst to the next
Stagger	Eliminate blind speeds in MTI systems	Several stable PRIs switched on a pulse-to-pulse basis in a periodic pattern
Sliding	Provide constant altitude coverage during elevation scanning or avoid eclipsing	Maximum PRI (at minimum elevation) usually less than six times minimum PRI at maximum elevation
Scheduled	Used in electronic scan, multiple function (search and track) computer-controlled systems	Numerous complex patterns (some periodic) may adapt to target situation
Periodic variations	Missiles guidance; avoidance of eclipsing, ranging	Nearly sinusoidal variations up to 5% of average PRI. Rates up to 50 Hz or more
Pulse groups	May improve range or velocity resolution	Also used for IFF and beacon interrogation

serve no useful purpose. If the peak PRI variations exceed 1% of the mean PRI, the variations may serve some particular function. Occasionally, circuit malfunctions or design peculiarities may cause large PRI variations that serve no useful purpose. These may also be placed in the "constant" category. For constant PRI radar signals, the parameters of interest include the mean PRI, the peak-to-peak incidental PRI variations, the amount of PRI drift, and the time period over which the drift was measured.

If the average PRI value is to be used to estimate the radar's maximum unambiguous range, determining the PRI to an accuracy of about 1% is clearly sufficient. On the other hand, very high accuracy may be needed if it is desired to investigate, for example, the variation in the mean frequency of the crystal oscillator controlling the PRI.

The incidental variations in the pulse interval are useful in determining the characteristics of the radar's trigger generating circuitry. For example, MTI systems that use a delay line for trigger generation and cancellation of returns from stationary targets require precisely controlled pulse intervals. The limit on the MTI improvement factor due to PRI jitter is

$$I_{jitter} \text{ (dB)} = 20 \log \frac{PD}{2 \cdot PRIJ} \tag{12.1}$$

where:

I_{jitter} = limitation on the MTI improvement factor due to jitter

PRIJ = RMS PRI jitter

PD = pulse duration

This means that for 30-dB MTI improvement, a radar system using 1-μs pulse duration can tolerate an rms PRI jitter of no more than 15.8 ns. Measuring individual PRIs to such accuracy is difficult. It requires both a high SNR and a scheme for avoiding apparent PRI changes due to changes in pulse amplitude as the radar scans. A good rule to follow is to try to obtain jitter measurements to about 1% of the pulse duration. However, practical limitations prevent meaningful jitter measurements of smaller than 10 to 20 ns. Clearly, whatever the SNR and other limitations faced by the analyst, the *maximum* jitter due to the radar can be estimated from the received signal. The radar obviously has incidental jitter of no more than the observed value. For that jitter, if (12.1) yields a useful MTI improvement factor, the radar could have MTI capability.

The PRI drift characteristic is also of interest because it can indicate the type of likely PRI stabilization in use. (For example, a trigger generator in an oven compared to one with no oven or even a proportional oven compared to an on/off thermostatically controlled oven.) Drift is expressed as a fraction of the average PRI over some time period. The procedure is to determine the average PRI over a short time interval (such as the number of pulses received during one scan of the radar main beam past the ELINT receiver antenna). These short time averages are then compared to the average of all the observed PRIs. An example is shown in

Figure 12.1. The maximum change in the short-term PRI average is divided by the long-term average PRI to obtain a fraction. In Figure 12.1, the maximum change is 12 μs in a 3,000-μs PRI. The fractional change is, therefore, 6×10^{-3} in 3 minutes. The time interval is usually at least several minutes but may extend over hours or even days.

It is normal for constant PRI radars to have several different PRIs that may be selected manually or by software control. This may be done to resolve range and/or Doppler ambiguities or to avoid eclipsing, as described in Chapter 2. In this case, the ELINT analyst must look for transmissions using all of the PRIs and perform the same jitter and drift analysis for each one.

12.2.2 Jittered PRIs

Intentional PRI variations are used for a variety of radar purposes, as indicated in Table 12.1. The use of such variations implies radar capabilities lacking in constant PRI radar. Intentional jitter (random PRI variations of large amounts) is used for electronic protection (EP) from certain types of jamming and would therefore be of interest to the ELINT analyst. The amount and type of jitter can also aid in identifying the type of radar transmission being received. The parameters of interest are the same as those for constant PRI radars, but analysis of the jitter waveform is emphasized. Generally, individual intervals should be measured to an accuracy of a fraction of a percent to give a good representation of the jitter waveform and its statistics.

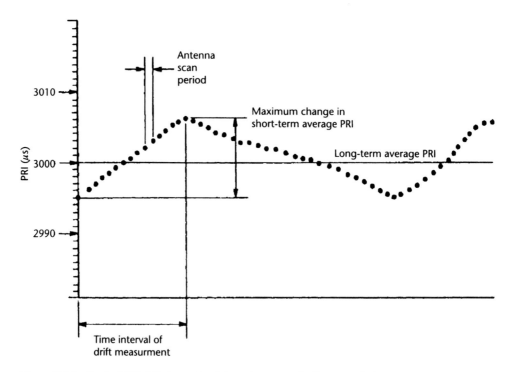

Figure 12.1 Typical PRI drift for a circuit in a thermostatically controlled oven.

Some radar sets may use jitter consisting of random selections among discrete interval values. Others may have intervals that can take on any value within a certain range. In the case of discrete intervals, the ELINT analyst should determine what values are used. For either discrete or continuous PRI variations, the shape of the distribution as well as the overall range over which the PRI varies must be determined.

12.2.3 Dwell and Switch PRI

In this type of radar, several (perhaps many) different PRIs may be selected for use. However, switching among them is done automatically and rapidly to perform certain radar functions. The parameters of interest include all of those used to characterize nominally constant PRIs, as well as additional information about how many PRIs are used, what switching patterns are observed, and how long the emitter transmits a fixed PRI (i.e., the dwell time). This technique can be used to resolve range ambiguities (in short PRI pulse Doppler radars) and velocity ambiguities or to eliminate target eclipsing (blind ranges) or blind speeds. Some range-tracking radars with short PRIs may adjust the PRI to maintain the target return at a more or less fixed time after the transmitted pulse. PRIs less than about 100 to 125 μs are switched among several values to eliminate eclipsing or resolve range ambiguities. PRIs in this range are intended to minimize Doppler ambiguities in pulsed Doppler radar. Having selected such a PRI, it then becomes necessary to use dwell and switch to eliminate eclipsing and resolve range ambiguities. In other words, that is the *result*, not the *objective* of using short-PRI (usually called high-PRF) radar. Once the ELINT analyst determines the set of PRIs used, the unambiguous range can be determined using methods described in the radar literature [2]. If the sequence of PRIs used is periodic, that tends to indicate that range ambiguity elimination or similar purposes are the reasons for the switching. A PRI sequence that is not periodic but has a pattern tends to indicate a process that adapts to the target situation. An apparently random sequence could be related to EP.

It is useful to determine the time required to switch PRIs. Any time should be characterized along with the dwell time for each PRI.

12.2.4 PRI Stagger

PRI stagger is the use of two or more PRIs selected in a fixed sequence. The sequence may contain more than one of the several intervals before it repeats. The sequence is described by the number of "positions" used to make up one period of the sequence and the number of different intervals used to make up one period of the sequence. A common stagger sequence consists of alternating long and short PRIs. This has two intervals and two positions. A sequence using the same two intervals but having a sequence of long, long, short has two intervals and three positions. Generally, stagger is used to eliminate blind speeds in MTI radar systems [3]. The use of stagger is a strong indication that MTI capability exists and that highly accurate PRI measurements are necessary to determine performance limitations due to incidental jitter. Of course, the intervals must be measured with sufficient accuracy to notice the presence of the stagger. This normally requires accuracies in the 0.1% range or better.

In the past, manually selected stagger was used to enable a radar operator to check to see if a target was beyond the unambiguous range. A target "blip" becomes two blips if stagger is used and if the target is beyond the maximum unambiguous range. Stable PRIs are not required for this technique. Incidental jitter measurements usually can be used to distinguish whether this or MTI is the reason for the stagger.

12.2.5 Sliding PRIs

A sliding PRI is characterized by monotonic increases or decreases in the PRI followed by a rapid switch upon reaching one extreme limit to return to the other extreme limit. This may be done to eliminate eclipsing (blind ranges) as in some dwell-switch PRI schemes. But, in this case, the variation is continuous. A sliding PRI may also be used to optimize elevation scanning by providing constant altitude coverage. In this application the sliding PRI pattern is usually periodic. The idea is to decrease the maximum unambiguous range as the elevation angle increases because targets of interest fly no higher than a certain altitude. Another application is to maintain a constant SNR in a mapping radar system.

The ratio of the minimum PRI to the maximum PRI is roughly the ratio of the minimum range to the maximum range in elevation scanning systems. For example, if the elevation scan covers about 0° to 30° and the altitude of interest is 15 km, then the minimum range is 30 km; whereas the horizon-limited distance at an elevation angle of 0° is about 500 km—a range ratio of nearly 17:1. A typical height-finder range limit is 150 to 200 km for a range ratio of 5 or 6 to 1. The ELINT analyst should determine the shape of the PRI "sweep" waveform, as well as the upper and lower limits and the time to complete the sweep. Short PRIs (100 to 125 μs) and more limited PRI variations in an unpredictable pattern may indicate target tracking in which the PRI is varied to eliminate eclipsing.

12.2.6 Scheduled PRIs

Scheduled PRIs occur in computer-controlled electronic scan radars that interleave search and track functions, often in three dimensions. The PRI variations can include some features of all of the other types as determined by the controlling program. While considerable variability and responsiveness to the environment is the rule, the complexity of the real-time software is such that wholesale changes rarely occur. The number of pulse interval patterns seen by an ELINT receiver depends on the number of targets being tracked and their locations. If no targets are present, a periodic raster-type search may occur with constant altitude coverage. It is important to discover the number of tracking pulses used (i.e., how many targets can be tracked at once) and how frequently the radar illuminates a tracked target. The ELINT analyst should determine all of the intervals used and some of the typical sequences. It is very useful to correlate pulse interval values with the beam angle, target range, or target velocity. Generally, interval accuracies of 0.1% are adequate. This type of PRI control is the rule for newer multifunction and/or solid-state radar systems.

12.2.7   Periodic PRI Variations

Periodic PRI modulation is a nearly sinusoidal variation over a more limited range than sliding PRI. It can also be used to avoid eclipsing or for ranging, but is more often used in conjunction with conical scan tracking systems as a missile guidance technique. When conical scan radar tracks a target, its boresight along the center of the cone is aimed at the target. If a missile is fired along the beam, the conical scan amplitude modulation will vanish if the missile is headed toward the target. To make course corrections, the missile needs a reference signal from the conical scanner. One method to do this is to sinusoidally modulate the radar PRI in synchronization with the conical scanning.

A deception-type electronic attack or countermeasure to defeat this system needs information on the sinusoidal PRI modulation. PRI modulation amounts of 5% at rates up to 50 Hz can be expected. Correlation with the conical scan envelope is a key analysis technique. The average PRI and maximum and minimum values should be determined to about 0.1%.

12.2.8   Pulse Groups

Radars may transmit groups of closely spaced pulses, possibly as part of a "scheduled" sequence, separated by longer times between pulse groups. These PGRIs can obviously be categorized just as any PRI can be. Pulse groups can be used for basic radar functions such as increased range and velocity resolution. (The range resolution is set by the duration of a single pulse within the group, whereas the Doppler resolution is set by the duration of the whole group.) Pulse groups can also be used to eliminate blind speeds in MTI systems. (Two pulses per group are used in the latter application.) Radar applications of pulse groups generally make use of a fixed pulse group. In other applications (e.g., where telemetry data are to be transmitted) pulse position modulation can be expected. IFF systems may also make use of pulse groups that can be changed from time to time, but the pattern may remain fixed for hours or days. The ELINT analyst should determine the pulse durations and intervals within the group if the group is fixed. If the pulse patterns within the group change, the analyst should determine the number of pulses (maximum and minimum) that can be included and any repeated synchronizing pattern that may be present. Such patterns are often used for data transmission. Pulse code modulation and pulse position modulation are representative of typical data transmission schemes.

12.3   Time Interval Measurements

Measurement of the radar pulse's TOA at an ELINT receiver is similar to the problem faced by the radar receiver in determining the range to a target. However, there are some differences.

1. The radar receiver is matched (more or less) to the pulse being received. The ELINT receiver is not usually matched to the pulses it receives.

2. The radar range delay uncertainty for a single pulse usually need not be much smaller than the pulse duration. Also, the target range can be determined based on the integration of a number of pulses. Some ELINT analyses may require a very small uncertainty in measuring the TOA of a single pulse, possibly a fraction of the pulse rise time (e.g., emitter location via TDOA).

3. In the radar, the time interval of interest is from a known time of the pulse transmission to the unknown time of the target echo return. The ELINT parameter of interest is usually the time between *two* pulses. There is uncertainty about the time of occurrence of the pulse at each end of the interval.

The effect of the ELINT receiver being poorly matched to the radar signal usually means that the ELINT receiver bandwidth is wider than the reciprocal of the pulse duration. This reduces the available SNR but also permits the pulse shape and rise time of the transmitted pulse to be the limiting factor in the timing measurement. A matched ELINT receiver would produce (approximately) a triangular video waveform of twice the pulse duration at the base. The "excessive" bandwidth of the ELINT receiver is the means by which the TOA of a single pulse can be determined to a fraction of its rise time. The other important factor is a high SNR in ELINT, which is sometimes available because of the R^{-2} versus R^{-4} propagation advantage of ELINT.

12.3.1   SNR Limitations

The problem of the TOA uncertainty due to noise has been analyzed thoroughly for the high SNR situation. If a fixed amplitude threshold is used and a pulse with a linear leading edge is applied, the rms variation of the measured time of arrival is [4]

$$\delta_T = \frac{1.25 t_R}{\sqrt{2\,\text{SNR}}} \tag{12.2}$$

where:

δ_T = rms variation of measured TOA due to noise

t_R = pulse time from the 10% to 90% points on the leading edge of the video pulse

SNR = signal-to-noise ratio at the receiver output, assumed much greater than 1

Notice that the average slope of the leading edge of a pulse of amplitude (A) is $0.8\,A/t_R$. The factor of $1.25 = 1/0.8$ is required because the rise time t_R is measured from the 10% point to the 90% point. The amplitude change in time t_R is then 0.8A. Note that the actual slope changes during the rise of the pulse and is generally steepest at the 50% point. Thus, the actual disturbance of the TOA due to noise depends on the threshold setting relative to the peak amplitude of the pulse. (The effects of bandwidth on rise time are discussed in Chapter 11.)

The SNR to be used in (12.4) is the SNR at the receiver output when a strong signal is present. In other words, the analysis is valid only when the pulse is clearly distinguishable above the noise. Next, consider that the PRI is the difference between the times of arrival of two pulses:

$$\text{PRI}_{i,j} = T_i - T_j \tag{12.3}$$

where:

$\text{PRI}_{i,j}$ = pulse repetition interval between pulse i and pulse j

T_i = time of arrival of pulse i

T_j = time of arrival of pulse j

The probability distribution of the values of $\text{PRI}_{i,j}$ is obtained by convolving the distributions for T_i and T_j. For a high SNR, the noise when the signal is present is a Gaussian random variable. Therefore, the distributions for T_i and T_j are Gaussian. Because T_i and T_j are separated in time by many times the reciprocal of the video bandwidth, the receiver noise at time T_i is independent of that at time T_j. Under these conditions, the probability distribution of $\text{PRI}_{i,j}$ is also Gaussian, with the standard deviation given by

$$\sigma_{\text{PRI}}^2 = \sigma_{T_i}^2 + \sigma_{T_j}^2 \tag{12.4}$$

If the rise times and amplitudes (i.e., SNR) for the two pulses are the same, then

$$\sigma_{\text{PRI}} = \sqrt{2} t_R / 0.8 \sqrt{2\,\text{SNR}} = t_r / 0.8 \sqrt{\text{SNR}} \tag{12.5}$$

Figure 12.2 shows a graph of this relationship between the rms pulse interval variation and the SNR for various rise times. As shown in Chapter 11, the measured rise time can be no less than that of the receiver system. For example, if a 2-MHz video bandwidth is used (typically the predetection bandwidth is then 4 MHz), the minimum pulse rise time is about 175 ns. If the available SNR is 30 dB, then the rms variation of the PRI values due to noise would be at least 13.7 ns according to (12.5). Note that, if the SNR is less than about 12 dB (shown by the shaded area), Figure 12.2 is not valid.

12.3.2   Limitations Due to Pulse Amplitude Changes

In the usual ELINT situation, the radar beam is moving past the ELINT position causing pulse-to-pulse amplitude variations. If the TOA of a pulse is determined by the crossing of a fixed threshold, the nonzero rise time combined with the amplitude variation causes TOA errors. If the amplitude is increasing, each successive pulse crosses the threshold a little earlier. If the amplitude is decreasing, the opposite is true. For a trapezoidal pulse the situation is shown in Figure 12.3. The change in pulse interval due to a change in pulse amplitude is

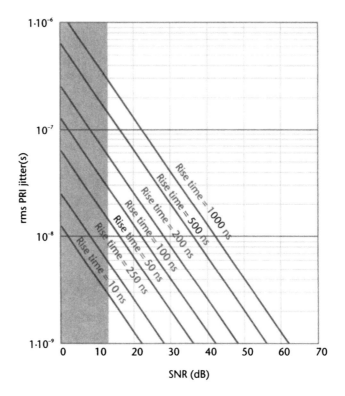

Figure 12.2   Pulse interval variation due to noise.

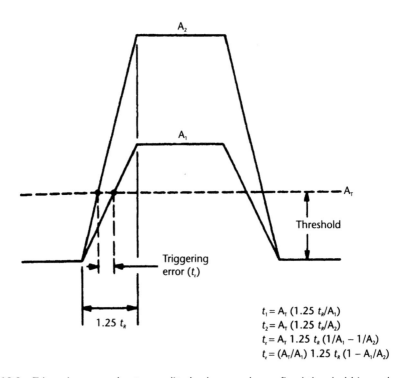

$$t_1 = A_T \left(1.25 \, t_R / A_1\right)$$
$$t_2 = A_T \left(1.25 \, t_R / A_2\right)$$
$$t_r = A_T \, 1.25 \, t_R \left(1/A_1 - 1/A_2\right)$$
$$t_r = \left(A_T / A_1\right) 1.25 \, t_R \left(1 - A_1 / A_2\right)$$

Figure 12.3   Triggering error due to amplitude change when a fixed threshold is used.

$$t_\epsilon = \frac{1.25 A_T (t_R)(1 - A_1/A_2)}{A_1} \qquad (12.6)$$

where:

t_ϵ = pulse interval change due to fixed amplitude threshold

A_T = threshold amplitude

A_1, A_2 = amplitude of pulses at beginning and ending of the pulse interval

t_R = rise time of pulse (10% to 90% points)

Figure 12.4 shows a graph of t_ϵ/t_R for various pulse-to-pulse amplitude ratios in decibels and for several values of A_T/A_1. If the first pulse just crosses the threshold, $A_T/A_1 = 1$. If the first pulse is two or five times the threshold, $A_T/A_1 = 0.5$ or 0.2, respectively. For pulses near the threshold, even 1 dB of pulse-to-pulse amplitude variation produces an apparent PRI variation of more than 15% of the rise time. Reducing this variation in PRI due to pulse amplitude changes can be accomplished through the use of a more sophisticated thresholding technique. For example, if the threshold is placed at the 50% amplitude point, the TOA is not affected by pulse amplitude variations when the leading edge rises linearly with time. Of course, finding the 50% point implies inserting a delay at least as long as the rise time of the pulse. If the pulse is digitized, the 50% points can be found in non-real time. An additional advantage of setting the threshold to the 50% point is that the slope of the leading edge is increased to $(1.0$ to $1.5)A/t_r$, compared to the $0.8A/t_r$ averaged over the leading edge, improving the accuracy of measurement for a given SNR.

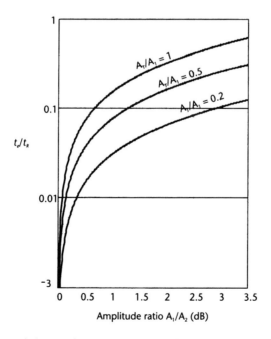

Figure 12.4   Pulse interval change due to pulse amplitude change for a fixed threshold.

A good test of the effectiveness of various thresholding schemes is to first measure the PRI using a constant amplitude pulse test signal to determine the residual jitter and then use alternating high- and low-amplitude pulses with a known rise time and amplitude ratio to measure the jitter again. The extent to which the measured jitter for the amplitude modulated signal is less than that expected when using a constant amplitude threshold indicates the effectiveness of the thresholding technique. Even if the threshold technique is effective in choosing a fixed amplitude point on the rise of the pulse, pulse-to-pulse shape variations due to multipath and transmitter anomalies contribute jitter that is not due to the radar's timing generator.

12.3.3 Improving Interval Measurements

The jitter due to noise of the output pulse is proportional to the rise time, as shown in (12.5). Likewise, the jitter due to amplitude variations and imperfect thresholding is proportional to the rise time, as shown by (12.6). However, the SNR is improved by reducing the receiver bandwidth, which in turn reduces the jitter but may increase the rise time. If the SNR is large, the jitter due to noise will be negligible and a bandwidth sufficient to pass the transmitted pulse without increasing its rise time should be used. If the bandwidth is narrowed in subsequent processing, TOA variation due to amplitude changes and noise may become important, even though the pulses at the receiver output had a very short rise time. The contribution of the receiver to the rise time of the output pulse is inversely proportional to the bandwidth, and the noise is directly proportional to the bandwidth. Jitter due to noise is reduced only in proportion to the square root of the SNR; therefore, reducing the bandwidth will not always reduce the jitter due to noise. Once the bandwidth is narrowed to the point at which the rise time becomes inversely proportional to the bandwidth, further bandwidth reductions actually increase the jitter due to noise. Using the expression for rise time from Chapter 11 gives

$$\sigma_{\text{PRI}}/t_{ri} \approx \frac{1}{0.8}\sqrt{1 + \left(\frac{0.7}{B \times t_{ri}}\right)^2} \frac{\sqrt{kT(\text{NF})B}}{\sqrt{S}} \tag{12.7}$$

where t_{ri} is the rise time of the input pulse.

A plot of (12.7) is shown in Figure 12.5. The optimum bandwidth occurs at $B = 0.7/t_{ri}$ (or $B_v = 0.35/t_{ri}$).

In processing signals for PRI analysis, frequent use is made of lowpass, notch, and highpass video filters. As noted above, the lowpass video filter is set for a minimum bandwidth so that the pulse rise time is not appreciably affected. Highpass and notch filters are used to eliminate portions of the frequency band where interference may be found. These include power line frequencies and interference from other analysis equipment.

Nonlinear processing such as amplitude clipping or limiting and slicing is used to remove unwanted portions of the signal's dynamic range For example, if the threshold is set well above the noise, the amplitude below that could be sliced off for a better view of the signal of interest. Likewise, if the scanning of the antenna

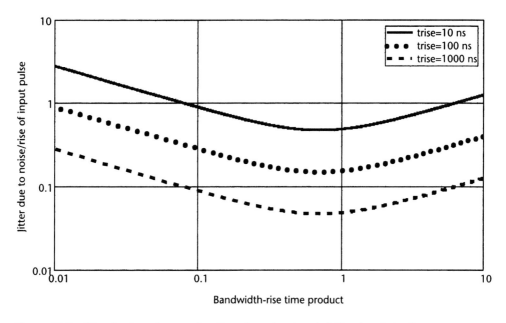

Figure 12.5 Jitter due to noise as a function of receiver bandwidth × input rise time.

produces very high amplitude pulses as well as other lower amplitude pulses of interest, the amplitude may be limited or clipped or compressed logarithmically to more easily view the sidelobes.

At the heart of digital time interval measurement systems is a timing reference oscillator used to generate a discrete time scale against which the time between the radar pulses is measured. There are two kinds of limitations. The first is due to the instabilities of the reference oscillator, which introduces its inaccuracies and drift into the measured PRI values. The second is the quantization error due to the noncoherence between the pulse arrival times and the reference oscillator.

Reference oscillators used in ELINT applications are typically the quartz type, although rubidium gas cell and cesium atomic beam oscillators are sometimes used. The latter is a primary standard requiring no calibration, whereas quartz and rubidium oscillators require periodic calibration to a primary standard to achieve the best long-term accuracy. Quartz crystal oscillators have excellent short-term stability and are usually used within the other types of frequency references. When an atomic standard is used, the quartz oscillator is usually phase locked to the atomic standard. Then the atomic standard determines the long-term stability while the short-term stability is determined by the quartz oscillator.

The long-term stability of the reference affects drift measurements and the ability to compare measurements at different times (e.g., hours or days apart). The short-term stability of the reference oscillator affects the observed pulse interval–to–pulse interval variations that make up the incidental jitter. If the reference oscillator can be characterized by its rms fractional frequency variation (or its Allan variance) over a time interval of one pulse interval,

$$\sigma_y \, (\text{PRI}) = (\text{Allan Variance})^{0.5} \tag{12.8}$$

then the rms variation of pulse interval measurements of an ideal stable pulse train would be

$$\sigma_{PRI} = \sigma_y \, (PRI) \cdot (PRI) \qquad (12.9)$$

Quite often the more important error is the quantization that results when counting the number of cycles of the reference oscillator between the radar pulses. PRIs from an ideally stable pulse train measured by counting cycles of a reference oscillator differ by one count, depending on the relative phase of the radar pulses and the cycles of the reference oscillator. For a given stable pulse interval and stable reference, the count can change by only one count for different phases; however, if a single pulse interval is measured, it is not known whether the count might increase by one or decrease by one for a different phase relationship. For this reason, PRI values have an ambiguity range from plus one count to minus one count even though the quantizing error is at most only one count. Because the counting rate of digital circuits exceeds hundreds of megahertz, the limitations due to the available SNR and the amplitude fluctuations are generally the limiting factors in making PRI measurements for ELINT purposes.

12.3.4 Digital Thresholding

The usual method to determine the TOA of a pulse is by the use of thresholds. However, high-speed A/D converters make it possible to use various digital signal processing algorithms to define the TOA. The A/D conversion rate (f_s) and quantization error are important factors in determining the accuracy with which the TOA can be determined. The minimum conversion rate required for reconstructing the signal is determined by the video bandwidth (or signal rise time). To simplify the threshold determining process, it would be useful to have a number of samples, h, within the rise time of the pulse:

$$f_s \approx h/t_r \qquad (12.10)$$

On the other hand, a large number of samples along the rising edge of the pulse may be beyond the capability of available A/D converters. As long as the sampling rate is greater than twice the highest frequency present in the video signal, interpolation between samples (signal reconstruction) can be used to reconstruct the pulse to interpolate to find the TOA.

The effect of the quantizing error of the A/D converter on the measured TOA must also be considered. Generally, the number of bits of amplitude accurately obtained from the converter decreases as the signal bandwidth (and conversion rate) increases. The reason is that there is a finite duration sampling window during which the A/D converter acquires the signal level.

The maximum SNR from an A/D converter with a given number of bits of resolution and for a full scale sine wave input is given by [5]

$$SNR_{max} = 6 \, (\# \, bits) + 1.76 \, dB \qquad (12.11)$$

The significance of (12.11) is that for a given SNR capability, the A/D converter must have at least that number of bits of resolution. For example, for preserving a 40-dB SNR through the conversion process, the A/D converter resolution required is in excess of 6.3 bits. Note that a high-speed low-resolution A/D converter can achieve a higher SNR by digitally lowpass-filtering the output. If the sampling rate is higher than required for the signal of interest, some of the noise can be eliminated by filtering. The improvement in SNR is given by

$$10 \, \log(F_{\text{Nyquist}}/B_{\text{Filter}}) \; \text{dB}$$

As can be seen, oversampling by a factor of four yields a 6-dB SNR improvement, equivalent to one additional bit of resolution.

Once the signal is in digital form, elaborate thresholding schemes can be envisioned. However, a high-speed resolution A/D converter provides a high data rate. (For example, a 1-GHz sampling rate at 8 bits per sample produces 10^9 bytes/sec.) Either a very large storage area is required or real-time thresholding must be done to reduce the quantity of data stored. In the latter case, only simple threshold algorithms are practical. If the computational resources are available, the calculation of the pulse centroid location can be considered as a definition of the TOA. A more precise location of the 50% amplitude point could be performed or a number of other schemes could be implemented. One scheme is as follows:

1. Find the peak amplitude of the pulse.
2. Define the amplitude as the median of the samples above 50% of the peak amplitude.
3. Interpolate to find the time at which the pulse rise crosses the 50% amplitude point.

12.4 PRI Analysis Techniques

PRI analysis is concerned with quantitatively describing the variations observed in the PRI, relating these variations to other aspects of the radar's signal, and permitting intelligent speculation about their functional use by the radar and the technique used in generating the PRI.

Historically, a frequently used manual technique for accurately determining the average PRI is to match the PRI of a synthetically generated pulse train to that of the unknown signal. The synthesized signal triggers an oscilloscope, and the signal to be analyzed is applied to the oscilloscope's vertical input. The sweep duration is set to be slightly less than the PRI so that the position of a single pulse on the screen is determined by the difference in time between occurrence of the trigger and the TOA of the pulse. If the PRI of the synthesized signal is different from the signal of interest, the horizontal position of the pulse on the screen will move—to the right if the synthesized PRI is too long and to the left if the synthesized PRI is too short. The synthesizer is adjusted until the pulse position is fixed. The synthesized PRI is then accurately measured using a counter. (It is easy to measure because it is not affected by scanning or by thermal noise.) This same idea is used in the raster displays discussed below.

The accuracy with which the signal PRI can be determined using this approach is limited by various factors. Of course, the synthesized PRI can be measured to very high accuracy, and this was sometimes mistakenly reported as the accuracy with which the unknown PRI was measured. The actual error involved in matching the reference PRI to the unknown PRI is greatly reduced compared to that involved in making a single PRI measurement because a large number of pulses are, in effect, averaged during the measurement process. First, consider the effects of noise. If the synthesized PRI exactly matched the unknown PRI, the observer would see a large number of overlaid pulses with noise and varying amplitudes. The noise would create some uncertainty about the precise location of the pulse. Usually there is a limited time available to make the measurement. The observer would be uncertain as to whether the pulse location moved due to noise disturbing the edges or whether the synthesizer was improperly adjusted. Thus, the measurement time and SNR both affect the accuracy of this method. Assume that the noise is the limiting factor in judging the position of the pulses on the screen. The rms variation in the location of the pulse due to noise would be approximately given by (12.4). The observer can make use of more information than the time at which the pulse crosses a threshold, in which case the error predicted by (12.4) could be the error over the entire averaging time.

Time-domain analysis techniques (originally based on the oscilloscope) are among the most useful and widely used. Incidental jitter and some stagger types can be analyzed with an oscilloscope having a dual time base (A delayed by B). The delay is needed because the jitter is usually very small compared to the PRI. A normal oscilloscope display showing a single PRI would permit rough peak-to-peak jitter measurements down to, for example, 1% of the PRI. With the delayed time base in use, the full width of the screen can be adjusted to display a small fraction of the PRI, making it much easier to analyze the extent of the jitter. The analyst should adjust the delay successively through several multiples of the PRI to examine the jitter of these intervals. For a random disturbance of a nominally constant PRI, the jitter should slowly increase as the delay increases. If the jitter is independent of the delay, the determining factor of the jitter may be related to threshold crossing variations due to noise rather than the properties of the radar's trigger generator.

The same procedure is also useful in analyzing stagger. One common problem is to distinguish staggered PRIs from discrete random jitter having only a few possible interval values. Because stagger is periodic, it is easy to tell when the delay selected is equal to that period [6].

12.4.1 Raster Displays

The raster display is an oscilloscope technique that uses z-axis modulation to indicate the presence of a pulse. The x-axis is a conventional time sweep. The y-axis is also time related such that each x-sweep is displaced slightly. The scheme in the most common use starts with the trace at the top of the screen and each sweep is displaced downward; hence the name "falling raster." The raster display makes it convenient to observe the variations of the PRI over a much larger number of intervals compared to the conventional oscilloscope display. This type of display is very useful and permits the analyst to observe at a glance the PRI variations of hundreds or thousands of pulses at once. With the advent of time base digitizing, it became possible to generate this type of display digitally. This resulted in the development of specialized processors dedicated to pulse sorting and deinterleaving as well as PRI measurements. These became available in the United States in the late 1970s. These units provided for time adjustment across the screen in steps as small as 1 ns and storage on the order of 16,000 events. Of course, if a separate time base digitizer is available to capture the data, a general-purpose computer and software can be used to create the raster-type displays for analysis. (Software was available that simulates common PRI variations and interleaving and also models the functions of these raster display units on the PC [7].)

Nearly all raster display units designed for pulse analysis also display pulse amplitude versus time. Simultaneously displaying time versus time and amplitude versus time shows at a glance correlation between the scan and the PRI. Often, cursors or markers are included to allow selecting those pulses that make up a portion of the display for further analysis.

There are several ways to control the triggering of raster displays. In one mode, the signal controls the start of the trace. After a trace is completed, the next trace occurs only after another pulse is received. In the free-running mode, the next trace starts immediately after the preceding one. Then trace duration must be adjusted to be equal to the average PRI to form a vertical line of pulses. In this mode the average PRI is determined by PRI synthesis. A zoom effect can be created by using a delay between the end of one trace and the start of the next to magnify the horizontal distance between pulses.

12.4.2 PRI Sounds

Listening to the sound of the pulse train using a loudspeaker or headphones is the oldest technique of PRI analysis. It is still useful today. The low duty cycle of radar signals makes pulse stretch circuitry an important aid. Also, because widely varying amplitudes are confusing to the listener, constant amplitude pulses may be used.

The simplest technique is to listen to an audio oscillator simultaneously with the radar pulse train. The analyst matches the tone of the generator to that of the pulse train by listening for beats as in tuning a musical instrument. An inexperienced analyst may set the audio oscillator to a harmonic or subharmonic of the PRI, but this error is rarely made after some practice. The analyst increases the sound level until the beat note is heard. The frequency of the beat note is equal to the difference between the frequency of the audio oscillator and the PRF = 1/PRI. The analyst tunes the audio oscillator until the beat note frequency is zero (i.e., the beat disappears). The error under the best of circumstances is about ±20 Hz, because this is the lower limit of human hearing. Additional error can be caused due to scanning, which makes the beat note more difficult to hear.

Modern ELINT systems are designed to produce audible sounds even when the PRF is well above the range of human hearing. This is done by nonlinearly mapping the true PRF to a synthetic PRF sound. For example, PRFs up to, say,

1 kHz are reproduced as they are. Then PRFs from 1 to 200 kHz could be mapped into the range from 1 to 20 kHz.

12.5 PRI Analysis Theory and Practice

Hardware is available to measure the TOA of individual pulses (or, equivalently, the time from the previous pulse) on a pulse-to-pulse basis. This may be done using counters and a high-quality quartz crystal time base or using other digital records of the signals of interest. The resulting digital TOA values can be analyzed (in non-real time) using a variety of computer-based techniques. As an example of the kind of analysis that can be done, consider the problem of estimating the PRI of an observed sequence of N pulses. The average PRI is usually estimated by dividing the time between the initial pulse (t_1) and the last pulse (t_N) by the number of pulse intervals:

$$\text{PRI}_{\text{ave}} = \frac{t_N - t_1}{N - 1} \qquad (12.12)$$

(This can also be done using a counter in the period averaging mode.) However, this calculation makes no use of the TOAs of the intermediate pulses. It might be better to find the pulse interval of the ideal constant PRI pulse train that most closely fits the observed TOA values. One scheme would be to find the pulse interval that minimizes the squared error between the observed TOAs and those of the idealized pulse train TOAs. The arrival times for the ideal pulse train would be

$$t_{o,n} = n \cdot I + \phi \qquad (12.13)$$

where:

$t_{o,n}$ = arrival time of the nth pulse from the idealized pulse train

n = an integer between 1 and N, the number of pulses observed

ϕ = time between the first pulse and the first pulse of the idealized pulse train

I = interval between pulses of idealized pulse train

The squared error is

$$\epsilon^2 = \sum (t_n - t_{o,n})^2 \qquad (12.14)$$

Substitute (12.13) into (12.14), differentiate ϵ^2 with respect to I and ϕ. Then set the results equal to zero, and solve for the value of I, the estimated PRI, and ϕ. With little loss of generality, the first pulse of the idealized pulse train can be made to coincide with the first actual pulse. Then ϕ is zero and the PRI estimate is found to be

$$I = \left[\sum_{n=1}^{N} n \cdot t_n\right] \frac{6}{(N-1)N(N+1)} \qquad (12.15)$$

The question of whether (12.12) or (12.15) is better for estimating the PRI was investigated in [8]. The best estimator depends on the type of disturbance perturbing the pulse TOAs. In [8], two types of jitter were considered. For noncumulative jitter (NCJ), the TOA of each pulse is perturbed independently from any other pulse. In other words, one adds a random number to the TOAs in (12.13) to get the perturbed TOAs. The second case is cumulative jitter (CJ). In the CJ model, the TOA of the current pulse is based on the TOA of the previous pulse plus the PRI plus a random number.

For NCJ the equation for modeling the TOA is

$$t_{o,n} = n \cdot I + \phi + v_n \qquad (12.16)$$

where v_n is the jitter due to measurement.

For CJ, the equation for modeling the TOAs is

$$t_{o,n} = t_{o,n-1} + I + \phi + u_n \qquad (12.17)$$

or

$$t_{o,n} = n \cdot I + \phi + \sum_{i=0}^{n} u_i \qquad (12.18)$$

In [8], the maximum likelihood estimators are shown to be PRI_{ave} for CJ and I for NCJ. Furthermore, the Cramer-Rao bounds (CRB) for estimating the PRI for each type of jitter are found and are the same as (12.12) and (12.15). Of course, it is unlikely that the type of jitter would be known in advance for an ELINT signal. This leads to the question of what happens when the nonoptimum estimator is used. The results are summarized in Table 12.2. The mean squared error (MSE) is the minimum for the estimators given in (12.12) and (12.15).

If the incorrect estimator is chosen, the MSE penalty is of interest. Comparing the MSE to the CRB gives insight into the best strategy. If the estimator I is used when CJ is present, the penalty is

Table 12.2 Estimator Performance for Both Jitter Models

Estimator	MSE for NCJ	MSE for CJ
I	$\dfrac{12\sigma^2}{N(N^2-1)}$	$\dfrac{6(N^2+1)\sigma^2}{5N(N^2-1)}$
PRI_{ave}	$\dfrac{2\sigma^2}{(N^2-2)}$	$\dfrac{\sigma^2}{(N-1)}$
CRB	$\dfrac{12\sigma^2}{N(N^2-1)}$	$\dfrac{\sigma^2}{(N-1)}$

$$\frac{6(N^2 + 1)}{5N(N + 1)}$$

This approaches 6/5 when N is large. On the other hand, if the PRI_{ave} is computed when there is NCJ present, the penalty is

$$\frac{N}{6}$$

This grows in proportion to N. It is easy to see that if there are many pulses (say, $N \sim 8$), it is best to use the NCJ estimator if the type of jitter is not known. In practice, both CJ and NCJ will be present to some extent.

There are many procedures that could be used to improve the PRI estimates. For instance, a weighting function based on pulse amplitude or SNR could be incorporated into the procedure so that those TOAs with better SNRs influence the estimate more than those with poorer SNRs. This type of analysis falls into the general category of fitting functional models to the observed data. Similar techniques can be developed for other functions (such as computing drift rates). In doing this, the analyst first assumes a functional model for the data and then uses the computer to calculate the best fit possible to that function and the resulting error. The analyst's judgment is still the key ingredient in the process. The dataset to be used must also be selected carefully. This requires looking at the raw data through a variety of display techniques.

12.5.1 Statistical Techniques

The TOA sequence can be analyzed through the computation of histograms and by analyzing the mean and standard deviations. A key choice faced by the analyst is in determining how much data to include in such computation—the window through which the data is to be viewed. Histograms and statistics, like the mean and standard deviations, are not affected by the order of the data. This means that such statistics are not useful for analyzing PRI variations in which the sequence of the intervals is important. The variation of statistics with time can be useful but this depends on computing them for different segments of the data, including overlapping segments. For example, to analyze drift, the mean PRI could be computed for several segments of the data separated in time.

Histograms are helpful in determining the overall statistics of a PRI sequence. A histogram is obtained by first dividing the expected parameter range into intervals (called bins). Then a count is made of the number of occurrences of the parameter value within each bin. For a random process, as the bin size approaches zero and as the number of samples approaches infinity, the histogram approaches the probability distribution of the random process. If the number of samples is very large but the bin size is fixed, the histogram level in a particular bin is proportional to the integral of the probability density function over the parameter range included in that bin. There is a delicate compromise between the amount of data available and the bin size. If the bin size is too small, then the average number of occurrences per bin will be very low and the histogram will consist of a large number of empty bins interspersed with bins with one or two counts. This is not useful to the analyst. Likewise, if the bin size is too large, all of the data samples fall into one or two bins, and once again the analyst is unable to see the shape of the probability distribution. The usual situation requires the generation of several histograms with different bin sizes. The analyst then decides which bin size is most appropriate for the quality and quantity of data available.

Figure 12.6 shows a histogram with many points per bin. The probability distribution is shown as well. The effect caused by reducing the number of pulses by a factor of 10 and 100 is shown in Figures 12.7 and 12.8. When there is limited data, wider bins improve reliability but sacrifice resolution, as illustrated in Figure 12.9.

The histogram of a PRI with sinusoidal jitter plus a small amount of random jitter is shown in Figure 12.10. The typical bucket shape is caused by the sine wave spending more time near its peak values than it does at its average value. Figures 12.11 and 12.12 are the same signal but the number of pulses is 1/10 and 1/100 of that used in Figure 12.10.

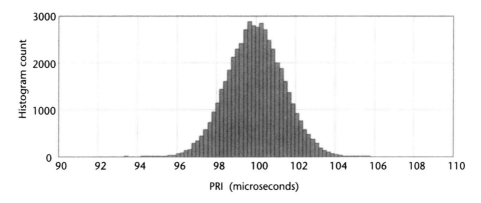

N = 5 × 10⁴ pulses   Width × 10⁶ = 0.2 microsecond   Jitnc = 1 microsecond

Figure 12.6   Histogram of PRI with random jitter (50,000 pulses).

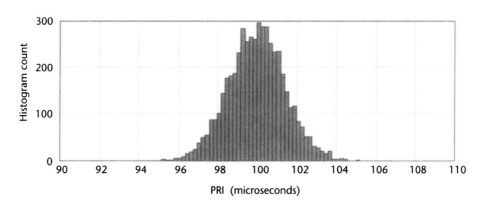

N = 5 × 10³ pulses   Width × 10⁶ = 0.2 microsecond   Jitnc = 1 microsecond

Figure 12.7   Histogram of PRI with random jitter (5,000 pulses).

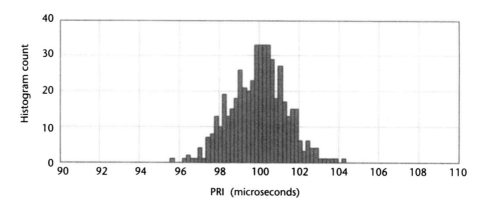

N = 500 pulses   Width × 10^6 = 0.2 microsecond   Jitnc = 1 microsecond

Figure 12.8   Histogram of PRI with random jitter (500 pulses).

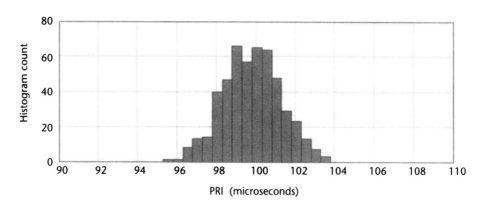

N = 500 pulses   Width × 10^6 = 0.5 microsecond   Jitnc = 1 microsecond

Figure 12.9   Histogram of PRI with random jitter—bin width increased (500 pulses).

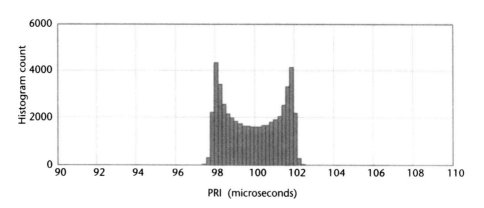

N = 5 × 10^4 pulses   Width × 10^6 = 0.2 microsecond   Jitnc = 0.1 microsecond

Figure 12.10   Histogram of sinusoidal PRI with random jitter (50,000 pulses).

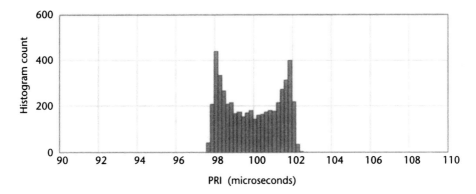

N = 5 × 10³ pulses   Width × 10⁶ = 0.2 microsecond   Jitnc = 0.1 microsecond
Jitsin = 2 microseconds

Figure 12.11   Histogram of sinusoidal PRI with random jitter (5,000 pulses).

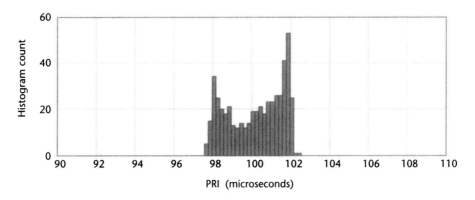

N = 500 pulses   Width × 10⁶ = 0.2 microsecond   Jitnc = 0.1 microsecond
Jitsin = 2 microseconds

Figure 12.12   Histogram of sinusoidal PRI with random jitter (500 pulses).

The histogram for a sliding PRI is shown in Figure 12.13. Note that the top is not quite flat. This is because at shorter PRI values there are more pulses per second than at the longer PRI values. Hence, there are more short PRI values than long ones in a fixed length record. Figure 12.14 shows the effect of adding 100 ns of random jitter.

This obscures the slope of the histogram. Figure 12.15 shows how 10 times as much data helps to show the slope of the histogram in spite of the random jitter. Table 12.3 summarizes the histogram shape characteristics of common PRI variations.

12.5.2   Delta-T Histogram

The delta-T histogram[1] is a histogram of pulse intervals made by considering the intervals between all pulse pairs (usually some maximum interval is specified). In

---

1.   The author believes that this histogram was first conceived by Dr. Peter J. Knoke in 1964.

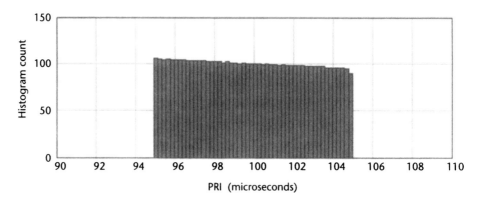

N = 5 × 10³ pulses   Width × 10⁶ = 0.2 microsecond   Jitnc = 0.1 microsecond
slidejit = 5 microseconds

Figure 12.13   Histogram of a sliding PRI.

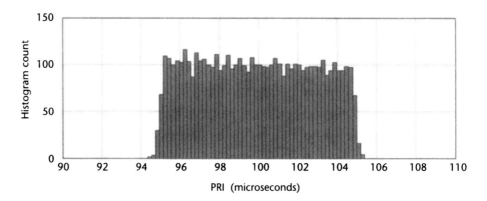

N = 5 × 10³ pulses   Width × 10⁶ = 0.2 microsecond   Jitnc = 0.1 microsecond
slidejit = 5 microseconds

Figure 12.14   Histogram of a sliding PRI (random jitter added).

other words, the intervals from pulse 1 to pulses 2, 3, 4, and so forth are considered, then the intervals from pulse 2 to pulses 3, 4, 5, and so forth are considered. The total number of intervals is given by the number of pulse pairs in a set of N pulses, which is

$$\binom{N}{2} = \frac{N(N-1)}{2} \tag{12.19}$$

The value of the delta-T histogram is clear if the effects of noise or interference pulses are considered. By examining the time between all pulse pairs, the true PRI (and its multiples) will be found and these interval values will accumulate in peaks in the histogram, whereas other pulse pairs will have intervals distributed across a number of histogram bins. Formally, the delta-T histogram can be shown to be equal to the integral over each histogram bin of the autocorrelation of a function

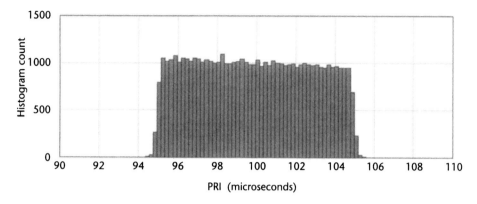

N = 5 × 10⁴ pulses   Width × 10⁶ = 0.2 microsecond   Jitnc = 0.1 microsecond

slidejit = 5 microseconds

Figure 12.15   Histogram of sliding PRI (10 times as many pulses as in Figure 12.16).

Table 12.3   Typical Pulse Interval Histogram Shapes

			Distribution			
Signal Type	*Spikes*	*Flat*	*Bell*	*Convex*	*Trapezoid*	*Ramp*
Discrete PRI jitter or stagger	X					
Scheduled	X					
Switched	X					
Gaussian random			X			
Uniform random		X				
Nonuniform random					X	
Sinusoidal				X		
Sliding						X

that consists of impulses located at the pulse TOAs. If the arrival times are denoted by t_n, this function is

$$f(t) = \sum \delta(t - t_n) \tag{12.20}$$

where $\delta(t)$ is the unit impulse.

The autocorrelation of $f(t)$ is

$$h(\tau) = \int_{-\infty}^{\infty} f(t) f(t - \tau) \, dt \tag{12.21}$$

$$= \int_{-\infty}^{\infty} \left[\sum_n \delta(t - t_n) \right] \left[\sum_k \delta(t - t_k - \tau) \right] dt \tag{12.22}$$

The integrand of (12.22) has a value only if $t - t_n = 0$ and $t - tk - \tau = 0$ simultaneously. This occurs when $t_n = t_k - \tau$. Thus,

$$h(\tau) = \sum_n \sum_k \delta(t_n - t_k - \tau) \, dt \qquad (12.23)$$

The integral of $h(\tau)$ from τ_1 to τ_2 is

$$\int_{\tau_1}^{\tau_2} h(\tau) \, d\tau = \int_{\tau_1}^{\tau_2} \sum_n \sum_k \delta(t_n - t_k - \tau) \, d\tau \qquad (12.24)$$

The integrand of (12.24) has a value for any pair of arrival times such that

$$\tau_1 < t_n - t_k \le \tau_2 \qquad (12.25)$$

Thus, a count of the number of pulse pairs, such that their arrival times are separated by an amount between τ_1 and τ_2, is equal to the integral of the autocorrelation function $h(\tau)$ over the same range of delay values—exactly the value of the delta-T histogram for the bin from τ_1 to τ_2.

As an autocorrelation function, the delta-T histogram emphasizes periodicities. Figure 12.16 shows the delta-T histogram from a three-position stagger. The intervals are 95, 100, and 105 μs.

For stagger, the Delta-T histogram peak occurs at the so-called stable sum. If one adds adjacent PRIs, eventually the period of the stagger is reached. It makes no difference on which pulse the sum starts; the stable sum is reached when the number of adjacent PRIs added is equal to the number of positions in the stagger sequence. Because the stable sum and the number of intervals and positions are often used to recognize radar threats, it is important for the ELINT analyst to correctly determine the stagger period and sequence.

The delta-T histogram is also useful in analyzing interleaved pulse trains. Consider a simple case of two signals having nearly the same PRI. Clearly, when an interval begins on a pulse of one pulse train and ends on a pulse of the other pulse train, the interval is somewhat random. When the interval begins and ends on pulses of the same pulse train, the interval is a multiple of one of the true PRIs.

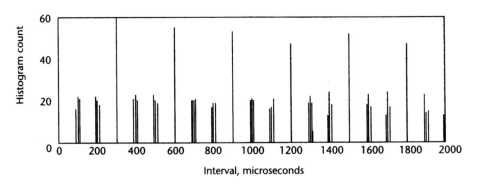

N = 100 3-interval 3-position stagger Width 1×10^{-7}

Figure 12.16 Delta-T histogram for three-interval three-position stagger.

Figure 12.17 shows a histogram for interleaved pulses from two emitters that have 5-μs (rms) NCJ and average PRIs of 100 and 110 μs. This is a puzzling situation for most analysis techniques. The histogram shown has a bin size of 1 μs. The short intervals near the left are clearly noise intervals consisting of those that begin on a pulse of one pulse train and end shortly thereafter on a pulse of the other pulse train. Near 100 to 110 μs, the large peak represents the jittered values of the both true PRIs. There is only one peak instead of two because the mean PRIs are separated by only the rms jitter times two. Near the 200- to 220-μs region, there is another large peak representing twice the true PRIs. Near the 300- to 330-μs region, there are what looks like two peaks. Near the 400- to 440-μs region, there are clearly two peaks. This leads to the conclusion that there are, in fact, two pulse trains present. Furthermore, the mean PRIs can be found by noting the positions of the fourth peak and dividing by four. The peaks of the multiple PRIs separate by more than the increase in the width of the peaks. This illustrates the fact that the jitter of the multiple intervals tends to increase no faster than the square root of the number of intervals added, while the peaks separate directly in proportion to the number of intervals added. If there is only one signal and the SNR is high (virtually no noise pulses), the delta-T histogram is no more useful than a regular histogram or time-domain displays of PRI versus time. When more than one signal is present or many noise pulses are present, the delta-T histogram can be a valuable aid to the analyst. Using the delta-T histogram as an aid in deinterleaving is covered in some detail in Chapter 13.

12.6 Interpreting the Results

While the major function of the ELINT analyst is to characterize the PRI waveform in detail, there is also the need to explain the waveforms in terms of rational radar design and operation. In terms of radar operation, it is common practice to synchronize the pulsing of different radar sets that are functionally related. A good example is a search radar (range and azimuth) controlling a height finder (range and elevation). When data from one radar is to be displayed with that from another, it is especially convenient to pulse the radars at the same time. To show the

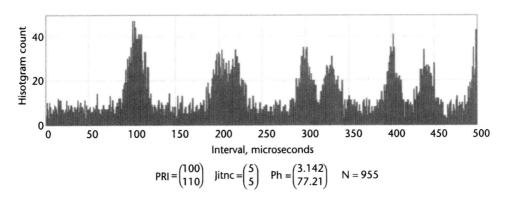

Figure 12.17 Delta-T histogram for two interleaved and jittered pulse trains (bin size = 1 μs).

synchronization, the analyst can use an oscilloscope triggered by one signal to observe the other (or a raster display can be used). Sometimes an overall synchronization exists but with one set operating at a multiple or submultiple of the controlling PRI. Sometimes the jitter characteristics of one transmitter's pulses will differ from that of the controlling PRI, especially if the controlled set transmits at the shorter PRI. Sometimes different radar PRIs will be simultaneously collected, possibly at different sites, for later analysis of the timing relationship (if any) among the signals.

Another relationship of great interest is the correlation of PRI changes to changes in other parameters (e.g., with pulse duration to maintain a constant duty cycle or with scan to maintain a constant number of pulses per beamwidth). Of still greater interest is the correlation of parameter changes with external events.

In addition to such relationships of the PRI to other radars, other parameters, and events, there is the fundamental relationship of the observed PRI to the circuit that generated it. After a microscopic look at PRI variations (drift, jitter, and the waveform used), the ELINT analyst is in a good position to infer what type of circuitry was the likely source of the PRI. The PRI generators in use as viewed from the ELINT perspective are described in the following sections.

12.6.1 Delay Line PRI Generators

Delay line PRI control has been used in conjunction with MTI systems that subtract the target returns from successive pulses to distinguish moving targets (whose echo changes) from stationary targets (whose echo is fixed). The delay line is used to store the return from the previous PRI as well as to control the pulse timing. A *pilot pulse* enters the delay line at the time the transmitter is pulsed. It emerges one PRI later; it is used to pulse the transmitter; and it is fed back through the delay line as the next pilot pulse. The PRI will be as stable as the delay line (usually very good). The delay line may use an oven to ensure stability. Close examination of the PRI drift can reveal whether or not an oven is used and whether a proportional or on/off-type thermostat is used. The fractional PRI stability expected from this type of generator is in the range of 1 part in 10^3 to 10^5. Use can be made of multiple delay lines to generate stagger.

The analog delay line techniques used were generally acoustic because delays of 1 ms or longer are often needed for search radars. Mercury columns, quartz blocks with multiple faces, and surface acoustic wave (SAW) devices are typical analog delay elements. The base band signals are generally impressed on an IF before being passed through the delay line. Of course, digital memories are also used to delay the echoes, followed by digital subtraction in real time.

12.6.2 Crystal Oscillators and Countdown Circuits

Quartz-crystal controlled oscillators have been used for many years to control radar PRI, and their use is increasing (the venerable SCR-584 used such an oscillator). One reason is that radar sets need an accurate timing reference for range measurement so that the oscillator is already available. Because crystal oscillators usually operate at frequencies well above the normal radar PRF values, digital or other types of frequency division or "countdown" are needed. Multiple position stagger, pseudo-random jitter, and PRI switching can be generated easily by programming the countdown circuitry.

While crystal oscillators are usually thought of as being fixed in frequency, small frequency changes (less than 0.1%) can be made by varying an adjustable capacitor in the resonant circuit of the oscillator. Temperature changes also cause changes in the mean PRI, but long-term drift is usually less than 1 part in 10^5. An accurate measurement of the mean PRI of such radar sets would show some variation from set to set because any group of crystal oscillators exhibits small differences in the mean frequency among its members.

The use of a crystal oscillator as a timing reference for the range delay often takes the form of range marks on the radar display. The time between the range marks spaced by a range interval ΔR is

$$\Delta T = 2\Delta R/c \tag{12.26}$$

where:

ΔT = time between the range marks (second)

ΔR = range spacing of range marks (meters)

c = speed of light (meters/second)

If each cycle of the crystal oscillator generates one range mark on the display, the crystal frequency required is

$$f_{xtal} = c/2\Delta R \tag{12.27}$$

The speed of light in a vacuum, c_v is 299.7925×10^6 m/s. In the atmosphere, the speed of light is reduced slightly. The speed depends on the density of the atmosphere, which is a function of altitude. A useful approximation to the speed of light in the atmosphere at an an altitude h is[2]:

$$c = \frac{c_v}{1 + 313 \cdot 10^{-6} \cdot e^{-h/h_0}} \tag{12.28}$$

where:

c = speed of light at altitude h

c_v = speed of light in vacuum

h = altitude (meters)

h_0 = reference altitude ($\approx$ 7,000m)

At sea level, $h = 0$ and the speed of light is approximately

2.   The author is grateful to Mr. David Barton for pointing out this relationship.

$$c(0) \approx c_v/1.000313 \qquad (12.29)$$

For radar set with 10-km range marks, the proper crystal frequency is 14,985 Hz.

The SCR-584 radar had 2,000-yd range marks. The proper crystal frequency (at sea level) is 81,939 Hz at sea level. The frequency for operation in a vacuum would be 81,964 Hz. The radar actually was specified to use a crystal frequency of 81,950 Hz, which would correspond to an assumed average altitude of about 4,110m throughout the radar beam's propagation path. When the SCR-584 radar was supplied to the Soviet Union during World War II, a crystal frequency of 74,919 Hz was substituted to make the range marks 2,000m apart instead of 2,000 yd apart. The crystal frequencies that correspond to common range marker spacings are given in Table 12.4.

Of course, range marks cannot be directly observed by the ELINT analyst. However, inside the radar circuitry, the PRI is sometimes obtained by selecting every Nth range marker pulse. This can be done in several ways, but the result is that the average pulse interval(s) multiplied by the crystal frequency (hertz) is the countdown integer N:

$$N = (\text{PRI})(f_{xtal}) \qquad (12.30)$$

When measured values for the PRI are available, and considering the possible range of crystal frequencies, the product is never exactly an integer. If N is not too large, it should be possible to determine N exactly. Obviously, if the measurement error of the PRI times the crystal frequency exceeds 0.5, there will be some uncertainty as to the value of N in use. Also, more than one common range mark crystal frequency may produce an integer value. It is usual to assume that the lowest crystal frequency that gives an integer result to (12.30) is correct.

For many radars, several PRIs are used. In this case, assuming that there is one crystal frequency, it is necessary to satisfy the relationship in (12.30) for all of the PRIs. This gives a set of countdown integers that are used by the PRI generator.

Countdown circuitry can be implemented in several ways. Some older sets used an analog timing circuit to select one of the range marks. Due to variations in the analog timing, any of several countdown integers might result. In this case, a

Table 12.4 Crystal Frequency for Common Range Marks

Range Marker Spacing	Crystal Frequency, Hz Sea Level Atmosphere	Vacuum
0.25 nmi	323,649	323,750
1.0 nmi	80,912	80,938
125m	1,198,795	1,199,170
500m	299,699	299,793
1,000m	149,849	149,896
2,000m	74,925	74,948
2,500m	59,940	59,959
2,000 yards	81,938	81,964

random discrete jitter is observed with the countdown values being a set of successive integers.

The more precise method is to use digital flip-flop networks. In times when such circuits were costly, it was common to make up the total countdown from a set of circuits, each of which divided by a relatively small (often prime) integer. If the total countdown can be factored into a set of smaller integers, these may represent the type of circuits in use. For example, a countdown value of 240 might be factored as $8 \times 3 \times 10$ or $2 \times 2 \times 2 \times 2 \times 3 \times 5$. A model of the countdown circuitry then consists of frequency division stages corresponding to each factor. Of course, this is purely speculation on the part of the ELINT analyst. However, if a PRI is observed (possibly a malfunction) that can be explained by a change in only one of the factors, more credence can be placed in the speculative model. Suppose the countdown integer changes from 240 to 216 but no other integer is observed. This could be explained by a divide-by-10 circuit malfunctioning to become a divide-by-9 circuit.

For the SCR-584 radar, a countdown of 48 is used, giving a PRI of 585.72 μs (a PRF of 1,707 pps). The metric range marker crystal of 74,919 Hz would have had a PRI of 640.69 μs (a PRF of 1,561 pps) if the countdown integer had remained 48. However, the actual countdown value was changed from 48 to 40 to give a PRI of 533.91 μs (a PRF of 1,873 pps). Countdown integers of 60 and 80 were also sometimes used in the SCR-584 radar.

The ELINT analyst should be aware that crystal-based PRI generators are very stable and any shift in PRI is usually in integral multiples of some basic interval. Finding this basic interval is equivalent to finding the greatest common divisor of all the observed PRIs. This value may represent the reciprocal of the frequency of the crystal oscillator that controls the PRI generation.

To see how to find the countdown integers from ELINT data, consider the following example. A radar pulse train is analyzed and the following intervals are found:

1. 2,440.8 μs;
2. 2,428.7 μs;
3. 2,465.3 μs;
4. 2,453.1 μs;
5. 2,562.9 μs.

The intervals occur in an unpredictable sequence; therefore, the conclusion can be drawn that this is discrete random jitter. The next step is to put these values in order from the smallest to the largest and find the differences as shown in Table 12.5.

The smallest difference is a candidate for the basic oscillator period from which all PRIs are derived. All other differences must be integer multiples of the basic period. In this example, the first three differences are nearly the same value. Bearing in mind that the resolution of the measurements is 0.1 μs, the best choice is to average the differences of 12.1, 12.2, and 12.3 μs to get a starting value of 12.2 μs. Note that the other difference value is eight times 12.2, which also points to using this value. Next, find the ratio of the observed intervals to the smallest

Table 12.5 Ordered Intervals
and Their Differences

Ordered Intervals	Difference
2,428.7 μs	
2,440.8 μs	12.1 μs
2,453.1 μs	12.3 μs
2,465.3 μs	12.2 μs
2,562.9 μs	97.6 μs

difference and determine if this ratio is close to an integer, as shown in Table 12.6. The average of the PRI values divided by the selected integers in the last column is a good estimate of the basic clock period; this is 12.20439 μs.

The reciprocal of this is 81.9377 kHz, which is close to the value for a 2,000-yd range mark from Table 12.4. Of course, it often takes many trial values before all of the ratios are close to an integer. If the value of 12.1 was selected in this example, the ratios become the values shown in Table 12.7.

The only problem is that the first column is not very close to an integer, and the resulting frequency is not very close to any of the common range mark values. In this example, the measurement resolution (0.1 μs) multiplied by the integers involved (about 200) is larger than the smallest PRI difference (about 12 μs). This means that it may be difficult to find the proper basic period even if it is concluded that a countdown sequence is in use by the radar. It is almost better to assume as a starting point the standard range mark period closest to the smallest PRI difference observed. When there is a large amount of data, it is useful to make histograms of the ratio of the PRI to the assumed basic period. This histogram can show at a glance how close the ratios are to various integers even if thousands of PRIs are being analyzed. The histogram also reveals if there are values between the integers in a pattern, which indicates that the basic oscillator period is actually the assumed

Table 12.6 Ratio of Observed Intervals to the
Smallest Difference Equal to 12.2 μs

PRI/12.2	Nearest Integer	PRI/Integer
199.073	199	12.20452 μs
200.065	200	12.20400 μs
201.073	201	12.20448 μs
202.074	202	12.20446 μs
210.074	210	12.20429 μs

Table 12.7 Ratio of Observed Intervals Integer
to the Smallest Difference of 12.1 μs

PRI/12.1	Nearest Integer	PRI/Integer
200.719	201	12.0831 μs
201.719	202	12.0831 μs
202.735	203	12.0842 μs
203.745	204	12.0848 μs
211.810	212	12.0892 μs
		12.0849 μs
	Average frequency	82.748 kHz

value divided by some integer. For example, if the histogram entries cluster near the integers and halfway between the integers, then divide the assumed basic period by 2.

It should be noted that radar designers may choose to use two crystal oscillators, one to control the PRI and the other to create the range marks. In this case, there is no need for the PRI increments to be related to a common range mark.

12.7   PRI and Range Velocity Ambiguities

As discussed in Chapter 2, in a modern pulse Doppler radar system, once the PRI sequences have been observed and categorized, it is natural to try to determine what purpose the sequence may have. One way to do this is to make a plot of the eclipsed ranges and blind velocities for each PRI and then overlay them to see how well the set of PRIs covers the expected ranges and velocities. If the PRIs do not cover the ranges and velocities well, it may be that the sequence has some other purpose.

The same basic method is used to resolve both range and velocity ambiguities through the use of different PRIs. An example showing how to resolve range ambiguity actually illustrates the principles for both. Suppose a radar has a PRI of 40 μs and there is a target echo with a delay of 12 μs from the last transmitted pulse. The maximum unambiguous range is 6 km. The indicated range is 1.8 km. After a few pulses, the PRI is switched to 30 μs and the target echo is observed at 2 μs. The maximum unambiguous range is now 4.5 km and the indicated range is 300m. The fact that the indicated range changed when the PRI was switched shows that the true range is different from the indicated range. A simple trial and error approach can be used to find the true range. The true range delay is some integer times 40 μs plus 12 μs. It is also some other integer times 30 μs plus 2 μs. Consider the information in Table 12.8.

The condition is satisfied when both values in the last two columns are equal. Therefore, the true range delay is 92 μs (or a multiple of 92). For this kind of radar, the maximum unambiguous range is determined by the least common multiple of the PRI values—in this case, 120 μs. It should be noted that the use of multiple PRIs is complicated considerably if there are multiple targets in the same angular cell.

12.8   MTI Radar Blind Speeds

MTI radar makes no attempt to measure target speed. Instead, the MTI feature eliminates large stationary echoes from the display. This is done by subtracting the

Table 12.8   Integers Versus Range Delay

Integer 1	Integer 2	Range Delay 1 (μs)	Range Delay 2 (μs)
1	1	52	32
1	2	52	62
2	2	92	62
2	3	92	92

echo phase from one pulse from that of the next pulse. If the phase does not change, the result is zero and the stationary targets produce nearly the same phase in each echo. However, if the echo phase changes by 360° (or a multiple thereof) during one PRI, this also produces nearly complete cancellation. Therefore, the blind speeds are those for which the target moves radially toward the radar any integer number of wavelengths during the PRI:

$$V_b = n \cdot c/[2(\text{PRI})(\text{RF})] \qquad (12.31)$$

where:

   V_b = blind speed (meters/second)

   n = an integer

   c = speed of light $(3 \times 10^8 \text{ m/s})$

PRI = pulse repetition interval (seconds)

RF = radio frequency (hertz)

   For example, a radar operating at 6 GHz with a 2,500-μs PRI has blind speeds at multiples of 10 m/s or 36 km/h. If the radar adds stagger so that a 2,500-μs PRI alternates with a 3,000-μs PRI, the blind speeds for this second interval if used alone would be multiples of 8.333 m/s. The blind speed of a radar with the average PRI of 2,750 μs would be 9.0909 m/s. The blind speed of the staggered PRIs is the lowest speed that is some integer times each blind speed—in this case, 50 m/s, which is 5×10 and 6×8.333. Another way to look at stagger is to reduce the PRIs to a ratio of integers (i.e., 3,000:2,500 = 6:5). These are the integers that multiply the blind speeds of the individual PRIs to obtain the blind speeds with stagger. Alternately, the blind speed for the average PRI can be found and multiplied by the average of the two integers (in this case $9.0909 \times 5.5 = 50$ m/s). If the PRIs are close together, the true blind speed can be made quite large; however, there will still be deep nulls in the MTI response (dim speeds).

12.9   Moving Target Detection

Moving target detection (MTD) is a combination of MTI and Doppler filtering. The velocity itself is not measured; additional clutter rejection is achieved by adding narrowband filters to further bring the signals out of the noise. From the ELINT viewpoint, the radar has a low PRF but the intervals have a pattern similar to dwell switch. To permit filtering after the canceller, the PRI must remain constant for enough intervals to "fill" the canceller and FFT processor. Typically, this may require 10 PRIs—two for the canceller and eight for the FFT. This allows reducing the bandwidth of the final filters to the PRF/8. In the FAA ASR radar, a three-pulse canceller alone achieves a 25-dB improvement factor. When the eight-pulse FFT is added, the improvement factor increases to 45 dB [9]. For this type of radar, the number of PRIs with each value should be on the order of 10; however, the problem of blind speeds is still the same as before. Therefore, there must still be a sequence of interval values to make sure that all targets are detectable regardless of their speed.

References

[1]   Vyse, B., and H. Levinson, "The Stability of Magnetrons Under Short Pulse Conditions," *IEEE Trans. on Microwave Theory and Techniques,* Vol. MTT-29, No. 7, July 1991.

[2]   Skolnik, M. I., *Radar Handbook*, New York: McGraw-Hill, 1970, p. 16–19.

[3]   Skolnik, M. I., *Radar Handbook*, New York: McGraw-Hill, 1970, p. 17–38.

[4]   Skolnik, M. I., *Introduction to Radar Systems*, New York: McGraw-Hill, 1980, pp. 401–402.

[5]   Tsui, J. B.Y., *Digital Techniques for Microwave Receivers*, Norwood, MA: Artech House, 1995, p. 167.

[6]   Wiley, R. G., *Electronic Intelligence: The Analysis of Radar Signals*, 2nd ed., Norwood, MA: Artech House, 1993, pp. 179–180.

[7]   Szymanski, M. B., and R. G. Wiley, *Pulse Analysis Using Personal Computers*, Norwood, MA: Artech House, 1986.

[8]   Gray, D. A., B. J. Slocomb, and S. D. Elton, "Parameter Estimation for Periodic Discrete Event Processes," *IEEE Proc. ICASSP-'94,* 1994.

[9]   Skolnik, M. I., *Introduction to Radar Systems*, New York: McGraw-Hill, 1980, pp. 127–128.

CHAPTER 13

Deinterleaving Pulse Trains

ELINT analysis often involves interleaved pulse trains. This means that several pulsed signals are present and that successive pulses may not come from the same emitter. The first problem the analyst faces, then, is to separate the interleaved pulses into groups of pulses from the same emitter. This process is called *deinterleaving*. The first step is usually to make use of information derived form individual pulses, often referred to as pulse descriptor words (PDWs). PDWs consist of parametric information such as pulse duration (PD), pulse amplitude (PA), pulse radio frequency (RF), angle of arrival (AOA), and time of arrival (TOA). Other on-pulse parameters such as descriptions of intrapulse modulation and polarization may also be available. If fine frequency measurements are made, the RF parameter may be thought of as the frequency of arrival (FOA) because the RF originally transmitted has been altered by the Doppler shift due to relative motion of the ELINT collector and the emitter.

13.1 Pulse Sorting

In many cases, deinterleaving can be accomplished using single pulse parameters as reported in the PDWs. This is also called *pulse sorting*. If there is no frequency agility, RF is a very useful first sort. AOA is a very useful sorting parameter because it is determined by the emitter's location and not by its signal design. If there are no reflected signals to cause confusion, a constant AOA will be present over rather long periods of time even when the platforms are moving. To have the same AOA, the emitters must be on the same line of bearing. Pulse duration can also be used; however, it is a less reliable sorting parameter due to multipath and thresholding problems associated with changing pulse amplitudes. When PDWs from more than one collector are available, emitter location also produces fixed values for the *difference* of the TOA and/or the *difference* of the FOA. These can also be used for pulse sorting once it is determined that the PDW from one collector is describing the same pulse as the PDW from another collector. Pulse sorting is often done using histogram techniques; that is, one might choose to determine the distribution of the PDWs in terms of, say, RF and AOA in a two-dimensional histogram. In the absence of frequency agility and multipath, the histogram will contain a single peak at the AOA and RF of each emitter. If the RF and AOA cells are each 1% of the range of values covered by intercept system, there will be "space" for $100 \times 100 = 10,000$ cells in the histogram. Suppose a space of one or two cells in all directions is allowed in order to separate the peaks (a 3×3 or 5×5 cell box). This means that the histogram could separate about $10,000/25 = 400$ to $10,000/9 = 1,111$ emitters—which is probably sufficient for most situations. However, when several emitters of the same type are received from the same direction with nearly the same frequency, and/or when frequency agility is present, when RF or AOA data is missing, or when multipath is present, this kind of sorting can be inadequate.

In the sorting/clustering operation, it may be that the RF, PW, AOA windows for the emitters overlap. It may be that the overlapping of the clusters causes problems in the deinterleaver. For example, the cluster boundaries may overlap, but the actual pulse data may be separated. Consider two overlapping circles. The pulses from emitter 1 may all lie in circle 1 and those from emitter 2 in circle 2, but none of them is in the overlap region. Or it may be that all of the pulses from the two emitters lie in the overlap region of the two circles. If the data can be actually separated by using smaller parameter windows for RF, PW, and AOA, the deinterleaving process can be aided by doing this ahead of the deinterleaver.

Use of smaller parameter windows may erroneously break up a cluster into smaller clusters. This might increase the variance of the parameter estimates later in the process. Another problem is that by imposing smaller limits on the cluster boundaries, artificial structure might be introduced that the deinterleaver subsequently detects and reports.

By definition, all of the pulses in the cluster have RF values within some "window." Clearly, the actual RF values can be examined and their spread determined. If the spread is significantly smaller than the original RF window, the window width can be reduced to see if the overlap condition is removed. The same thing can be done with the other parameters of pulse width and AOA. If the overlap conditions can be removed in the sorting/clustering, then deinterleaving is already accomplished. Even if the sorting is only partial, deinterleaving should be easier.

Of course, the sorting process has a downside as well. Suppose a cluster is erroneously divided into subclusters, and as a result two emitters appear where there should be only one. For example, consider a frequency agile signal where only a limited number of pulses was intercepted. The underlying distribution might be uniform, but a few samples might appear to cluster.

The deinterleaver may also create reports of more emitters than are actually present. For example: a stagger could erroneously be called two or three constant PRI pulse trains instead of one.

Generally, the clustering window widths would be chosen a priori to get good clustering results, and these widths would be based on variations expected in the measured parameter values for actual emitters. The values reported for the cluster to a subsequent deinterleaver are usually the means of the parameters for those pulses in the cluster. Going back to the original data values and examining the distribution of the measured values relative to the original window widths used to define the cluster is one way to possibly improve the sorting of the pulses.

These sorting processes are key to the operation of Radar Warning Receivers (RWRs) and Electronic Warfare Support Receivers (ES). These generally make use of relatively short snapshots of the environment, whereas ELINT operations often make use of much longer collection periods. Nevertheless, today's RWRs often can make the same quality of measurements as ELINT receivers. The difference is in the length of time allowed for both collection and signal analysis.

13.2 PRI-Based Gating

One of the earliest (and still useful) deinterleaving techniques is a gating device. A block diagram is shown in Figure 13.1.

These devices operate by having the first pulse arrive at the input to start a delay. At the end of the delay, an acceptance interval begins. If the next pulse arrives during the acceptance interval, it restarts the delay and the process is repeated. If no pulse is found in the acceptance interval, the next pulse to arrive restarts the delay and the process is repeated. In this way, if there is a signal present having a PRI value between the delay and the delay plus the acceptance interval, the device "locks on" to that pulse train and only those pulses appear at the output. This process was usually done in real time; however, it may also be applied to digitized or recorded signals. These devices can be used in series in the reject mode. (This is similar to the ideas in [1].) In this way, more than one interfering pulse train can be rejected. In a series connection, only the last device in the chain can be in the accept mode. The number of rejection devices that can be successfully used in a series connection is limited. If the fraction of time that the rejection gates taken together block the input is significant, then the probability increases that one of the desired pulses will also be blocked. When this happens the deinterleaving device loses lock and the process must start over. In the worst case, so many of the desired pulses will be missing that the final deinterleaver cannot lock on to the desired pulse train at all. Deinterleavers are also available to continue when one or more pulses are missing.

The acceptance gate width can be selected to allow for uncertainty in the PRI value or to allow for PRI modulation such as jitter or stagger. The delay interval

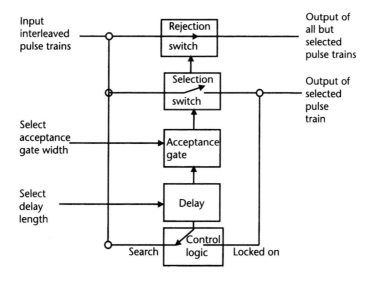

Figure 13.1 Simple PRI accept/reject device.

should be selected to allow passing the shortest PRI of the signal of interest, and the acceptance gate width plus the delay length should be equal to the longest PRI in the signal of interest. Choosing these values usually requires a preliminary analysis of some kind. Various start-up methods are also in use in computerized deinterleaving. More sophisticated tracking devices adapt the values of the acceptance gate width and the delay length to match the observed signal values seen at the output. Such tracking can follow the pulse train of interest even if its PRI is drifting.

It is important to note that if the sorter is set to any multiple of the PRI, the sorter will lock on to that multiple. Therefore, it is best to try short PRI values first.

13.3   Deinterleaving Algorithms

If pulse train data is reduced to digital form (such as PDWs), a variety of computer algorithms can be used to assist in deinterleaving. At the outset, the analyst should realize that, in principle, there is insufficient information in the pulse arrival times alone to be certain about how the pulses should be sorted into groups representing different emitters. In practice, it is possible to do a good job with TOA data alone if there are not too many emitters present at once.

Given that such pulse sorting will not be 100% correct and that the actual emitter configuration that produced a dataset is unknown, how is the analyst to judge the results of deinterleaving? Suppose that the analyst can try a number of sorting algorithms, each having a number of variable parameters (e.g., acceptance gate widths, the number of pulses allowed to be missing, and the range of acceptable pulse intervals, and so on). Each of these choices of algorithms and parameters provides a slightly different sorting result. How can the analyst decide which is the best? This calls for the analyst's judgment, based on experience. In general, though, a deinterleaving algorithm should:

1. Assign as many pulses as possible to some emitter;
2. Result in as few different emitters as possible;
3. Have the least PRI variation associated with each emitter.

The end result of the deinterleaving process can be thought of as a model of the environment that produced the interleaved pulse train. In other words, it is a theory that is thought to explain the observed data. Generally, the best theory is one that accounts for (predicts) the most pulses correctly with the least complexity. For example, it could be said that each pulse came from a separate, independent transmitter that transmitted one pulse (i.e., the number of pulse trains is equal to the number of pulses). This would require explaining any periodicities observed as being due to chance. This model should be rejected as having too much complexity and requiring improbable assumptions. At the other extreme, it could be said that all of the pulses came from a single emitter that has a very complex PRI, pulse duration, and pulse amplitude generation sequence. This model accounts for all of the pulses and has as few emitters as possible, but the PRI, pulse duration, and amplitude variations that must be assumed for that emitter are too great and this model must be rejected. These extreme cases illustrate the need for the three criteria listed. The key to successful deinterleaving is the analyst's judgment, both in selecting algorithms and the parameter limits used, as well as in determining when the results are realistic. The analyst must also judge when, in spite of the application of all of the best techniques, the pulses cannot be successfully deinterleaved. In other words, it is necessary to know when to quit and move on to a new analysis task!

In an environment of signals from emitters that scan in angle, the pulses from an emitter tend to arrive in bursts. Frequently there will be at least a few pulses that are not interleaved with those from another signal. A good approach is to try to work with those pulse trains that are at least partially deinterleaved to begin with. The following algorithm works well in this case.

Step 1. Look for a few pulses separated by equal intervals. The number of pulses must be at least three, but four or five may be better choices. Based on typical radar target illumination times, 10 pulses would probably be too many. The tolerance to which the intervals are required to match must be larger than the digitizers quantizing noise and also larger than the jitter expected from the radar's pulse interval generator's design.

Step 2. Upon finding such a set of pulses equally spaced pulses, try to extend the pulse train both forward and backward in time to account for more pulses. Extension techniques are discussed below. The extension is terminated when a certain number of "missing pulses" occurs. Those pulses associated into a group from a single emitter are deleted from the interleaved pulse train and the process is repeated. Because removal of a set of pulses may result in another pulse train for which four or five pulses are no longer interleaved with any other pulse train, the process should be repeated by returning to Step 1. This algorithm is very reliable in that it rarely locates pulse trains not actually present. However, it may be ineffective in very dense environments because it is unable to get started.

The method of extending the group of associated pulses can range from very simple to very complex. Some simple techniques are to compute the average pulse interval and hypothesize pulses at times spaced by that amount. A window of tolerance is used that can be based either on the observed variability of the few available intervals or on the tolerance used initially. Another simple method is to base the next pulse interval on the immediately adjacent pulse interval. This allows for the tracking of a drifting pulse interval (but also can cause locking on to the wrong pulse train). One procedure for estimating the "correct" pulse interval for a group of pulses is that for the least mean square error. (This technique is discussed in Chapter 12.) Whatever method is selected, when the algorithm can extend the associated group of pulses no further, it is fruitful to take all those pulses defined as being associated, recalculate the predicted pulse locations based on all the data, and search again to see if any additional pulses are located. The loop terminates when no pulses are added to the group. Note that successful algorithms of this type allow for some fraction of pulses to be missing so that the associated groups are not broken into small segments by occasional missing pulses. In computing the radar's pulse interval, the missing pulses must be considered as present.

There will be situations where there are no segments of any pulse train free of interleaving and the technique just discussed cannot get started. For example, perhaps 75% of the pulses are successfully deinterleaved, but a residue of 25% remains. Another start-up technique should then be applied. A useful one is to select one pulse (possibly near the center of the data being analyzed) and assume that it is paired with the pulse next to it. The interval between those two pulses is used in an attempt to project forward and backward in time to see if that is the interval of a pulse train. If it is, the same pulse association extension techniques can be employed. If that interval does not work, the same starting pulse is paired with the next closest pulse and that interval is tried. This process continues until the trial interval reaches some maximum or until a pulse train burst is found. If no pulses are associated, another pulse is selected as the starting point. The process continues until every pulse is either included in a burst or has been tried as a starting point. By considering every pair of pulses as defining a potential PRI, the process resembles the delta-T histogram described in Chapter 12. Various methods of using histogram techniques to obtain PRI values to start the deinterleaving are described in [2, 3].

Staggered pulse intervals cause two kinds of problems for these pulse-sorting algorithms. In one case, the acceptance window is so wide that all intervals of the stagger are accepted into a burst. Now, however, the average pulse interval in the burst will not be related to any of the stagger values. A technique that may overcome this difficulty is the following. For each interval in the stagger sequence a different delay function is available that attempts to locate a pulse within an acceptance gate (i.e., one staggered PRI interval plus or minus a specified tolerance). If a pulse is found, the next different delay in the stagger sequence is started. This chaining process continues until a pulse cannot be found within an acceptance gate. A second problem occurs if the tolerance window is so small that several constant PRI pulse trains with a multiple of the average pulse interval are found instead of one staggered pulse train. This is a case of erroneous deinterleaving. If two bursts overlap in time and have the same average pulse interval, they may be pulses from the same pulse train. If the pulses are truly from more than one emitter, their relative phase will change slowly. One way to detect this is to compute the mean and variance of the time between the pulses of one pulse train and the pulses of another pulse train. If they are from a staggered emitter or from emitters that are synchronized in their pulsing, the variance will be small. Signals having multiple pulse characteristics can cause similar kinds of problems and require extra attention. Usually, a preliminary analysis of the data by the analyst can reveal the multiple pulse groups.

13.4 Delta-T Histogram Applied to Deinterleaving

The delta-T histogram described in Chapter 12 can be used to help find intervals that may be PRIs. These intervals may then be used as starting values for other algorithms. The more interleaved pulse trains, the more difficult the problem becomes.

The delta-T histogram can give gives some insight into how many pulse trains can be successfully deinterleaved. One of the outstanding problems of ELINT is determining how many pulse trains can be deinterleaved under various conditions. The following gives one way to make an estimate of the number of pulse trains that can be deinterleaved. For the purposes of this analysis, assume that all of the pulse trains have nominally constant PRIs. For a segment of time, T, there are a total number of pulses, N, present. These are made up of n_k pulses from each of k pulse trains. Each pulse train has a nominal PRF.

The following relationships are needed for this analysis:

$$N = \sum_{k=1}^{K} n_k = T \sum_{k=1}^{K} PRF_k$$

$$n_k = PRF_k T = \frac{T}{PRI_k}$$

Next consider that some of the intervals in the histogram are made up of the difference in the TOA between two pulses from the same pulse train, while others are made up of the difference in the TOA of pulses from different pulse trains. Denote the former as *pure* intervals and the latter as *impure* intervals. Then the ratio of the number of pure to impure intervals is given by the purity ratio, PR:

$$PR = \frac{\sum\limits_{k=1}^{K} n_k(n_k - 1)}{\left(\sum\limits_{k=1}^{K} n_k\right)\left(\sum\limits_{k=1}^{K} n_k - 1\right)}$$

If all of the pulse trains have about the same PRF, the number of pulses from each will be about the same. For this special case, the purity ratio becomes

$$PR = \frac{n - 1}{Kn - 1} \approx \frac{1}{K} \text{ (if } Kn \gg 1)$$

This makes clear the fact that the ratio of pure to impure intervals decreases as the number of interleaved pulse trains increases.

Next, consider the heights of the histogram bins that result. Suppose the impure intervals are spread uniformly among all of the histogram bins and the pure intervals fall into sets of adjacent bins spaced by the PRI. (Usually, the size of the bins in the histogram is on the order of the incidental jitter of the individual pulse trains.) The number of bins, B, will be T/b, where b is the width of the individual bins. The average height of the impure intervals is given by

$$\frac{1}{2}\left[\left(\sum_{k=1}^{K} n_k\right)^2 - \sum_{k=1}^{K} n_k^2\right]\frac{b}{T}$$

The pure intervals are placed in sets of a few adjacent bins near multiples of the PRI. There are $n = T/(\text{PRI})$ sets of such bins. In each set, the number of bins into which intervals are placed is approximately J/b, where J is the PRI jitter. This makes the average height of the bins due to the pure intervals equal to

$$(n_k - 1)\left(\frac{b}{J}\right)$$

Note the fact that the bin height at multiples of the PRI decreases slightly has been ignored.

For the special (but interesting) case where all the PRIs are equal, the ratio of the height of the pure to impure intervals in a bin is given by

$$R = \frac{n-1}{Kn(Kn-n)}\frac{T}{2J} \approx \frac{\text{PRI}}{2K^2J}$$

Consider the case illustrated in Figure 13.2. The average height of the first 400 bins (all impure intervals) is 27.7. The height of the tallest peak is 94, and the observed ratio, R, is then 2.4. Applying the equation above gives a theoretical value of R of 2.9 when the mean PRI is 115.4 μs and the total jitter is 200 ns.

The bin heights due to the pure intervals should be larger than those of the impure intervals in order to attach much significance to that bin. It is clear that the number of pulse trains that can be reasonably deinterleaved using interval-only techniques is on the order of

$$K = \left(\frac{\text{PRI}}{2RJ}\right)^{1/2}$$

In this equation, R is the ratio of pure to impure intervals required in a bin to achieve sufficient height for the analyst to believe it is a potential PRI value.

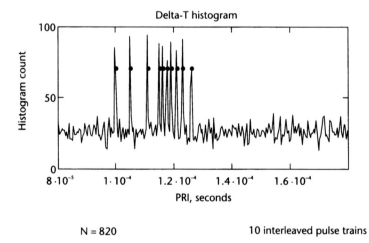

Figure 13.2   Delta-T histogram for 10 interleaved pulse trains (bin size = 400 ns).

This relationship is a quantitative expression of the intuitive idea that the ability to deinterleave pulse trains improves if they have quite stable PRIs. In a typical application, the typical PRI is 1 ms and there may be jitter on the order of 1 μs. This analysis indicates that for a ratio $R = 3$, the number of pulse trains that could theoretically be distinguished is about 13. On the other hand, for 100 μs of jitter and $R = 2$, the number of distinguishable pulse trains may be less than two. Clearly, the presence of a number of intentionally jittered signals can cause problems for interval-only deinterleavers.

The use of *cumulative difference* (CDIF) histograms and *sequential difference* (SDIF) histograms in deinterleaving has been described [2, 3]. The first CDIF consists of a histogram of the first differences of the TOAs. This is examined for peaks. Peaks will occur if there are regions in the data record where there is no interleaving present. If no peaks are found, a histogram of the second differences of the TOAs is added to the histogram of the first differences and the cumulative histogram is examined for peaks, and so on. (This creates the delta-T histogram as it accumulates.) The advantage is that a noninterleaved pulse train can be found immediately. This greatly reduces the number of differences needed to get started (approximately n instead of n^2) and may also reduce the confusion caused by multiples of the PRIs. The SDIF histogram approach makes use of the first differences of the TOAs in the same way as the first step of the CDIF approach. However, the second differences of the TOAs are used to form a separate histogram, rather than being added to the histogram of the first differences. Each higher order difference is used to form its own histogram, whose peaks are then used to infer interval values in conjunction with the results from the other lower order difference histograms.

The threshold function to be used to determine which histogram bins represent possible PRIs is considered in [3]. Clearly, if the number of pulses available is fixed, the number of intervals present is inversely proportional to the length of the interval in the SDIF histogram. The impure intervals, which represent the time between the pulses of different pulse trains, are represented as random Poisson points [3]. This leads to a threshold function of the form,

$$\text{threshold } (PRI) = x(N - d)e^{-(PRI)/gB}$$

where N is the total number of pulses, B is the total number of bins, d is the difference level, and the constants x and g are experimentally determined. Using simulated data, the exponential threshold is shown to be better than the reciprocal of the interval or its square root [3]. (The parameter x depends on the percentage of missed pulses in the data.) In the full delta-r histogram (or CDIF histogram with all differences considered), the bin height falls slowly as a function of the length of the time differences—at least in the region of greatest interest, which is a relatively small integer multiple of the typical PRI.

Another algorithm based on finding the period of any complex pulse train has been developed that is capable of finding the period of any single pulse train of the dwell-switch or staggered type [private communication with R. B. Millette, Delata Consultants, Baie O'urfe, Quebec, Canada, July 1992]. Once this period is found, one can then predict when the next pulse will occur regardless of the complexity of the PRI sequence. This algorithm is not designed for deinterleaving and (so far) does not specify the type of PRI sequence present. However, it can be used to drive an interval-based deinterleaver once the true period has been found.

One of the major problems with interpreting the delta-T histogram is the peaks at multiples of the true PRIs. It would be quite helpful to have a method for making histograms without peaks at multiples of the PRI. One method for doing this is to add a complex value to the histogram bin for each interval found rather than to add one count. Dr. Doug Nelson described this technique to the author around 1983. The technique was conceived independently by K. Nishiguchi in Japan at about the same time [4]. The excellent analysis in [4] is covers nearly all aspects of this type of histogram. Note that in [4], the complex delta-T histogram is called the *PRI transform*.

The complex value to be added to the appropriate histogram bin can be thought of as the phase of the TOA of the pulse at one end of the interval with respect to a constant PRI pulse train. The PRI is equal to the particular bin into which that TOA difference falls. This means that for each pair of pulses, one computes the difference of their TOAs as usual. However, instead of increasing the count of the proper histogram bin by unity, it is increased by a complex number given by

$$\exp[j(\text{phase}_{kn})]$$

where:

$$I_{kn} = T_n - T_{n-k}$$
$$\text{phase}_{kn} = 2\pi(T_n \bmod I_{kn})/I_{kn}$$

After this complex histogram has been formed, the absolute value of the contents of the bins is examined. Consider the bin contents for a constant PRI. If the interval is a true PRI and not a multiple, then

$$T_n = nI + T_o$$
$$\text{phase}_{kn} = 2\pi(T_o)/I = \text{Constant}$$

For intervals that are multiples of the PRI, the phase values are spaced around the unit circle like spokes in a wheel, with the number of spokes equal to the integer multiple of the PRI under consideration. When these complex quantities are added, the result is nearly zero. Therefore, the only bin with a significant accumulation is that bin corresponding to the true PRI. Figure 13.3 compares the delta-T histogram with the complex delta-T histogram. The suppression of the peaks at multiples of the PRIs is clearly shown when there is a small amount of jitter. Note that the same 10 PRIs were used in Figure 13.3 as were used in Figure 13.2. The dots show the true PRIs.

Next, consider the effect of jitter on this kind of histogram. Once the number of PRIs considered exceeds I/J, where I is the PRI and J is the jitter, the phase value will begin taking on random values around the unit circle. Therefore, the height of the bin will be somewhat limited by the jitter. Without jitter, the height

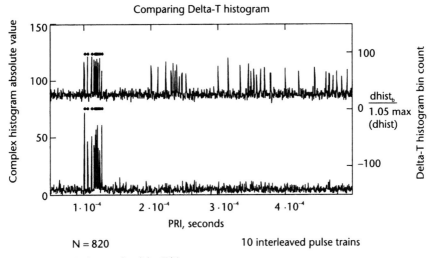

Figure 13.3   Comparison of the delta-T and complex delta-T histograms.

of the bin representing the fundamental PRI is equal to the number of pulse intervals which is one less than the number of pulses in the record:

$$n = (T/I)$$

With jitter, the bin height is approximately limited to

$$(I/J)$$

Consider an example with $T = 0.1$ second, $I = 1$ ms, and $J = 1$ μs. There will be about 1,000 pulses. With no jitter, the bin height will be 999. With 1 μs of jitter, the bin height is about the same. If the jitter is increased to 100 μs, the bin height decreases to less than 10.

The situation modeled in Figure 13.3 resulted an observed noise level of about 13 for the delta-T histogram and about 3.2 for the complex delta-T histogram. Because the phases of the impure intervals are random, the resulting amplitude of the sum might be expected to be approximately the square root of the number of intervals in that bin. The height of the bins where PRIs occur remains about the same because in the absence of jitter, the phase remains constant. In [5], a Poisson model is used to compute the expected noise level of the complex histogram. The result obtained is that the expected noise level (ENL) is given by

$$\text{ENL} = \sqrt{n^2 b/T}$$

where:

n = the number of pulses in the record

b = histogram bin size

T = duration on the data record

Applying this result to the situation in Figure 13.3 gives a calculated noise level of 3.7, which compares well to the observed noise level of 3.2.

Figures 13.4 to 13.6 show the effect of increasing jitter on the regular and complex delta-T histogram. The noise increases and the ability to resolve the peaks is degraded because the bin width was increased to 0.5 μs. With 5 μs of jitter and 10 emitters near 100 μs PRI, the utility of the histograms is limited.

In [4] an improved complex histogram is described and analyzed in detail. In it, the fact that jitter causes larger and larger phase errors for intervals far from the time origin is partially overcome by shifting the time origin according to simple rules. The second improvement is to use histogram bins that are wide enough to accommodate the jitter but whose spacing is determined by the desired PRI resolution. As shown in Figure 13.7, this provides good resolution of PRIs along with maintaining the harmonic suppression properties of the complex delta-T histogram.

When these kinds of histograms have sufficient data and the peaks are clearly formed, automated techniques can be used to find PRI values. The following method can be considered as an example [5].

Step 1. Arrange the histogram bins in order of amplitude from largest to smallest.

Step 2. Find those bins larger than a threshold. The threshold can be based on the observed average histogram level.

Step 3. Arrange the PRI values corresponding to the bins above the threshold from shortest to longest.

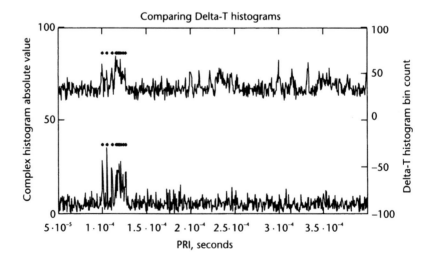

Jitnc$_0$ = 0.5   Jitcum$_0$ = 0.5   N = 820   width = 5 × 10^{-7} 10 interleaved pulse trains

Figure 13.4   Effect of jitter on delta-T histograms (jitter = 1 μs).

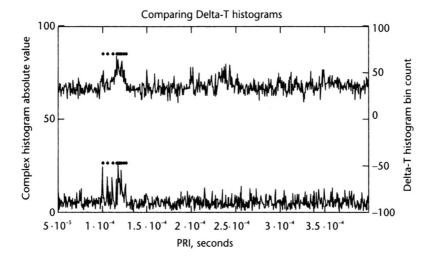

Jitnc$_0$ = 1 Jitcum$_0$ = 1 N = 820 width = 5 × 10^{-7} 10 interleaved pulse trains

Figure 13.5 Effect of jitter on delta-T histograms (jitter = 2 μs).

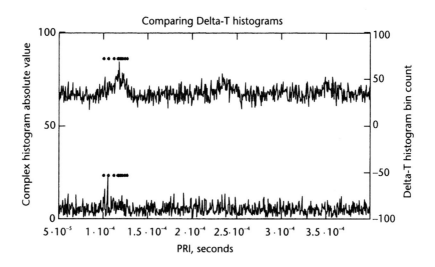

Jitnc$_0$ = 2.5 Jitcum$_0$ = 2.5 N = 820 width = 5 × 10^{-7} 10 interleaved pulse trains

Figure 13.6 Effect of jitter on delta-T histograms (jitter = 5 μs).

Step 4. Eliminate adjacent bins.

Step 5. Eliminate PRIs that are more than twice the smallest (these could be PRI multiples).

These values may be used to start a deinterleaving process or may be considered directly as the PRIs present in the data. For the example shown in Figure 13.3. The results can be very good, as shown in Figure 13.8.

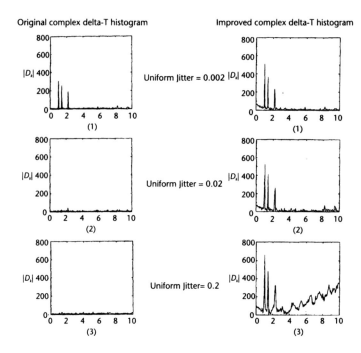

Original complex delta-T histogram   **Improved complex delta-T histogram**

Uniform Jitter = 0.002

Uniform Jitter = 0.02

Uniform Jitter= 0.2

Figure 13.7 Complex delta-T histogram: original and improved (PRI values 1, $\sqrt{2}$, and $\sqrt{5}$). (*From:* [5]. © 2000 IEEE. Reprinted with permission.)

	Delta-T hist.		Complex Delta-T		Input PRI values
	0		0		0
0	$1 \cdot 10^{-4}$	0	$1 \cdot 10^{-4}$	0	$1 \cdot 10^{-4}$
1	$1.048 \cdot 10^{-4}$	1	$1.05 \cdot 10^{-4}$	1	$1.05 \cdot 10^{-4}$
2	$1.11 \cdot 10^{-4}$	2	$1.11 \cdot 10^{-4}$	2	$1.11 \cdot 10^{-4}$
3	$1.15 \cdot 10^{-4}$	3	$1.15 \cdot 10^{-4}$	3	$1.15 \cdot 10^{-4}$
4	$1.162 \cdot 10^{-4}$	4	$1.164 \cdot 10^{-4}$	4	$1.163 \cdot 10^{-4}$
5	$1.176 \cdot 10^{-4}$	5	$1.178 \cdot 10^{-4}$	5	$1.177 \cdot 10^{-4}$
6	$1.19 \cdot 10^{-4}$	6	$1.192 \cdot 10^{-4}$	6	$1.191 \cdot 10^{-4}$
7	$1.21 \cdot 10^{-4}$	7	$1.21 \cdot 10^{-4}$	7	$1.21 \cdot 10^{-4}$
8	$1.23 \cdot 10^{-4}$	8	$1.23 \cdot 10^{-4}$	8	$1.23 \cdot 10^{-4}$
9	$1.26 \cdot 10^{-4}$	9	$1.26 \cdot 10^{-4}$	9	$1.26 \cdot 10^{-4}$
10	0	10	0		
11	0	11	0		
12	0	12	0		
13	0	13	0		
14	0	14	0		
15	0	15	0		

pk = (column 1), pkc = (column 2), PRI · 10^{-6} = (column 3)

Jitter = 10 ns cumulative and 10 ns noncumulative.
Histogram bin size 200 ns.

Figure 13.8 Example of automated peak processing algorithm results.

13.5   The Pulse Train Spectrum

Another technique available for finding the PRFs in interleaved data is the pulse train spectrum [6]. This provides a frequency domain view of the interleaved data and the peaks occur at the PRF = 1/PRI of the pulse trains. One motivation for this technique is to reduce the computational load of the delta-T histogram type of auto correlation calculations The number operations required is approximately equal to the number of pairs of pulses or is approximately equal to the square of the number of pulses, while the FFT can be computed with $n(\log n)$ operations. In this calculation, the pulse TOAs are converted to phase values equal to

$$\text{Phase} = 2\pi\left(\frac{\text{TOA}}{T}\right)$$

These complex values spaced around the unit circle. Their FFT is computed and its absolute value is the pulse train spectrum. Figure 13.9 shows the pulse train spectrum for the same example of 10 interleaved pulse trains used above.

The number of pulses used in this pulse train spectrum is 8,705—about 10 times the number used in the delta-T histograms above. This provides 10-Hz resolution. For a PRI of 100 μs, this corresponds to a PRI resolution of 100 ns. The equation below relates PRI resolution to PRF resolution for a constant PRI and when PRI = 1/PRF.

$$d\,\text{PRI} = \frac{-d\,\text{PRF}}{\text{PRF}^2}$$

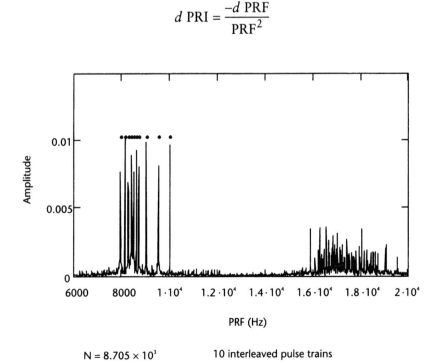

N = 8.705 × 10³   10 interleaved pulse trains

Figure 13.9   Pulse train spectrum of 10 interleaved pulse trains (PRF resolution = 10 Hz).

To achieve the same resolution as the histograms, it is necessary to use a relatively long record. Such a long record is unlikely to be available from radars in a tactical environment. When the record is shortened to the same number of pulses (870) used in the delta-T histograms, the resolution of the pulse train spectrum is insufficient. This is shown in Figure 13.10.

13.6 Combining Pulse Bursts

After deinterleaving, many algorithms produce bursts of pulses with the same PRI but separated in time. One example is pulses from a scanning emitter. The bursts may consist of the main beam and the stronger sidelobes as the antenna bema moves past the intercept receiver. It is necessary to group these bursts together into one emitter pulse train. This can be done using the burst parameters of average PD, PRI, AOA, and so on. Another example is that of a coherent pulse Doppler radar with high PRF bursts with different PRIs transmitted during a single coherent processing interval. Many airborne multimode radars are of this type. In this case, the PRI is not the same from one CPI to the next. But the bursts are close in time and probably have approximately the same CPI duration. This is explored in [7]. A typical signal of this type is described in Table 13.1.

The deinterleaver proposed in [7] consists of two stages. The first extracts the PRI bursts and the second extracts sequences of packets that share a coherent timing structure. A coherent structure is one in which the event times (in this case, the packet times) have the same phase with respect to a periodic reference signal. It is this use of coherence that sets this approach apart from the other in the references. Clearly, any information known about the structure of the signal should be used to help deinterleave and identify it, including the coherence of the signal from CPI to CPI. In [7], the method is used to separate two emitters having the same PRI sequences and overlapping CPIs.

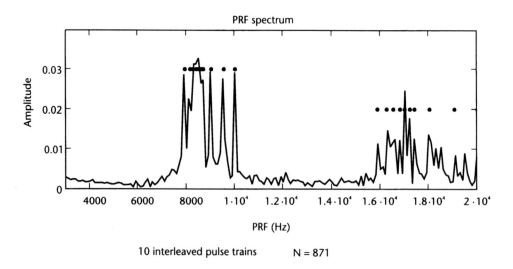

Figure 13.10 Pulse train spectrum of a shorter record.

Table 13.1 Typical Pulse Doppler Signal

CPI	PRI(μs)	Number of Pulses	CPI Duration
1	25.8	121	3.4848
2	45.6	76	3.456
3	32.0	109	3.4880
4	50.4	69	3.4776
5	35.2	99	3.4848
6	48.8	71	3.4648
7	33.6	104	3.4944
8	47.2	74	3.4928
9	30.4	115	3.4960
10	44.0	79	3.4760

Source: [7].

13.7 Raster Displays and Deinterleaving

The raster displays described in Chapter 12 could also be used for deinterleaving. When the time across the raster horizontally is not equal to the PRI of any signal present, constant PRI signals form diagonal patterns of straight lines at various angles in the raster. The use of the Hough transform is suggested in [8] to detect these straight lines and the angles at which they occur. These angles are related to the radar PRI. Then the pulses making up each linear pattern can be grouped together and hence are deinterleaved. Another way to process such data may be to use the fractional Fourier transform [8].

13.8 Measuring Deinterleaver Performance

There are at least two aspects to deinterleaver performance. One is the processing complexity: how much processing power is needed to apply the deinterleaver to a batch of data? This aspect is often addressed analytically in the references. The second is the accuracy of the deinterleaving results. This aspect is often addressed by showing the accuracy when the deinterleaver is applied to specific data sets. Generally, simulated data must be used because in that way the accuracy of the results can be verified by comparing the known sources of the pulses in a burst formed by the deinterleaver. Consider the effects of jitter on accuracy. Performance could be determined by examining each burst it formed. Then one can call correct those pulses having the most prevalent source in that burst. The erroneous pulses are the rest in that burst. Now a simple ratio of incorrect pulses to correct pulses gives the error rate. Several techniques can be applied to the same data and the technique with the lowest error rate can be determined.

In a fixed gate deinterleaver, such as that described in Section 13.1, the time tolerance is an important parameter entered to start the process. The score is a function of the choice of the time tolerance. Clearly, if the time tolerance is less than the time jitter of the PRI, many pulses will be missed. If the time tolerance is much larger than the jitter, incorrect pulses may be included in the bursts formed. For a given scenario, there will be a minimum error rate for some particular value of the time tolerance. Improvement of performance when additional parameters are added to deinterleaving process can also be explored in this way.

References

[1] Costas, J. B., "Residual Spectrum Analysis—A Search and Destroy Approach to Signal Processing," Technical Information Series No. R80EMH8, Syracuse, NY: General Electric Company, Military Electronic Systems Operation, Box 4840 CSP, 4-18.

[2] Mardia, H. K., "New Techniques for Deinterleaving of Repetitive Sequences," *IEE Proc. F, Communication, Radar, and Signal Processing*, Vol. 136, No. 4, 1989, pp. 149–154.

[3] Milojevec, D. J., and B. M. Popovic, "Improved Algorithm for the Deinterleaving of Radar Pulses," *IEE Proc. F, Communications, Radar, and Signal Processing*, Vol. 139, No. 1, February 1992, pp. 98–104.

[4] Nishiguchi, K., and M. Korbyashi, "Improved Algorithm for Estimating Pulse Repetition Intervals," *IEEE Trans. on Aerospace and Electronic Systems*, Vol. 36, No. 2, April 2000.

[5] Frankpitt, B., J. Baras, and A. Tse, "A New Approach to Deinterleaving for Radar Intercept Receivers," *Proc. of SPIE*, Vol. 5077, 2003, pp. 175–186.

[6] Orsi, R., J. Moore, and R. Mahony, "Interleaved Pulse Train Spectrum Estimation," *Intl. Symp. on Signal Processing and Its Applications, ISSPA*, Gold Coast, Australia, August 25–30, 1996.

[7] Perkins, J., and I. Coat, "Pulse Train Deinterleaving Via the Hough Transform," *Proc. IEEE 1994 Intl. Conf. on Acoustics, Speech, and Signal Processing*, Vol. 3, April 1994, pp. III-197–III-200.

[8] Ray, P. S., "Defining Signal Descriptors by Fractional Fourier Transform," *5th Intl. Symp. on Signal Processing and Its Applications ISSPA '99*, Brisbane Australia, August 22–25, 1999, pp. 247–249.

CHAPTER 14

Measurement and Analysis of Carrier Frequency

14.1 Pulsed Signal Carrier Frequency

The determination of the carrier frequency of a particular pulsed signal is largely a function of the receiving system. The typical manual receiver has a calibrated frequency display and a tuning aid (e.g., a spectrum analyzer display or discriminator), which are used to center the signal in the IF bandpass of the receiver. The typical automatic system searches the band with certain center frequencies and receiver bandwidths specified in advance. In that case, the signal may not be centered in the receiver bandwidth and the tuning steps may be small enough so that if the signal is near the band edge on one frequency setting, it will be near the center of the band on the next setting. Then the processing must recognize the fact that the same signal (with distortion) has been received on adjacent tuning positions. The receiver reports the center frequency indicated by the local oscillator setting. The actual RF for a signal requires determining the offset from the center of the band using frequency discrimination techniques. A manual process may make use of a signal generator and frequency counter. The operator attempts to match the frequency of the signal generator to that of the incoming signal. This procedure can give rise to reported frequency values with many significant digits (e.g., 1-Hz resolution). However, the accuracy of the reported value is subject to many variables, including:

1. The accuracy of the available frequency standard (e.g., the counter time base or synthesizer reference oscillator calibration);
2. The available SNR for signal pulses and the ability of the receiver system to integrate a number of pulses;
3. Doppler shifts due to the motion of the emitter or the receiver.

14.1.1 Frequency Measurement Accuracies

Counter or synthesizer-based measurement systems now in common use are generally accurate to less than 1 part in 10^6 to 10^7 per month. At 10 GHz, such a frequency measurement would be accurate to 1 to 10 kHz. (In the past, the calibration of mechanically tuned receivers was accurate to less than 1% or 2% of the band covered, or 10 to 20 MHz at 1 GHz.) Note that accurately determining the exact center of the receiver bandpass is only the first step in measuring the signal frequency. Analysis of predetection data or discriminator outputs is required to find the true frequency.

Excellent measurements of pulsed signal frequencies (if the SNR is sufficient) are achieved by accumulating pulses until the total duration of the pulses is equal to a selected time. For example, if 1-μs pulses occur at a PRI of 1,000 μs and the time selected is 1 ms, it will take 1,000 pulses to fill the 1-ms time requirement, this will take 1 second of real time if every radar pulse is received. The accuracy of RF measurements made by counting the total number of cycles in the accumulated pulses over the selected time is approximately:

$$\sigma_{\text{pul}\sin g} = \frac{1}{\text{PD}\sqrt{\text{NP}}} \tag{14.1}$$

where:

PD = pulse duration

NP = number of pulses or selected time per PD

For this example,

$$\sigma_{\text{pulse}} = \frac{1}{10^{-6}\sqrt{10^3}} = 3.16 \times 10^4 \text{ Hz} \tag{14.2}$$

For a scanning radar at 12 seconds per revolution and 40 pulses per beamwidth, accumulating 1,000 pulses requires 25 scans or 5 minutes of data. Of course, if such accuracy is not needed, a shorter time may be selected. Often, the effective pulse duration must be reduced by 30 to 40 ns from its observed value to account for cycles lost at the leading and trailing edge. This becomes an important consideration for short pulses.

Achieving this accuracy also requires a reasonably good SNR so that correct cycle counting can occur. The number of independent noise samples per second is approximately twice the receiver bandwidth. The time the receiver is susceptible to noise triggering is the time required to accumulate the number of pulses needed. If the probability that each pulse triggers the counter is P_D and the probability that a single noise sample triggers the counter is P_{FA}, then the probability of detecting that number NP of consecutive pulses is

$$P_{D,\text{NP}} = (P_D)^{\text{NP}} \tag{14.3}$$

Note that this requires constant signal amplitude over NP pulses.

Likewise, the probability of detecting no false triggers in the time required for NP pulses is

$$P_{\text{FA},BT} = 1 - (1 - P_{\text{FA}})^{BT} \tag{14.4}$$

where B is the noise bandwidth and T is the time over which the pulses occur.

For example, the required single pulse SNR can be estimated from P_D and P_{FA} for NP = 1,000, B = 20 MHz, T = 1 second, $P_{P,D}$ = 0.9, and $P_{FA,BT}$ = 10^{-4}; then,

$$P_D = 0.99989$$

and

$$P_{FA} = 5(10^{-12})$$

Then using tabulated functions for P_D, P_{FA}, and SNR, it is clear that the required single pulse SNR is about 18 dB for this example.

14.1.2   Doppler Shifts

All measured frequency values are affected slightly by the relative motion between the emitter and the intercept receiver. The Doppler shift in the RF is given by

$$f_D = \frac{V_R}{c} f_C \qquad (14.5)$$

where:

  f_e = carrier frequency

  V_R = radial velocity (positive for the movement of the transmitter and receiver toward each other; negative if they are moving apart)

  c = speed of light

  f_D = Doppler shift

At 300-m/s radial velocity, the Doppler shift is about 1 part in 10^6 or 1,000 Hz/GHz. Obviously, this is not important for nominal RF carrier frequency reporting purposes. However, it becomes an important parameter in keeping a reference oscillator phase locked to the receiver signal. This is done when using the differential Doppler shift (or FDOA) between receivers at different locations and different motions to help determine flight paths and trajectories of moving emitters, as discussed in Chapter 6. Doppler shifts also limit efforts to measure the coherence or short-term stability of signals.

14.1.3   Drift Measurement

Any RF measurements must be made over some time period. Frequency changes that occur over time periods that are considerably longer than the measurement time are considered drift. The analyst may infer drift from reported RF values. For example, if an RF value is reported for the same emitter on successive hours, the drift may be inferred as the difference in RF values divided by the elapsed time between the intercepts. Generally, drifts over time periods of minutes to hours are of most interest in helping to determine the transmitter type.

Notice that very small drifts over very short intervals may be impossible to measure. Determining a small frequency change requires very accurate frequency measurements. For a pulsed signal from a scanning emitter, measurements to accuracies of tens of kilohertz may require several minutes of data. Obviously, a drift of a few kilohertz per minute could not be measured unless it continued for a long time in the same direction.

Finally, it may not be possible to distinguish small drifts from Doppler shifts without information about the transmitter's motion.

14.1.4 FM Ranging in Radar

Intentionally changing the frequency slightly from pulse to pulse is a common technique used in high PRF radar to determine range. The frequency shifts used are such that the difference frequency between the transmitted frequency and the received echoes is less than the PRF. The radar determines range by using knowledge of the frequency versus time slope and by measuring this difference frequency. The FM ranging computation must be competed during the time the radar beam looks at the target—generally a few milliseconds. This type of pulse-to-pulse frequency change can be observed by ELINT using the same techniques as for drift and Doppler shift.

Note that the SNR may be too low to allow measurement of the frequency of each and every pulse with sufficient accuracy to determine the characteristics of the FM ranging waveform. It may be necessary to process in a manner similar to the radar by multiplying delayed and undelayed versions of the signal together to obtain the difference frequency and then determining the difference frequency by spectral analysis. The delay used is nominally a multiple of the PRI so that the delayed and undelayed pulses are aligned. Obviously, a delay of a multiple of the PRI plus half of the PRI would not normally result in any overlapping pulses in the product. This is not as good as the processing in the radar, where the undelayed version can be noise free. The ELINT processor must multiply two noisy signals together.

14.2   Intrapulse Frequency or Phase Modulation

As discussed in Chapter 11, intentional frequency modulation on pulse can indicate the pulse-compression capabilities of that type of radar. Unintentional FMOP can indicate the type of radar transmitter in use. Wideband radar signals can be demodulated at the collection site or, if high-speed digitization is used, demodulation may occur during subsequent analysis.

Historical Note.   In the past, the analyst obtained this type of data in the form of wideband tape recordings of the output of a discriminator at the intercept site. Signals that did not have such wide bandwidths were recorded using predetection techniques. These two kinds of recordings are quite different. If the intercept site recorded the output of a frequency discriminator, the analyst must deal with a postdetection recording. In this case, at least three tracks were generally recorded:

1. Track 1, with an AM detector or video signal;
2. Track 2, with a precision time reference tone;
3. Track 3, with a discriminator output (FM recording process required).

Ideally, the tape included a calibration signal for both the AM detector (e.g., the amplitude versus the intercept signal level in 10-dB steps) and the FM detector (e.g., the discriminator output versus the frequency shift in convenient frequency steps). To preserve the precise amplitude of the discriminator output, an FM tape recording process was used if sufficient FM recording bandwidth was available.

The AM track was necessary to determine the times at which the discriminator output is valid. The calibration run indicated the time delay between the AM signal and the discriminator output. Because a limiter-discriminator has an output even if there is no signal, the AM video signal was used to gate the FM video signal prior to analysis. This was often done visually using a dual-trace oscilloscope or an actual gating device was used. The AM signal was also used to control the oscilloscope triggering and the z-axis intensification input of the oscilloscope.

For determining pulse-to-pulse variation such as drift, the average discriminator amplitude for each pulse was traced on the oscilloscope in much the same way an antenna scan pattern was made. The discriminator output was observed using a very slow sweep speed and the AM output was used for z-axis intensification.

As noted in Chapter 11 and in Appendix B, the ability of a frequency discriminator to indicate an instantaneous frequency is a function of the RF band covered, the output FM video bandwidth, and the available SNR. The standard deviation of a frequency discriminator's noise output (measured in hertz) is (B.50):

$$\sigma_f \, (\text{rms}) = B_V \sqrt{\frac{1}{3\,\text{SNR}} \frac{B_V}{B_R}} \qquad (14.6)$$

where:

σ_F = standard deviation of the discriminator output (Hz) = discriminator
   rms frequency noise, f_n

B_V = discriminator video output bandwidth (Hz) = maximum video
   frequency, f_V

B_R = discriminator input RF bandwidth (Hz) = input noise bandwidth, B_n

SNR = available input signal-to-noise ratio

As an example, consider a 40-MHz discriminator input bandwidth and a 5-MHz video or recording bandwidth. At an SNR of 20 dB (100 to 1), the discriminator noise will have a standard deviation of at least 105 kHz (see Appendix B). For determining FMOP parameters, the same limitations apply. Intentional FMOP would have a large frequency deviation and the limitation due to noise would be less important. The extent of the frequency variation on the pulse is usually a good indicator of the bandwidth of the RF spectrum. Some care must be exercised in making this bandwidth estimate because the bandwidth of an FM signal also depends on the waveform of the modulating signal. For instance, if a carrier is frequency modulated by a sine wave, the bandwidth of the spectrum is approximately the peak-to-peak frequency deviation or twice the frequency of the modulating sine wave, whichever is larger. (Also see the discussion of Carson's rule in Appendix B.) For FMOP used for pulse compression, a linear FM or some similar waveform is used. In this case, the linear modulating function has a bandwidth comparable to the reciprocal of the pulse duration; however, the bandwidth of the resulting linear FM will be much larger—tens or hundreds of times the reciprocal of the pulse duration. Achieving a large time bandwidth product requires that the frequency deviation be the dominant factor in determining the overall width of the radar spectrum. Woodward's theorem [1] then requires that the RF spectrum shape be approximately the probability density function of the modulating waveform. For wideband linear FM, the RF spectrum is approximately uniform between the limits set by the peak deviation. For FMOP, the analyst tries to determine the shape of the modulating function and the extent of the FM.

Phase reversal signals or other phase modulation on pulse waveforms can also be analyzed using the output of a frequency discriminator. The instantaneous frequency of a signal is the derivative (with respect to time) of the instantaneous phase. Therefore, a phase reversal type of signal will produce a series of positive and negative impulses at the output of a frequency discriminator. The impulses will represent the impulse response of the frequency discriminator and their duration will be approximately the reciprocal of the video bandwidth. The chip rate can be determined by noting the separation of those impulses closest together in time. The 3-dB bandwidth of a BPSK-coded signal is approximately equal to the chip rate. This is illustrated in Figure 14.1. If the phase shifts are other than 180°, the amplitudes of the impulses may show this. A quadraphase signal should show four discrete, equally separated impulse amplitudes.

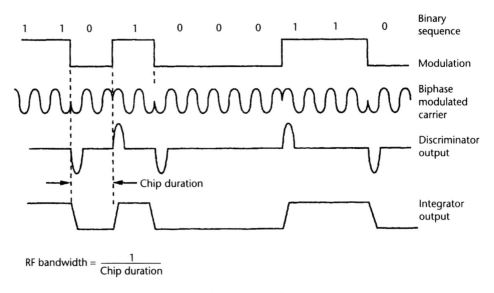

Figure 14.1 Discriminator response to a biphase modulation signal.

An integrator may be beneficial in analyzing the output of a frequency discriminator when dealing with phase-modulated signals. The integrator should be reset after the end of each pulse.

Historical Note.   Analog predetection recordings required heterodyning the signal down to a low intermediate frequency so that it could be recorded directly onto tape. This was done using a single tape channel with the IF centered in the tape recorder bandpass. For example, if the tape recorder could handle a 5-MHz bandwidth, the selected IF would be 2.5 MHz and the signal band could extend ±2.5 MHz on each side of the IF. Practical limitations prevented the use of the full bandwidth of the tape recorder. For a tape recorder bandwidth of B, a good rule of thumb was that the center frequency of the predetection recording should be at $0.6B$ instead of $0.5B$ and that the signal band should extend from $0.2B$ to B for a total bandwidth of $0.8B$ instead of B. The tapes had the precise center frequency used in making the recording marked on the accompanying log sheet. It was generally thought that about five cycles or more of the IF should be available from each pulse for predetection recording. This meant, for example, that for a 5-μs pulse, at least a 1-MHz predetection center frequency should be used. Another predetection recording process was to use a zero frequency IF and provide two channels of recording: one for the in-phase (I) component and the other for the quadrature (Q) component. These processes are shown in Figure 14.2.

The single-channel approach is simpler compared to the I and Q approach, but the total bandwidth required of the single channel is twice that required for

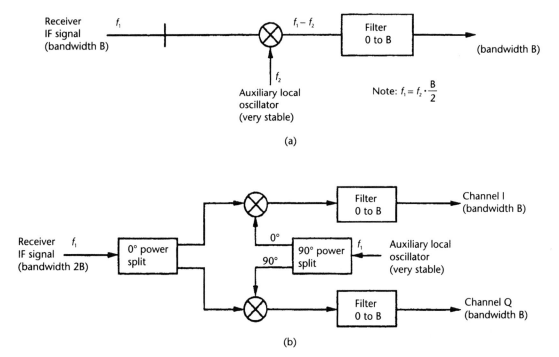

Figure 14.2   Predetection processing: (a) low IF process, and (b) zero-IF I and Q process.

each of the I and Q channels. Both processes can make use of digital recording. An A/D conversion rate of at least $2B$ is required for each channel. For pulsed signals, a high-speed A/D conversion can be followed by a low-speed digital recording as long as the average digital recording rate is in excess of the peak A/D conversion rate of all channels added together and then multiplied by the duty factor (PD/PRI) of the pulsed signal.

14.2.1 Analysis of Predetection Data

If single-channel predetection recording data is available, it must be demodulated to proceed with the analysis. An envelope detector is used to produce the AM video pulses. This AM detection can be done using a conventional envelope detector and a lowpass filter or by squaring the digital samples[1] and then digitally smoothing them using a lowpass filter. The lowpass filter must have a cutoff frequency below $B/2$.

Frequency demodulation of the channel data can be accomplished using any of several frequency discriminator approaches. One good discriminator makes use of the zero crossings of the signal. A standard amplitude and duration pulse is generated at each zero crossing and then this pulse train is lowpass filtered at a frequency of less than $B/2$. The result is a voltage proportional to the frequency. Accurate demodulation over the full band of zero to B is possible using the zero crossings. The time between the zero crossings is carefully measured. The reciprocal of twice this time represents the average frequency during that interval. This stair-case-like waveform is often a sufficiently accurate demodulation for most purposes. If large changes in the average frequency occur from one pair of zero crossings to the next, reconstructing the true instantaneous frequency may require more processing using an iteration [2]. Another FM demodulation approach is to convert the single-channel data to I and Q form using the process shown in Figure 14.2(b), with $f_1 = B/2$ and the filter cutoff frequency equal to $B/2$.

If two-channel I and Q data is available, the AM and FM can be found using the techniques discussed in Chapter 12.

Note that the instantaneous phase can be observed on an x-y display on which the I signal drives the x-axis and the Q signal drives the y-axis. The angle of the displayed spot at each instant represents the instantaneous phase of the signal.

14.3 Coherence (Short-Term RF Stability)

The term "coherence" implies an ability to predict the future waveform of a signal based on a past observation of it. In other words, if the RF and phase of a carrier frequency can be determined during one period of time, an ideal coherent waveform would be completely predictable for all times in the future. In practice, one cannot

---

1. Care must be exercised in performing nonlinear processing on digitized data to avoid aliasing. Squaring the samples will produce frequency components at twice the frequencies present before the squaring operation. If these components exceed the Nyquist rate, they will be aliased into the band of interest. The sampling rate must be increased prior to the nonlinear operation and can be decreased after filtering has ensured that the Nyquist criterion has once again been met at the original sampling rate.

determine the frequency and phase with complete accuracy. In addition, these quantities vary with time and ambient conditions. Therefore, complete coherence is an abstraction that never exists in practice. Curry [3] was one of the first to suggest a quantitative (and elegant) measure for short-term stability.

Coherence is a very important property of some radar signals. This is true for radars designed to measure or to discriminate among targets on the basis of target velocity. The reflected signal (as Doppler shifted by the target's motion) must be analyzed by the radar. If the transmitted signal is not coherent (stable), its spectrum will have sidebands of energy near the carrier frequency. These sidebands, when reflected by large stationary targets can produce strong returns that appear in the same part of the frequency spectrum as that of the Doppler-shifted target echo signal.

As an example, consider a radar observing an aircraft flying near a mountain. This mountain has a huge radar cross-section and reflects a large signal back to the receiver. The aircraft has a small radar cross-section, but its radial motion produces a frequency shift of the reflected signal. If the transmitted spectrum is down 40 dB (at the frequency separated from the carrier that is equal to the Doppler shift), then the smallest radar cross-section moving target that can be detected by that radar will be one whose radar cross section is 1/10,000 that of the mountain range. In practice, it is larger than this by the required output SNR.

Coherence is another way of saying that the signal frequency must be like an ideal sinusoid with a constant frequency (or linear phase). Therefore, to determine the degree of coherence, one must determine the degree to which the signal frequency fluctuates over the time periods of importance to the operation of the system transmitting the signal. For a radar, this time is that for which the target echoes are integrated coherently—typically no more than the time that the beam dwells on (or illuminates) a single target. Because these times are on the order of milliseconds, the frequency stability to be measured is called short-term stability. (Long-term stability could extend over days, months, or years.) Those using frequency stability measures realized that there were two kinds of measurements needed for different applications and situations. On the one hand, frequency stability could be determined by examining the narrowness of the Fourier frequency spectrum and its sidebands. On the other hand, it could be expressed in terms of the variation of the instantaneous frequency as a function of time. Note that the word "frequency" is used in two different ways: one is the independent variable in the frequency domain, which is independent of time (the Fourier frequency); and the other is the time-dependent frequency of the oscillator whose stability is to be measured. Since the 1960s, these stability measures have been developed in theory and applied in practice [4].

These two frequency stability measures are referred to as frequency-domain measures and time-domain measures. Attempts were made to use the variance of the instantaneous frequency as the time-domain measure of stability; however, for real oscillators, as the measurement time decreases, the estimate of the variance increases without a limit. Therefore, it was necessary to introduce measurements that are based on the changes in the average of the instantaneous frequency over some averaging time.

One of the earliest ways to measure frequency stability was spectrum analysis. To see how the frequency spectrum relates to the fractional frequency stability, note that (from FM theory) for a small sinusoidal frequency modulation, the ratio of the carrier level to the sidelobe level is related to the index of modulation by

$$\beta = \frac{\Delta f}{f_m} = \frac{V_{SB}}{V_C} \tag{14.7}$$

where:

β = index of modulation

Δf = peak frequency deviation

f_m = modulation frequency

V_{SB} = spectrum sidelobe level (V)

V_C = carrier level in spectrum (V)

If the fractional frequency variation is of interest, one can obtain a high-resolution spectrum of the signal and measure the sidelobe to carrier ratio at a given frequency f_m away from the carrier. Then the fractional frequency stability due to disturbances in a bandwidth equal to the spectrum analyzer resolution at a frequency f_m away from the carrier is

$$\frac{\Delta f}{f_c} = \frac{V_{SB}}{V_c} \frac{f_m}{f_c} \tag{14.8}$$

The fractional stability is used because it is independent of frequency multiplication or division.

For example, if the spectrum level of a sideband 400 Hz away from the carrier is 40 dB below that of a 4-GHz carrier, then the fractional frequency stability is

$$\frac{\Delta f}{f_c} = \frac{0.01}{1} \frac{400}{4 \cdot 10^9} = 10^{-8} \tag{14.9}$$

There are several problems when trying to apply this to pulsed signals. Because of the short duration of the pulses (typically 1 to 10 μs), the spectrum is rather wide (the order of megahertz). The separation of the PRF lines is typically less than 1 kHz. Thus, there may be several hundred spectral lines in the vicinity of the carrier line and all of them have nearly the same amplitude. Spectral lines other than the carrier line are affected by both PRI jitter and RF instability; therefore, it is necessary to locate the carrier line if there is significant PRI jitter. The portion of the spectrum of interest extends on each side of the carrier line out to one-half of the PRF. RF disturbances that occur more rapidly than one-half of the PRF cannot be distinguished from slower ones due to aliasing. This is caused by the fact that the RF carrier is, in effect, sampled at the rate of the PRF.

For a pulsed signal, disturbances of interest all have modulating frequencies in the range $0 < f_m < \text{PRF}/2$. Regardless of what measurement processes are used, disturbances that occur more rapidly will be identified as disturbances at rates less than one-half of the PRF.

This discussion of spectral analysis shows that the character of the power spectrum is related to the time-varying frequency of the sinusoid. In the discussion that follows, this relationship is explored in detail for three time-domain measures of frequency stability:

1. The rms differences of the phase values measured at intervals τ apart;
2. The rms differences of the frequency averaged over intervals of duration D spaced τ apart[2];
3. The average fraction of the signal remaining when one segment is subtracted from another segment separated by an interval τ.[3]

While these measures offer different views of the frequency stability, they are related mathematically to each other and to the frequency-domain measure, which is the spectral density of the fractional frequency variations. Note that the desired spectrum is that of the variations in the fractional frequency. This is not the same as the power spectrum, which also includes the effects of changes in the signal amplitude. If the variations in signal amplitude are negligible or somehow eliminated, then the power spectrum and the spectrum of the frequency variations are related.

Although the recommended stability measures are the Allan variance in the time domain and the spectrum of the fractional frequency variations in the frequency domain, other measures (as previously listed) plus the frequency-domain measure of the carrier to the sideband ratio, are sometimes used in ELINT. This is the reason for including them here.

In the discussion that follows, the relationships between the measures are obtained by using the autocorrelation function of the instantaneous phase. This is a mathematical convenience and is only a means to allow the relationships to be derived. It is not necessary to actually measure the autocorrelation of the phase.

The following important symbols are used:

$\phi(t)$ = instantaneous phase variation in excess of the ideal phase, which is 2π times the carrier frequency times the elapsed time

$R_\phi(\tau)$ = autocorrelation of the phase variation $\phi(\tau)$ at a lag of τ

σ_x^2 = variance of the subscripted variable

$S_x(f)$ = spectral density of the subscripted variable

f = Fourier frequency

2. When $D = \tau$, this is the Allan variance, named for D. W. Allan of the National Bureau of Standards [5]. This is the recommended time-domain measure.
3. The subtraction of one pulse from another and comparing the residue energy to the energy of one pulse has been called the *cancellation ratio*. This ELINT usage is not the same as the use of this term in MTI radar.

For example,

$$\sigma_\phi^2 = \text{variance of } \phi(t)$$

$$\sigma_{\Delta\phi}^2(\tau) = \text{variance of } \phi(t) - \phi(t - \tau) = \Delta\phi$$

$$\sigma_{\Delta^2\phi}^2(\tau) = \text{variance of } \Delta\phi(t) - \Delta\phi(t - \tau)$$

$\sigma_{\Delta f}^2(D, \tau_0) = $ variance of the difference between the average frequencies measured for a duration D and separated by T_o

$\sigma_y^2(\tau_0) = $ Allan variance $= \sigma_{\Delta f}^2(\tau_0, \tau_0)$ divided by the average or carrier frequency.

14.3.1   RMS Phase Fluctuation

If a measurement of a sinusoid phase at specific times is made, measures of the stability of the sinusoid can be derived. For example, the differences in the phases at times separated by T can be measured and the variance of these differences can be used as a stability measurement.

Let the total phase of the sinusoid be represented by

$$\theta(t) = \omega t + \phi(t) \tag{14.10}$$

where ω is 2π times the nominal carrier frequency and $\phi(t)$ is the variation of the phase.

The autocorrelation of the phase disturbance $\phi(t)$ is given by

$$R_\phi(\tau) \equiv \int_{-\infty}^{\infty} \phi(t)\,\phi(t - \tau)\,dt \tag{14.11}$$

Note that

$$R_\phi(0) \equiv \int_{-\infty}^{\infty} \phi^2(t)\,dt = \sigma_\phi^2 \tag{14.12}$$

Useful stability measures can be obtained by sampling the phase variation periodically. Suppose the phase is sampled at particular times separated by τ_0. Then sequence of values obtained is

$$\phi(0), \phi(\tau_0), \phi(2\tau_0), \ldots, \phi(n\tau_0) \tag{14.13}$$

The first differences of this sequence are

$$\phi(\tau_0) - \phi(0), \phi(2\tau_0) - \phi(\tau_0), \ldots, \phi(n\tau_0) - \phi[(n - 1)\tau_0], \ldots \tag{14.14}$$

The variance of this sequence expressed in terms of the autocorrelation function is

$$\sigma^2_{\Delta\phi}(\tau_0) = \lim_{N\to\infty} \frac{1}{N} \sum_{n=1}^{N} \{\phi(n\tau_0) - \phi[(n-1)\tau_0]\}^2$$

$$= \lim_{N\to\infty} \frac{1}{N} \sum_{n=1}^{N} \{\phi^2(n\tau_0) - 2\phi(n\tau_0)\phi[(n-1)\tau_0 \; \phi^2[(n-1)\tau_0]\} \quad (14.15)$$

$$= 2R_\phi(0) - 2R_\phi(\tau_0)$$

Similarly, higher order differences can be computed. The variance of the second phase differences is obtained by taking the differences of the first phase differences and averaging their squares:

$$\sigma^2_{\Delta^2\phi}(\tau_0) = \lim_{N\to\infty} \frac{1}{N} \sum_{n=1}^{N} (\{\phi[(n+1)\tau_0] - \phi(n\tau_0)\} - \{\phi(n\tau_0) - \phi[(n-1)\tau_0]\})^2$$

$$(14.16)$$

In terms of the autocorrelation function, this works out to

$$\sigma^2_{\Delta^2\phi}(\tau_0) = 6R_\phi(0) - 8R_\phi(\tau_0) + 2R_\phi(2\tau_0) \quad (14.17)$$

These measurements can be performed on both pulsed and CW signals. In the case of the pulsed signal, the time separating the phase measurements must be a multiple of the PRI. For a pulsed signal, the first phase differences are a measure of the pulse-to-pulse phase variation. For such a signal, one would be interested in the phase change over the pulses separated in time by more than one PRI because, in a coherent radar, coherent integration would take place over a number of pulses. Therefore, one would measure not only $\sigma^2_{\Delta\phi}$ (PRI), but also $\sigma^2_{\Delta\phi}$ ($K \times$ PRI) where K is an integer up to the order of the number of pulses to be coherently integrated. Similarly, for a CW radar signal, one would want to know the phase variation over time intervals that are comparable to the reciprocal of the bandwidth of the Doppler filter being used for velocity measurements.

14.3.2 RMS Frequency Fluctuations

If the average frequency of a signal is measured over a time duration D, this is equivalent to measuring the phase change during that time and dividing by the time (see Figure 14.3).

Suppose the average frequency is measured during the two pulses of duration D and the frequency change over the interval τ_0 is obtained by subtracting

$$\Delta f = \frac{\phi(\tau_0 + D) - \phi(\tau_0)}{2\pi D} - \frac{\phi(D) - \phi(0)}{2\pi D} \quad (14.18)$$

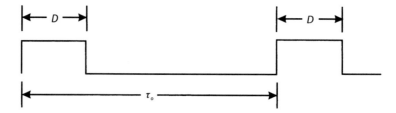

Figure 14.3 Measurement time and time between measurements.

The variance of this frequency difference is of interest because it can be related to the Allan variance as follows.

The variance of this frequency difference can be related to the autocorrelation of the phase disturbance to obtain

$$2(2\pi D)^2 \, \sigma_{\Delta f}^2(\tau_0, D) = 4R_\phi(0) - 4R_\phi(D) + 2R_\phi(\tau_0 - D) \qquad (14.19)$$
$$+ 4R_\phi(\tau_0) + 2R_\phi(\tau_0 + D)$$

In this expression, D is the measurement or averaging time for the individual frequency measurements (in Figure 14.3, it is the pulse duration) and τ_0 is the time between the frequency measurements.

For pulsed signals, if a single phase value is measured for each pulse, the average frequency over the entire pulse interval can be computed. In this case, $D = \tau_0$ and using (14.19) gives

$$2(2\pi\tau_0)^2 \, \sigma_{\Delta f}^2(\tau_0, \tau_0) = 6R_\phi(0) - 8R_\phi(\tau_0) + 2R_\phi(2\pi_0) = \sigma_{\Delta^2\phi}^2(\tau_0)$$
$$(14.20)$$

In other words, this variance of the frequency differences is equal to the variance of the second phase differences divided by $2(2\pi\tau_0)^2$. The frequency variance for the case where the measurement interval is equal to the interval between the measurements is called the Allan variance when it is expressed as a fraction of the average (or carrier) frequency. The Allan variance is denoted in the literature by $\sigma_y^2(\tau)$ and is related to the variance of the second phase difference by

$$\sigma_y^2(\tau) = \frac{\sigma_{\Delta^2\phi}^2(\tau)}{2(2\pi\tau f_c)^2} \qquad (14.21)$$

where:

$\sigma_\phi^2 \tau =$ Allan variance

$\tau =$ frequency measurement time and time between measurements

$f_c =$ average or carrier frequency

$\sigma_{\Delta^2\phi}^2(\tau) =$ variance of second phase differences

For stable pulsed signals, the Allan variance can be measured for τ equal to a multiple of the PRI if the phase can be sampled once per pulse. The rms fractional frequency variation can also be measured and reported when the measurement time is less than the time between measurements. This is of value when the instabilities are too great to allow meaningful phase difference measurements from pulse to pulse. In this case, a single-pulse frequency measurement can be made (by means of a discriminator, a counter, spectral analysis, or other techniques) and the variance of these frequency differences can be computed.

Consider the case where the phase $\phi(t)$ from one pulse to the next is a uniformly distributed random variable over the range $(-\pi, \pi)$. Then the variance of the phase measurements is given by

$$\sigma_\phi^2 = \frac{1}{2\pi} \int_{-\pi}^{\pi} \phi^2 \, d\phi = \frac{\pi^2}{3} \tag{14.22}$$

$$\sigma_\phi = \frac{\pi}{\sqrt{3}} = 103.9 \text{ deg} = \sqrt{R_\phi(0)} \tag{14.23}$$

Therefore, if the standard deviation of the phase measurements approaches 100°, the phase change from pulse to pulse is essentially random. In this case, the standard deviation of the first phase differences would be about 147°, while that of the second phase differences would be about 254°.

$$\sigma_\phi = \sqrt{R_\phi(0)} = \frac{\pi}{\sqrt{3}} \tag{14.24}$$

$$\sigma_{\Delta\phi} = \sqrt{2R_\phi(0)} = \frac{\sqrt{2}\,\pi}{\sqrt{3}} \tag{14.25}$$

$$\sigma_{\Delta^2\phi} = \sqrt{6R_\phi(0)} = \sqrt{2}\,\pi \tag{14.26}$$

In other words, this technique to obtain the Allan variance through phase differences is valid only for Allan variance values less than

$$\sigma_y(\tau)_{\max} = \frac{1}{\sqrt{2}\,\tau f_c} \tag{14.27}$$

For example, a 300-MHz signal with a 3,333.33-μs PRI would be required to have a square root of the Allan variance no larger than

$$\sigma_y(\tau)_{\max} = \frac{1}{\sqrt{2} \cdot 10^6} = 0.707 \cdot 10^{-6} = 0.707 \text{ ppm} \tag{14.28}$$

to make use of the phase measurement technique.

14.3.3 Signal Repeatability

Another method of measuring signal stability is to take a waveform section and overlay it on top of another waveform section and note the differences. This can be called *signal repeatability measurement*. The smaller the difference between the two waveform samples, the more repeatable it is. Again, this can be done with pulsed CW signals, but it is most appropriate for a pulsed signal. In this situation, a pulse is overlaid with a nearby pulse and the rms difference is measured. Indeed, the procedure would be to adjust the amplitude and phase of one of the pulses to minimize their difference. This is the procedure used in Chapter 11 to compare waveforms, which was defined there as the normalized Euclidean distance (NED). In Chapter 11, the application was to compare the demodulated envelopes of the AM and FM waveforms of signals received at different locations or times. Here the NED is used to compare the predetection waveforms of successive pulses from the same radar.

Consider the NED defined as

$$\text{NED} = \frac{\displaystyle\int_0^D [s_1(t) - s_2(t)]^2 \, dt}{\displaystyle\int_0^D s_1^2(t) \, dt + \int_0^D s_2^2(t) \, dt} \tag{14.29}$$

where

NED = normalized Euclidean distance

$s_1(t)$ = signal 1

$s_2(t)$ = signal 2

D = pulse duration

Usually, for coherence measurements,

$s_1(t)$ = one pulse of a radar signal

$s_2(t)$ = next pulse of a radar signal

Then $s_2(t) = s_1(t - \tau_0)$ where τ_0 is the separation of the pulses in time and $s_1(t)$ is assumed to be the radar's output signal for all time. Rewriting (14.29) this way gives

$$\text{NED} = \frac{\displaystyle\int_0^D [s_1(t) - s_1(t - \tau_0)]^2 \, dt}{\displaystyle\int_0^D s_1^2(t) \, dt + \int_0^D s_1^2(t - \tau_0) \, dt} \tag{14.30}$$

The denominator is simply the total energy of the two waveform samples (separated in time) that are being compared in the numerator. Usually, the denominator will be a constant and serve as a normalizing factor. The numerator is of more interest. Expanding it gives

$$\int_0^D s_1^2(t)\,dt + \int_0^D s_1^2(t - \tau_0)\,dt - 2\int_0^D s_1^2(t)s_1(t - \tau_0)\,dt \qquad (14.31)$$

so that the NED can be written as

$$\mathrm{NED}(\tau_0) = 1 - \frac{2\displaystyle\int_0^D s_1(t)s_1(t - \tau_0)\,dt}{\displaystyle\int_0^D s_1^2(t)\,dt \displaystyle\int_0^D s_1^2(t - \tau_0)\,dt} \qquad (14.32)$$

The numerator is the correlation coefficient at $\tau = \tau_0$. The denominator is actually the total energy in the signal over the interval from zero to $\tau_0 + D$, as is shown in Figure 14.4.

Suppose

$$s_1 = \sin(\omega t + \phi_1) \qquad (14.33)$$
$$s_2 = \sin(\omega t + \phi_2)$$

Then the NED becomes

$$\mathrm{NED}(\tau_0) = 1 - \cos(\phi_2 - \phi_1) \qquad (14.34)$$

Assume $\phi_2 - \phi_1$ is a small angle (i.e., that there is not much pulse-to-pulse phase change); then the NED is approximately

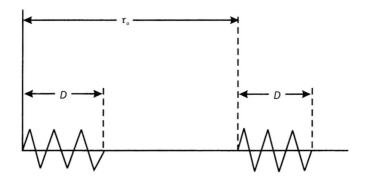

Figure 14.4 Timing relationships.

$$\text{NED}(\tau_0) \cong \frac{(\phi_2 - \phi_1)^2}{2} \tag{14.35}$$

The expected value of the NED can then be related to the autocorrelation function of the phase. Let $\phi_1 = \phi(t)$ and $\phi_2 = \phi(t + \tau_0)$. Then,

$$\text{NED}(\tau_0) \cong R_\phi(0) - R_\phi(\tau_0) = \frac{1}{2}\sigma_{\Delta\phi}^2 \tag{14.36}$$

All of these time-domain stability measures are now seen to be related. For example, the Allan variance can be expressed in terms of the NED ratio as

$$\sigma_y^2(\tau_0) = \frac{\langle 4\text{NED}(\tau_0)\rangle - \langle\text{NED}(2\tau_0)\rangle}{(2\pi f_c \tau_0)^2} \tag{14.37}$$

To derive this, note that

$$\sigma_{\Delta^2\phi}^2 = 4\sigma_{\Delta\phi}^2(\tau_0) - \sigma_{\Delta\phi}^2(2\tau_0) \tag{14.38}$$

These relationships are summarized in Table 14.1.

14.3.4 Effects of Variations in τ_0

When making these measurements, the value of τ_0 may fluctuate. As long as this fluctuation is small (10% or so), the effect will be to obtain the average of the stability measure over the range of the measurement separation times represented by the fluctuation of τ_0. Because, typically, the stability measure changes roughly in proportion to the measurement separation time, small variations in τ_0 are not

Table 14.1   Relationship Among Three Time-Domain Frequency Stability Measures

	Allan Variance	NED	RMS Phase Difference
In terms of Allan variance	—	$\langle\text{NED}(\tau_0)\rangle = (1/2)(2\pi\tau_c f_c)^2$	$\sigma_{\Delta\phi}^2(\tau_0) = (2\pi\tau_0 f_c)^2$ $\sum_{n=8}^{\infty}(1/2)^n\sigma_y^2(2^n\tau_0)$
In terms of NED	$\sigma_y^2(\tau_0) = \dfrac{4\langle\text{NED}(\tau_0)\rangle - \langle\text{NED}(2\tau_0)\rangle}{(2\pi f_c \tau_0)^2}$	—	$\sigma_{\Delta\phi}^2 = 2\langle\text{NED}(\tau_0)\rangle$
In terms of rms phase difference	$\sigma_y^2 = \dfrac{\sigma_{\Delta\phi}^2(\tau_0)}{2(2\pi\tau_0 f_c)^2}$ $= \dfrac{2\sigma_{\Delta\phi}^2(\tau_0) - \sigma_{\Delta\phi}^2(2\tau_0)}{2(2\pi\tau_0 f_c)^2}$	$\langle\text{NED}\rangle = (1/2)[\sigma_{\Delta\phi}^2(\tau_0)]$	—

troublesome. Of course, such variation may make the measurement process more difficult to carry out.

14.3.5   Frequency-Domain Stability Measures

The spectral density of the phase fluctuation is given by the Fourier transform of the autocorrelation function of the phase disturbance, or

$$s_\phi(f) = 4 \int_0^\infty R_\phi(\tau) \cos 2\pi f \tau \, d\tau \qquad (14.39)$$

Because the instantaneous frequency fluctuation is the derivative of the instantaneous phase fluctuation divided by 2π, the spectral density of the frequency fluctuation is related to the spectral density of the phase fluctuation by

$$s_f(f) = f^2 s_\phi(f) \qquad (14.40)$$

When divided by the square of the carrier frequency (or average frequency), the spectral density of the fractional frequency variations is obtained:

$$s_y(f) = \left(\frac{f}{f_c}\right)^2 s_\phi(f) \qquad (14.41)$$

Any or all of these spectra may be estimated to specify frequency stability. Barnes et al. [5] have defined $s_y(f)$ as the general frequency-domain measure of stability. It is clearly easy to obtain from $s_\phi(f)$. In practice, the ratio of the single-sideband phase noise level to the carrier level is often measured. Assuming that the phase modulation index is small, this is equal to $s_\phi(f)/2$. The phase noise is usually expressed in decibels below the carrier level per hertz of analysis filter bandwidth (dB/Hz). Typically, oscillator stabilities are specified by giving the maximum levels of the phase noise spectral density at different frequency separations from the carrier, as shown in Table 14.2. This type of frequency-domain stability measure is also valuable and in wide use in the radar community for specifying oscillators to be used in Doppler systems.

When measuring the phase spectral density stability measure for pulsed signals, it is first necessary to use high-quality filtering and down-conversion so that a

Table 14.2 Maximum Levels of Phase Noise Spectral Density at Various Frequency Separations

Offset from Carrier (Hz)	Maximum Phase Noise (dB/Hz)
0.1 to 1	−70
1 to 10	−100
10 to 100	−110
10 and up	−120

single line of the pulsed signal spectrum can be isolated for analysis through CW techniques. If the pulsing function itself is stable, any of the lines near the center of the spectrum will be acceptable. However, if the pulse interval jitters, only the center line of the spectrum will be acceptable. This is because the pulsing instabilities contribute to the spectral width of each line in proportion to their separation from the carrier frequency. The practical problems with measuring the phase noise spectra of pulsed microwave signals are first to mix down to the base band coherently, without any aliasing or spectrum fold-over, and second to locate the center line of the spectrum. Once this is done, the resulting single line can be analyzed in the same way as a CW signal.

The question of how well the frequency stability can be determined from noisy, time-limited, or band-limited data must be answered. For this reason, it is important to specify if the stability estimate is limited by the SNR or the measurement bandwidth. The next section gives these limitations for the Allan variance. Based on the relationships between the various measures, it is clear that the available SNR and bandwidth are important parameters no matter how the stability is measured.

In the general case, the Allan variance is determined by measuring the average of the instantaneous frequency over intervals of T sec and then computing the mean square of the difference of successive readings:

$$\sigma_y^2(\tau) = \left\langle \frac{(\bar{y}_{n+1} - \bar{y}_n)^2}{2} \right\rangle \tag{14.42}$$

where:

$$\bar{y} = \frac{1}{\tau} \int\limits_{n\tau}^{(n+1)\tau} y(t)\, dt$$

$\langle x \rangle$ = infinite time average of x

$y(t)$ = instantaneous fractional frequency

In the usual case, there are limitations in the measurement process due to bandwidth restrictions and thermal noise.

14.3.6 Bandwidth Limitations on Allan Variance Measurements

If the signal is processed through a narrowband filter before the Allan variance estimate is made, the variance cannot be computed over an infinite frequency range. In practice, there is always a restricted measurement bandwidth. The effect of this on the Allan variance estimate depends on the value of τ and on the spectral density, $s_y(f)$, of the fractional frequency variation.

Barnes [5] shows that the Allan variance may also be obtained from $s_y(f)$ through the relation

$$\sigma_y^2(\tau) = 2 \int_0^\infty s_y(f) \left[\frac{\sin^2 \pi \tau f}{(\pi \tau f)^2} - \frac{\sin^2 2 \pi \tau f}{(2 \pi \tau f)^2} \right] df \qquad (14.43)$$

$$= 2 \int_0^\infty s_y(f) \left[\frac{\sin^2 \pi \tau f}{(\pi \tau f)^2} \right] df$$

Note that this relates the Allan variance and all the other time-domain measures in Table 14.1 to the frequency-domain measure of stability. Suppose the fractional frequency variation spectrum is white so that

$$s_y(f) = s_0$$

then the true value of the Allan variance is found to be

$$\sigma_y^2(t) = 2 s_0 \tau$$

However, if the signal from the oscillator is passed through a filter of bandwidth $\pm f_\beta$ from the carrier, then the measured value of $\sigma_y^2(\tau)$ will be less than the true value. Substituting f_β for the upper limit of integration in (14.43) and integrating numerically gives the result shown in Figure 14.5. Here, the vertical scale is the

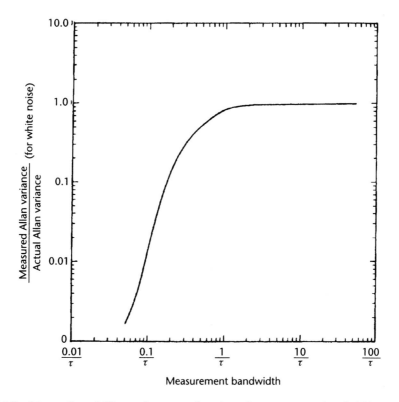

Figure 14.5 Attenuation of Allan variance as a function of measurement bandwidth.

ratio of the estimated Allan variance to the true Allan variance, and the horizontal scale is the measurement bandwidth, f_β, expressed in terms of $1/\tau$. As can be seen, a measurement bandwidth of $2/\tau$ provides better than 90% accuracy.

For short values of τ, it is important to ensure that sufficient measurement bandwidth is used.

14.3.7 Noise Limitations on Allan Variance Measurements

Suppose it is necessary to estimate the Allan variance in the presence of white noise. The rms frequency error, due to thermal noise, in making frequency measurements over a signal of duration τ is [6]

$$\sigma_f = \frac{1}{2\pi\tau\sqrt{\text{SNR}}} \qquad (14.44)$$

The fractional rms error is then

$$\sigma_y(\tau) = \frac{1}{2\pi f\tau\sqrt{\text{SNR}}} \qquad (14.45)$$

where:

SNR = signal-to-noise power ratio

f = frequency being measured

τ = time interval over which measurements are made

Notice that $(f\tau)$ is the number of cycles available within the pulse. Equation (14.45) is easily seen to be applicable to a device that measures the time required for $(f\tau)$ cycles.

If it is assumed that successive average frequency measurements are independent Gaussian-distributed random variables, as would be expected for noise-induced errors, then the Allan variance as given by (14.43) is equivalent to (14.45) because the variance of the difference of two identically independent Gaussian random variables is twice that of one of them. Equation (14.45) is plotted in Figure 14.6 for various SNRs as a function of the product of averaging time and frequency.

Of course, the measurement bandwidth has a direct effect on the available SNR. The SNR is given by

$$\text{SNR} = \frac{s}{kT(\text{NF})f_\beta} \qquad (14.46)$$

where:

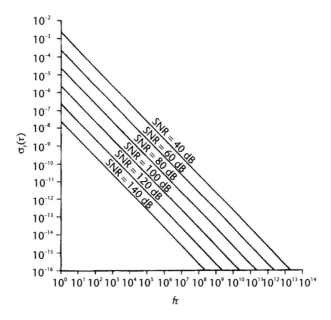

Figure 14.6   Noise floor for $\sigma_y(\tau)$.

s = signal power

k = Boltzmann's constant

T = absolute temperature

(NF) = system noise figure

f_β = measurement bandwidth

The optimum measurement process would use the smallest bandwidth commensurate with the averaging time being used. If this strategy is adopted, then

$$f_\beta \sim \frac{2}{\tau} \tag{14.47}$$

$$\mathrm{SNR} = \frac{s\tau}{2kT(\mathrm{NF})} \tag{14.48}$$

$$\sigma_y(\tau) = \frac{\sqrt{2kT(\mathrm{NF})}}{2\pi f\tau^{3/2}\sqrt{s}} \tag{14.49}$$

14.3.8   Frequency Stability Measures for Power Law Spectra

Figure 14.7 shows the spectral density of frequency fluctuations for five common types of instability observed in practice. These fluctuations are given names according to the exponent in

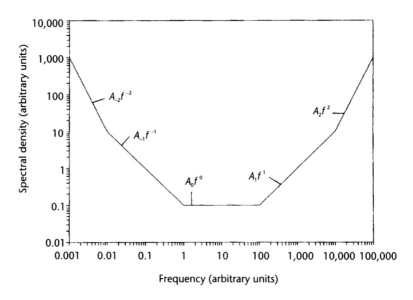

Figure 14.7 Spectral density of common frequency instabilities.

$$s_y(f) = \sum_{n=-2}^{+2} A_n f^n \qquad (14.50)$$

These names and the corresponding theoretical Allan variance are shown in Table 14.3. Here, f_β is the upper bandpass limit of the stability measuring system.

Because the Allan variance (as a function of the averaging time for the white phase noise and flicker phase noise) is virtually identical among these noise types, it cannot be used to distinguish among them. For this reason, a modified Allan variance was defined [7].

Table 14.3 Common Power Law Frequency Instabilities and Their Allan Variances

Noise	Spectral Density $S_y(f)$	Exponent	Allan Variance	
			Rectangular Lowpass	Single Pole Lowpass
White phase	$A_2 f^2$	−2 (−20 dB per decade)	$\dfrac{3 f_\beta A_2}{(2\pi)^2 \tau^2}$	$\dfrac{3 f_\beta A_2}{(2\pi)^2 \tau^2}$
Flicker phase	$A_2 f$	−1 (−10 dB per decade)	$\dfrac{[1.038 + 3\ln(f_\beta\tau)]A_1}{(2\pi)^2 \tau^2}$	$\dfrac{3\ln(f_\beta\tau)A_1}{(2\pi)^2 \tau^2}$
White frequency	A_0	0	$A_0/2\tau$	$A_0/2\tau$
Flicker frequency	$A_{-1}f^{-1}$	(+10 dB per decade)	$2[\ln(2)]A_{-1}$	$2[\ln(2)]A_{-1}$
Random walk frequency	$A_{-2}f^{-2}$	(+20 dB per decade)	$\dfrac{2\pi^2}{3}\tau A_{-2}$	$\dfrac{2\pi^2}{3}\tau A_{-2}$

14.3.9   Sinusoidal FM and Linear Frequency Drift

Often the frequency disturbance is periodic. In this common case, it is important to realize that the Allan variance approaches zero when the averaging time is equal to a multiple of the disturbance period. If it is not known in advance that the disturbance is periodic, different measurements of the Allan variance may take on widely differing values if slightly different averaging times are used by different observers. The Allan variance and spectral density for sinusoidal disturbances are given by [7]

$$S_y(f) = \frac{1}{2} \left(\frac{\Delta f}{f_c} \right) \delta(f - f_m) \tag{14.51}$$

$$\sigma_y(\tau) = \left(\frac{\Delta f}{f_c} \right)^2 \frac{\sin^4 (\pi f_m \tau)}{(\pi f_m \tau)^2} \tag{14.52}$$

If there is linear drift of frequency with time, this should be removed before attempting to make frequency-domain (spectral) stability measurements because there is no useful model for the spectrum in this case. If the fractional frequency is given by the product of the drift, d, and the time separation, τ, then

$$Y = d(\tau) \tag{14.53}$$

and direct calculation in the time domain gives the Allan variance as [7]

$$\sigma_y(\tau) = \frac{d \cdot \tau}{\sqrt{2}} \tag{14.54}$$

Note that the Allan variance contribution from a linear frequency drift has the same functional dependence on the averaging time as the contribution from random walk frequency noise ($n = 2$; see Table 14.3). For ELINT purposes, it is usually not fruitful to attempt to determine which type of frequency process is present (i.e., to find the value of the exponent) from the dependence of the Allan variance estimate on the averaging time.

14.3.10   Short-Look Problem

Most of the literature relating to short-term stability measurement assumes that the signal is available for an indefinitely long time. In ELINT, the signal may be observable at high SNR only during the relatively short times that the radar beam passes. This means that instead of an idealized line spectrum of a pulsed signal, the observable spectrum (as viewed through the duration "window" of the radar beam illumination time) appears as a series of sin $(x)/x$ functions, each located at a multiple of the PRF and having a width approximately equal to the reciprocal of the illumination time. Clearly, if the sidebands due to frequency variations are hidden beneath those sidebands of the spectrum caused by the limited duration of the pulse train, it will not be possible to measure the true carrier to sideband ratio. Generally speaking, if there are more than 30 pulses in a group available for analysis, the spectrum contributions due to the short-look problem will be small enough so that one can assume that the components located between the PRF lines are due to either thermal noise or frequency fluctuations. If there are fewer than 20 pulses, the spectral components due to the short-look problem will probably exceed those of the frequency fluctuations. Of course, the Allan variance can still be estimated, but again the limited number of samples will cause a greater uncertainty in the result.

Usually, there are many such short looks available—one is received each time the radar beam scans past the ELINT site. These additional short looks can help improve the estimates by averaging together more frequency values or more spectra. However, this will not solve the problem completely because the successive looks are so far apart in time that they are essentially independent insofar as their phase is concerned. Figure 14.8 shows this effect for 50, 30, 20, and 10 pulses received in a single scan. The resolution of these plots is approximately 0.2% of the PRF.

As some consolation, the ELINT analyst should note that the radar is limited in its ability to detect targets in clutter by this same problem.

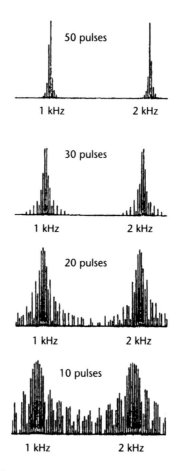

Figure 14.8   Short-look spectra.

14.4   Frequency Character of CW Signals

All of the techniques relating to the measurement of the carrier frequency characteristics of pulsed signals generally apply to CW signals. For CW signals, the spectrum analysis problems caused by the convolution of the carrier line spectrum with the video pulse spectrum are eliminated. On the other hand, there is usually some type of modulation present that may obscure the properties of the carrier. The modulation may include:

1. Antenna scanning;
2. Frequency sweeping;
3. Multiple channel subcarrier modulation using AM or FM;
4. Discrete modulation such as frequency shift keying (AM or FM) or phase reversals.

Measuring the carrier frequency of modulated CW signals may require long-time averaging in the case of FM sweeps. In this case, the average frequency (if it is of interest) can be found by counting over an integral numbers of complete sweep repetitions.

In the case of AM, a limiter can precede the frequency-measuring counter, and the carrier or center frequency can be properly measured. In the case of FM, the long-term average can be investigated through counting. With FM-FSK, the average frequency may never actually be transmitted.

The usual technique in analyzing the modulation details is to try various types of demodulators and to examine their outputs. These can include discriminators, filter banks, and envelope detectors. Once the type of modulation is determined, demodulation can usually be achieved. For example, if 180° phase reversals are involved, and, if the SNR is high, a frequency doubler followed by a divide-by-two circuit provides a means for recovering the carrier. This can be analyzed using any of the techniques described for pulsed signals. In addition, this recovered carrier can be used as one input to a phase detector to determine the code sequence being used.

In the case of many ECM signals that are CW in nature, a *look-through interval* is often present. The signal may have a duty cycle approaching one. However, the pulsing effects of the on-off look-through modulation must be considered. For frequency-multiplexed signals, spectrum analysis can usually locate the subcarriers. If FM multiplexing is used, the failure of a limiter to remove the modulation can indicate that FM is in use.

Tracking filters of narrow bandwidths can be used with CW signals to follow weak signals in the noise. These may be of the phase-locked loop type. In this case, the frequency variations of the incoming signals are represented by the variations of the voltage-controlled oscillator tuning the voltage within the phase-locked loop.

14.5   Pulsed Signal Example

Digitized predetection data is normally used for measuring intrapulse AM and PM. It can also be used for measuring pulse-to-pulse RF coherence if proper digitizing techniques are used. The key is to measure the instantaneous phase of many pulses at precisely known points in time. The collection system needed for interpulse coherence measurement differs from that needed for intrapulse unintentional modulation on pulse (UMOP) measurement by the addition of a counter that measures the number of clock pulses that elapse between the end of one pulse and the beginning of the next. In addition, the digitizing clock and all local oscillators must have good stability over the time duration of the coherence measurement (e.g., 100 ms).

Consider the following example:

$$\text{nominal pulse duration} = 1 \times 10^{-6} \text{ sec}$$

$$\text{nominal PRI} = 1,000 \times 10^{-6} \text{ sec}$$

$$\text{digitizing clock period, } T_s = 5 \times 10^{-9} \text{ sec}$$

The UMOP data will be contained in approximately 200 samples that are obtained during a single pulse. Then, for about 200,000 clock pulses, the signal is absent and the A/D clock pulses are simply counted. Then, the next pulse rises above the threshold and digitizing of about 200 more samples occurs, and so on. This data stream is sufficient to measure coherence, as is shown here:

$$\text{Phase of pulse 1, 1st sample} = \phi_{1,1}$$

$$\text{Phase of pulse 1, 2nd sample} = \phi_{1,1}\Delta\phi, \ \Delta\phi = 2\pi f_{IF} T_S$$

$$\text{Phase of pulse 1, 3rd sample} = \phi_{1,1} + 2\Delta\phi$$

$$\vdots$$

$$\text{Phase of pulse 1, 200th sample} = \phi_{1,1} + 199\Delta\phi$$

The number of missing samples between pulse 1 and pulse 2 is 199,800.

$$\text{Phase of pulse 2, 1st sample} = \phi_{2,1} = \phi_{1,1} + 200,000\Delta\phi$$

$$\text{Phase of pulse 2, 2nd sample} = \phi_{2,1} = \phi_{1,1} + 200,000\Delta\phi$$

and so on.

If the sampling rate is well above the Nyquist rate, $\Delta\phi$ will be less than 180°. Suppose the signal of interest is mixed down to an IF of 60 MHz and that the bandwidth is about 20 MHz (e.g., 50 to 70 MHz). Suppose the actual center frequency is 59.444 MHz. If the digitizing rate is 200 MHz, then $\Delta\phi$ is 107°. Arbitrarily assume $\phi_{1,1} = 0$ and the phase data become as shown in Table 14.4.

Note that the modulo-360° phase of the first sample of the nth pulse is given by $[(n - 1)(200,000)(107)]$ modulo 360 and is shown in Table 14.5. Coherence is indicated by the linear progression of the unwrapped phase, at least over the number of pulses used by the radar for coherent integration. Any deviation from a linear phase progression indicates instabilities of the signal relative to the A/D converter clock and system local oscillators.

Table 14.4   Phase Samples Versus Unwrapped Phase

Phase Sample (modulo 360°)		Unwrapped Phase
(First sample of Pulse 1)	$\phi_{1,1} = 0$	0
	$\phi_{1,2} = 107$	107
	$\phi_{1,3} = 214$	214
	$\phi_{1,4} = 321$	321
	$\phi_{1,5} = 68$	428
	$\phi_{1,6} = 175$	535
	$\phi_{1,7} = 282$	642
	$\phi_{1,8} = 29$	749
	$\phi_{1,9} = 135$	856
	$\phi_{1,10} = 243$	963
	$\phi_{1,11} = 350$	1,070
	$\phi_{1,12} = 97$	1,177
(Last sample of Pulse 1)	$\phi_{1,200} = 53$	21,293
(First sample of Pulse 2)	$\phi_{2,1} = 160$	21,400,000
(Second sample of Pulse 2)	$\phi_{2,2} = 267$	21,400,107
(Third sample of Pulse 2)	$\phi_{2,3} = 14$	21,400,214
	$\phi_{2,4} = 121$	21,400,321
	$\phi_{2,5} = 228$	21,400,428
	$\phi_{2,6} = 335$	21,400,535
	$\phi_{2,7} = 82$	21,400,642

Table 14.5   First Phase Sample Versus Unwrapped Phase

Pulse Number (n)	Phase of First Sample (°)	Unwrapped Phase
1	$\phi_{1,1} = 0$	0
2	$\phi_{2,1} = 160$	160
3	$\phi_{3,1} = 320$	320
4	$\phi_{4,1} = 120$	480
5	$\phi_{5,1} = 280$	640
6	$\phi_{6,1} = 80$	800
7	$\phi_{7,1} = 240$	940
8	$\phi_{8,1} = 0$	1,100
9	$\phi_{9,1} = 160$	1,260
10	$\phi_{10,1} = 320$	1,580
11	$\phi_{11,1} = 120$	1,700
etc.	etc.	etc.

Note that if the PRI is constant and the threshold is insensitive to pulse-to-pulse amplitude changes, the first samples of consecutive pulses may be separated by a fixed (constant) number of A/D clock periods. In this case, it is not necessary to count the number of A/D clock periods between the pulses, and the coherence of the signal can still be estimated.

So far, only the first phase sample of each pulse has been used to explore the pulse-to-pulse phase progression. However, when there are additional samples within the pulse, it is reasonable to also make use of these. Because the first sample of pulse (n), $\phi_{n,1}'$ and the first sample of pulse $(n + 1)$, $\phi_{n+1,1}$ are separated by exactly the same time as the second sample of pulse (n), $\phi_{n,2}$ and the second sample of pulse $(n + 1)$, $\phi_{n+1,2}$, it is clear that the phase change will be the same if the phases of the second samples of each pulse are subtracted—provided that there is the same FMOP or PMOP on each pulse. Notice that in this example:

$$\phi_{2,1} - \phi_{1,1} = 160 - 0 = 160$$

$$\phi_{2,2} - \phi_{1,2} = 267 - 107 = 160$$

$$\phi_{2,3} - \phi_{1,3} = 14 - 214 = -200 \text{ (or } +160)$$

$$\phi_{2,4} - \phi_{1,4} = 124 - 324 = -200 \text{ (or } +160)$$

$$\phi_{2,5} - \phi_{1,5} = 228 - 68 = 160$$

and so on

This indicates one way to use these additional intrapulse phase samples; namely, compute the difference in the phase between like numbered samples on successive pulses and find the average of the differences. This has the effect of reducing the effects of quantizing noise and thermal noise.

14.6   Measuring Coherence

Once the pulse-to-pulse phase change is determined, any of the methods described in this chapter may be used to determine a coherence estimate. For example, the best straight line may be fitted to the unwrapped phase data and then subtracted. The residues are then Fourier transformed to obtain the phase noise spectrum. The better the coherence, the lower the phase noise spectral components. Another method is to examine the changes in the average frequency to compute the Allan variance. The average frequency during a particular pulse interval is

$$f_n = \frac{\phi_n - \phi_{n-1}}{2\pi T_n}$$

where:

f_n = average frequency between pulse $n - 1$ and n

ϕ_n = phase of the nth pulse (rad)

ϕ_{n-1} = phase of the $(n - 1)$th pulse (rad)

T_n = time between pulse $(n - 1)$ and pulse n

In this example,

$$f_n = \frac{160°}{360°} \times \frac{1}{1 \times 10^{-3} \text{ sec}} = 444.44 \text{ Hz}$$

To compute the Allan variance, it is necessary to find the average of the sum of the squares of the frequency changes over many pulses:

$$\overline{\Delta f^2} = \frac{1}{N - 1} \sum_{n=1}^{N} (f_{n+1} - f_n)^2$$

In this example, $f_{n+1} = f_n$, and the Allan variance is zero because the signal is completely stable. Then this value is divided by the square of the carrier or RF:

$$\text{Allan variance} = \frac{\overline{\Delta f^2}}{(\text{RF})^2}$$

Notice that for a completely noncoherent signal, the individual phase values will be random numbers uniformly distributed over 360°. The difference of two such random numbers will also be random, but with a triangular distribution extending from −360° to +360°. An Allan variance value can be computed, but there is no real coherence present.

14.7 Effects of Drift

If the frequency of any of the local oscillators (including the digitizing clock) drifts, the effect is to create a quadratic (rather than linear) phase versus time plot. Then unwrapping the phase may become difficult and the pulse-to-pulse phase changes may also appear to be random. While in principle it is possible to remove the drift by fitting a quadratic to the unwrapped phase, this is of limited value unless it is possible to determine whether the drift is due to the intercept equipment or to the incoming signal. Making the local oscillators very stable is a prerequisite for a useful coherence measuring system.

References

[1] Blachman, N. M., and G. A. McAlpine, "The Spectrum of a High Index FM Waveform: Woodward's Theorem Revisited," *IEEE Trans. on Communications Technology*, Vol. COM 17, No. 2, April 1969, pp. 201–207.

[2] Wiley, R. G., et al., "Demodulation Procedure for Very Wideband FM," *IEEE Trans. on Communications*, Vol. COM 25, No. 3, March 1977.

[3] Curry, T. F., "Specification of Short-Term Frequency Stability by Maximum Likelihood Estimates," *IEEE-NASA Symposium on Short-Term Stability Proc.*, NASA Rep. Sp-80, November 1964, pp. 101–109.

[4] Allan, D. W., "Statistics of Atomic Frequency Standards," *Proc. IEEE*, Vol. 54, February 1966, pp. 221–230.

[5] Barnes, J. A., et al., "Characterization of Frequency Stability," *IEEE Trans. on Instrumentation and Measurement*, Vol. IM-20, May 1971, pp. 105–120.

[6] Skolnik, M. I., *Radar Handbook*, New York: McGraw-Hill, 1970, pp. 4–7.

[7] Rutman, J., and F. L. Walls, "Characterization of Frequency Stability in Precision Frequency Sources," *Proc. IEEE*, Vol. 79, June 1991, pp. 952–960.

CHAPTER 15

Determining ELINT Parameter Limits

15.1 Introduction

When faced with a number of parameter values measured by various methods in different locations at different times, the analyst must determine the likely range of the parameter involved for a given type of radar. One simple method is to take the largest and smallest observed values as the parametric limits (e.g., the RF range could be from the lowest reported RF value to the highest reported RF value). Unfortunately, this can place too much emphasis on extreme values that may actually represent errors.

One common practice is to report ELINT parameter limits as being those that include 95% of the data values. Selection of the data to be used requires care and judgment on the part of the analyst. It may be useful to eliminate values having a poor accuracy or that are too old. Generally, it is necessary to look at the range determined by the overall set of data values and then compare this to the ranges indicated when various subsets of the data are analyzed. For example, the parameter range indicated by 5 years of data values may be compared to those indicated by data from each year individually to see if changes with time are occurring.

15.2 Histograms Used to Determine Parameter Limits

A histogram gives the number of occurrences of a parameter value within a range of values. For example, a table showing the number of people from 0 to 10 years of age, from 10 to 20 years, from 20 to 40 years, from 40 to 60 years, and 60 years and older would be a histogram showing the distribution of the population by age groups.

Figure 15.1 shows a histogram of pulse duration values. The horizontal axis consists of 0.1-μs bins, and the vertical axis is the number of reported occurrences of each pulse duration value observed. This histogram was made from tens of thousands of pulse duration values, but it gives a very compact and convenient representation of this large body of data.

A more challenging histogram of pulse duration is shown in Figure 15.2. The amount of data available to build this histogram is smaller by a factor of about 400, which may account for the gaps and its generally ragged appearance. In this case, each individual intercept may require special scrutiny. For example, all of the points above 5 μs are accounted for by only nine intercepts. Examining these

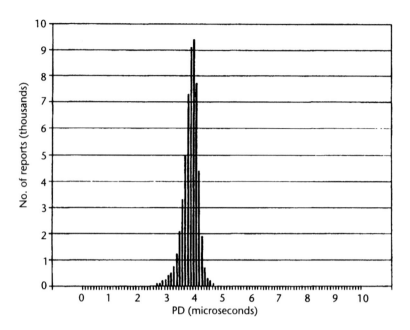

Figure 15.1   Histogram of pulse duration values.

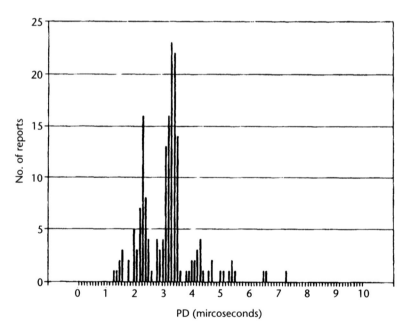

Figure 15.2   Another pulse duration histogram.

to determine where and when they took place, and whether or not they have been properly identified, may allow some or all of them to be deleted. Also, the gaps may represent different pulsewidth modes, in which case separate histograms for each mode's pulse durations are called for. Discovering radar modes often requires the use of conditional histograms, which are described in Section 15.3.

The histogram approximates the shape of the probability density function (PDF) of a random process. The nature of the approximation is that the histogram more closely resembles the PDF as (1) the bin size becomes very small and (2) the number of data values or occurrences becomes very large.

In practice, the number of data values is often fixed, and the smallest bin size that is practical depends on the resolution of the measuring device. For example, in Figures 15.1 and 15.2, if the pulse duration measuring system uses a 0.1-μs clock interval, the histogram bin size cannot be reduced.

To use histograms, it is frequently necessary to make several, each having different bin sizes. The analyst chooses the bin size range that seems to give the best approximation to the underlying distribution. This judgment is somewhat subjective; however, one looks for a range of bin sizes over which the essential shape or character of the histogram is the same. There are two extreme cases. When the bin size is very large, all of the values fall into the same bin. For example, if Figure 15.1 were redrawn with 5-μs bins (e.g., 0 to 5 μs and 5 to 10 μs), all of the data values would fall into the first bin. When the bin size is too small, there is insufficient data to provide a statistically significant number of points per bin. The histogram would typically have a single point in some bins and the others would be empty. Generally, the ratio of the number of points to the number of bins, which is the average number of points per bin, should be more than 10 and preferably more than 100.

The effect of insufficient data on the histogram is shown in Figure 15.3, which shows histograms of data values drawn from a uniform distribution and from a triangular distribution. With an average of only one point per bin, the character of the distribution is not apparent, whereas with 100 points per bin it is.

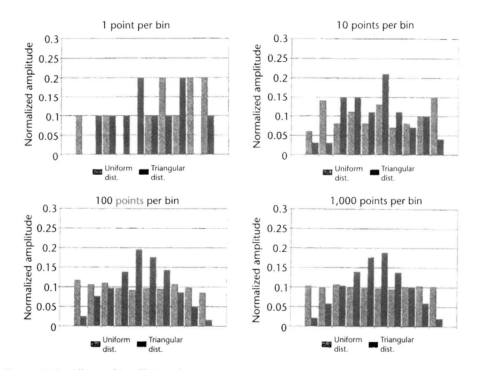

Figure 15.3   Effects of insufficient data.

Another property of histograms is that the bin size can be increased to any multiple of a smaller bin size by simply adding the number of occurrences in a number of adjacent bins. This is convenient when dealing with very large quantities of data because one can begin with a relatively small bin size, compute the histogram, and then create larger bin size histograms from the first one without examining the individual data values.

A second additive property is that several histograms with the same bins can be added bin by bin to create a composite. For example, a parameter might be studied for changes in time by making histograms of the data for each year for the five previous years. A composite (5-year histogram) could then be created by simply adding the occurrences in each bin for each of the five 1-year histograms.

15.3 Types of Histograms

In addition to the conventional histogram giving an approximation to the PDF, histograms also often show the cumulative distribution function (CDF). This is done by simply tabulating a total for the number of occurrences in all the bins below a given bin. The CDF makes it very easy to determine important parameters of interest. For example, the median is that bin for which half of the occurrences are above and half below. Using the CDF, the median can be easily located. Similarly, one can easily find the range of values within which some fraction (e.g., 95%) of the occurrences lies. The CDF is the integral of the PDF. A good histogram program provides both of these.

A second histogram type is the multiple-dimension histogram. For example, one might plot the number of intercepts reporting both a pulse duration in a certain bin and a PRI in a certain bin. This would be a three-dimensional plot above the PRI by pulse duration plane. Adding more dimensions makes the plotting problem more difficult (e.g., using the PRI, pulse duration and RF as coordinates). To avoid such problems, it is usual to make multiple histograms of, say, the pulse duration with the added condition that the PRI reported was also within some specified range. This is necessary to determine what modes of operation are possible for the emitter. For example, the overall pulse duration histogram may reveal two distinct values, and the overall PRI histogram may also reveal two distinct values, as shown in Table 15.1.

To find out whether (and how frequently) PRI_1 is observed with PD_1, conditional histograms are needed. Therefore, we count the occurrences of the pulse duration values for which the radar's PRI was in the range typical for PRI_1 and also the pulse duration values for which the radar's PRI was in the range typical of PRI_2. From the results shown in Figure 15.4, we see that PRI_1 is usually seen

Table 15.1 Pulse Duration Values
Versus PRI Values

Pulse Duration Values	PRI Values
PD_1 (near 3 μs)	PRI_1
PD_2 (near 5 μs)	PRI_1, PRI_2

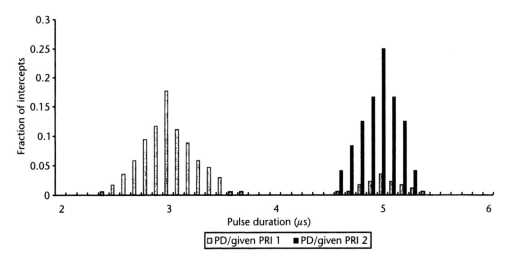

Figure 15.4   Conditional histograms.

with PD_1, but is sometimes observed with PRI_2. On the other hand, PD_2 is never observed with PRI_1.

The conditional histograms can be added (if the bin sizes are the same) to obtain a composite. Adding annual histograms over a 5-year period is a form of conditional histogram, where the condition is placed on the date of the reported intercept rather than on other parameters of the radar.

Note that simple histograms are completely independent of the order in which the intercepts are analyzed. To take a simple example, shuffling a deck of cards will have no effect on a histogram of the number of aces, twos, and threes. There will always be four of each.

A special histogram that is dependent on the order of the data values is the delta-τ histogram, which is described in Chapter 12. This histogram consists of the time intervals between all pairs of pulses and is related to the autocorrelation of the TOAs.

There are two parameters to be selected before making a histogram: the parameter range to be covered and the number of parameter subranges, or bins, into which the overall range of interest is to be divided. If the size of the parameter bins is too large, the histogram gives little information because all of the data are in one or two bins. If the bin size is too small and there is insufficient data, the histogram again fails to be useful because each bin contains only one or two measurements. Analyst judgment and experience must be used to select the proper number of bins or bin size so that there is enough data in enough bins to give a useful result. In this way, the 95% limits of the available data can be directly determined. The analyst simply notes the parameter range as being those bins that exclude the lower 2.5% and upper 2.5% of the data. For example, if there are 400 data values, the analyst would eliminate those bins containing the lowest 10 data points and the highest 10 data points. Then, one estimate of the 95% limits would be the limits of the parameter values represented by the remaining histogram bins. This can always be done regardless of the number of data values or the shape of the distribution. There is no doubt that 95% of the observed values are contained within limits found in this way. It is not certain that 95% of all of the measurements that might have been made would also fall in this range. Our confidence about this depends on the amount of data available.

To account for this lack of certainty about the limits of the parameter values that are possible as opposed to those values that have actually been observed, it is necessary to examine what is called the confidence level of the limits as determined by the analyst from the data. This can be done using the methods of statistics known as "parametric" [1]. It is possible to state that the limits obtained in the previous example of 400 data points by deleting the 10 lowest and 10 highest include 93.5% of the parameter values at a confidence level of 0.9. In other words, 9 times out of 10, 93.5% of a set of 400 data values would fall within these limits. We could also state that 99 times out of 100, 92% of the data points would fall within those same limits.

The amount of data to be deleted at each end of the histogram in order to specify parameter limits so that 95% of the data falls between the limits 19 times out of 20 (i.e., in 95% of all tests using that much data) varies with the amount of data, as shown in Figure 15.5. For fewer than 100 data values, none of the data should be deleted. The parameter range should be given as that covered by the extremes of the available data and then a notation made that the quantity of data

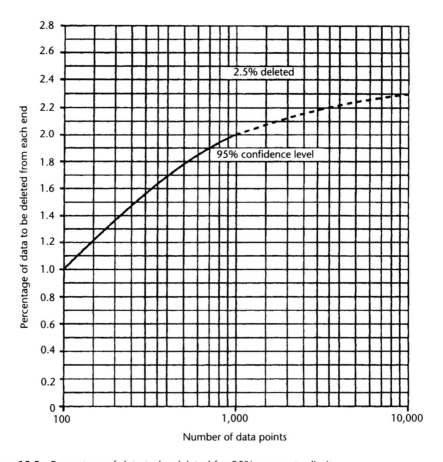

Figure 15.5 Percentage of data to be deleted for 95% parameter limits.

available is insufficient to specify the 95% limits to a 95% confidence level. Above 10,000 data points, 2.5% of the data at each end of the histogram can be deleted with virtually no loss of confidence. (If only 1,000 data points are available and 2.5% are deleted at each end, then for 9 out of 10 groups of 1,000 measurements, 95% of the data, will fall between the specified limits; that is, the confidence level drops from 95% to 90%.)

It should be noted that this technique requires no assumptions about the underlying distribution of the data, and the limits can be easily determined without computation. If the measurements are ordered from the smallest to the largest, the 95% range is determined as the extreme values of the data when the fraction of points given in Figure 15.5 is deleted from each end of the list. A histogram can also be used and the bins containing that fraction of the data are eliminated from each end to obtain the 95% limits of the parameter.

Note that the analyst may wish to break the parameter range up into several modes if the histogram shows the data points clustered around several peaks. The number of data values in each peak can be noted and individual intervals or parameter ranges about each peak can be specified using the same method. Note also that statistically it does not matter if equal numbers of data points are deleted from each end of the range of values. The total percentage of the data to be deleted is twice that shown in Figure 15.5. Different numbers of points can be deleted at each end of the range to achieve the desired total amount of data to be deleted.

15.4 Two-Sigma Parameter Limits

In addition to the histogram, another frequently used statistical approach involves computation of statistical parameters based on the data values. The most common ones are the sample mean and the sample standard deviation. The formulas are as follows:

$$m_s = \frac{\sum_{i=1}^{N} x_i}{N} \tag{15.1}$$

$$\sigma_s = \left[\frac{\sum_{i=1}^{N} (x_i - m_s)^2}{N - 1} \right]^{1/2} \tag{15.2}$$

where:

m_s = sample mean

σ_s = sample standard deviation

N = number of data values

x_i = the ith data value

The sample mean is at the center of gravity of the histogram of the data. The sample standard deviation is a measure of the spread or dispersion of the data. In many cases, it is common to assume that the parameter values being observed are distributed according to the normal distribution. This is a bell-shaped curve given by

$$\phi(x) = \frac{1}{\sqrt{2\pi}\sigma} e^{-(x-m)^2/2\sigma^2} \tag{15.3}$$

where $\phi(x)\, dx$ is the probability a data value will be between the values x and $(x + dx)$, σ is the standard deviation, and m is the mean value.

If the data values being observed are distributed according to the normal distribution, then 95% of them will lie between the mean minus twice the standard deviation and the mean plus the standard deviation:

$$\int_{m-2\sigma}^{m+2\sigma} \phi(x)\, dx \cong 0.95 \tag{15.4}$$

If the analyst chooses, these so-called "two-sigma" limits may be specified for a parameter. This means that the lower value is given as the sample mean minus twice the sample standard deviation, and the upper value is given as the sample mean plus twice the sample standard deviation. There are two problems with this approach. First, as with the histogram approach, there is no assurance that the data available is typical and that the sample mean and standard deviation are the same as the true mean and true standard deviation. Second, we cannot be certain that the data values are distributed normally.

The problems of the differences between the sample mean and true mean and the sample standard deviation and true standard deviation can be handled by the methods of statistics using the *student's* t *distribution* and the *chi-squared distribution*. Using these distributions and assuming that the true distribution of the data is normal, we can specify an interval in which the true mean or standard deviation will lie, 9 times out of 10 (a 90% confidence level). For example, 9 times out of 10 the true mean will fall within the range

$$m_s \pm \frac{\sigma_s}{\sqrt{N}} t_{(n,90)} \tag{15.5}$$

where $t_{(n,90)}$ is given by Table 15.2.

Any standard text or handbook [2] on statistical methods as applied to experiments can be consulted to learn more about these methods.

Quite often the true distribution of data values is not described by the normal distribution. For this reason, it is preferable to use the histogram approach to specify a parameter range. This is done very simply by deleting a few points at each end of the range and is valid for any underlying distribution of the data.

For example, if the data distribution is actually uniform between two limits, L_{min} and L_{max} (as in a radar set operating equally probably at any frequency within

Table 15.2   Student's *t* Multiplier Values

Number of Data Points	$t_{(n,90)}$	Number of Data Points	$t_{(n,90)}$
2	6.314	12	1.796
3	2.920	15	1.761
4	2.353	20	1.729
5	2.132	30	1.699
6	2.015	40	1.684
8	1.895	60	1.670
10	1.833	∞	1.645

an assigned band), the mean is at the center of the range between two limits and the standard deviation is the range divided by $\sqrt{3}$. Then the estimated two-sigma limits include about 15% more than the actual range:

$$\frac{4\sigma}{L_{max} - L_{min}} = \frac{2}{\sqrt{3}} \cong 1.15 \qquad (15.6)$$

Of course, even if the parameter values came from a uniformly distributed process, the inaccuracies present in measurement of the values would add additional variations, so that the distribution estimated from measured data would appear as shown in Figure 15.6. Mathematically, Figure 15.6 is obtained by convolving the distributions of the measurement errors and the population's radar signal values.

These measurement errors would increase the estimated standard deviation and push the two-sigma limits even farther out. In fact, the sample standard deviation can be significantly affected by just a few measurements that are grossly in error. For this reason, in order to get reasonable two-sigma limits, outlying points must be discarded before the sample standard deviation is calculated.

While statistical techniques can be useful to the analyst, judgment and experience must be combined with knowledge of statistical techniques to specify parameter limits.

15.5   Histogram Analysis Techniques

15.5.1   Parameter Limits Example

In many cases, the important results to be obtained from the analysis process are the parameter limits of the emitter (e.g., the RF range over which the radar transmits).

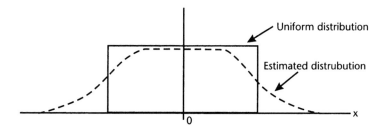

Figure 15.6   Uniform distribution and the effects of measurement errors.

The following hypothetical situation illustrates the complications caused by the use of different ELINT receivers at different sites.

Suppose there are 100 ELINT stations and that 90 of these are equipped with old receivers capable of ±50-MHz accuracy RF measurements while 10 are equipped with newer receivers capable of a ±5-MHz accuracy. Suppose each site reports 100 intercepts of a particular radar and a histogram of the RF values from 10,000 intercept reports. The result is shown in Table 15.3.

For this histogram, the mean $\pm 2\sigma$ limits include nearly all of the data. (These limits would include 95% of the data if it followed a normal distribution.) The technique shown in Figure 15.5 does not require the assumption of a normal distribution and would simply delete 250 data values (2.5%) from each end of the histogram and give the limits as the extreme values of the remaining 9,500 points. This would give limits of 2,945 to 3,085 MHz. For this bin size, the closest we can come to deleting 250 points is to delete one bin at the lower end and one bin at the upper end of the distribution. Note that the median is near 3,010 MHz, as indicated by the CDF.

A significant problem is to determine the accuracy with which we know the parameter limits. In a sense, the accuracy is ±50 MHz because that is the worst accuracy in the raw intercept report data. Certainly the accuracy is no worse than that; however, it does not take into account the benefits of having 10,000 intercepts to consider. Nevertheless, when parameter limits from histograms are being entered into a database, it seems more descriptive to the eventual user to enter the accuracy of the measurements themselves rather than try to compute a confidence interval for the lower and upper limits. One should comment on the histogram (e.g., give the bin size, number of data values, and overall shape) to assist the end user.

Sometimes the accuracy entered by the analyst may be the histogram bin size because this is one factor that determines his ability to determine where the 95%

Table 15.3 Histogram of RF Values from 10,000 Intercept Reports

Bin (MHz)	Number of Reports	Cumulative Number of Reports
2,940* (and below)	179	179
2,950**	268	447
2,960	453	900
2,970	1,057	1,927
2,980	723	2,680
2,990	631	3,311
3,000	1,059	4,370
3,010	716	5,086
3,020	627	5,713
3,030	1,067	6,780
3,040	731	7,511
3,050	619	8,130
3,060	1,066	9.196
3,070	444	9,640
3,080	175	9,815
3,090 (and above)	185	10,000

Note: mean = 3,015 MHz; σ = 38.7 MHz; mean -2σ = 2,937.5; mean $+2\sigma$ = 3,092.5.
*RF values below 2,945.
**RF values from 2,945 to 2,954.99.

limits are located. However, because the bin size can be selected arbitrarily, it is not as meaningful as the accuracy value associated with the measurement.

15.5.2 Intercepts Separated into Accuracy Classes

In the previous example, not all of the intercepts have a ±50-MHz accuracy. Ten percent (1,000) of the reports have a ±5-MHz accuracy. If separate histograms for the two types of intercepts are made (i.e., conditional histograms, with accuracy as the conditioning factor), the results are as shown in Table 15.4.

Using only the data from the newer receivers seems to indicate that the radar operates on four RF channels separated by about 30 MHz. The analyst needs to make four finer resolution histograms centered at 2,970, 3,000, 3,030, and 3,060 MHz to determine the distribution of the values for each channel.

The data from the older receivers is still of value. The older receivers may be the only ones available in many geographic areas. It is useful to compare frequency operation by geographic area and the data from the less accurate receivers would allow that. Figure 15.7 shows the data of Table 15.4 graphically.

This example also points out the different information needs of different users. Suppose the user was an intercept operator with one of the older receivers. One of his most important jobs is to identify the signal type. If he were told that this radar operates on one of only four specific frequencies, then he possibly could not identify the radar because his receiver might indicate some different frequency. On the other hand, a designer of a radar warning receiver that uses an IFM would be very interested in the channelization statistics to help identify the radar. The analyst should recognize these different information needs and provide as complete a picture of the data as possible.

Table 15.4 Results of Conditional Histograms

Bin (MHz)	New Receivers Only PDF	CDF	Old Receivers Only PDF	CDF
2,940* (and below)		0	179	179
2,950**		0	268	447
2,960		0	453	900
2,970		243	814	1,714
2,980		243	723	2,437
2,990		243	631	3,068
3,000	243	495	807	3,875
3,010		495	716	4,597
3,020		495	627	5,218
3,030	252	752	810	6,028
3,040		752	731	6,759
3,050		752	619	7,378
3,060	257	1,000	818	8,196
3,070		1,000	444	8,640
3,080		1,000	175	8,815
3,090 (and above)	248	1,000	185	9,000
	Mean = 3,015 σ = 33		Mean = 3,014 σ = 39	
	Mean -2σ = 2,948		Mean -2σ = 2,938	
	Mean $+2\sigma$ = 3,082		Mean $+2\sigma$ = 3,090	

*RF values below 2,945.
**RF values from 2,945 to 2,954.99.

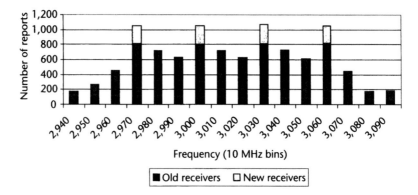

Figure 15.7 RF histogram example.

A better example of the effects of RF channelization is shown in Figure 15.8. Here each histogram peak should be examined in detail and their individual ranges and an overall range reported. A discussion of the apparently unused channels (e.g., near 2,140, 2,300, 2,360 MHz) would also be appropriate.

15.5.3 Most Probable Values

When the histogram of a parameter indicates a well-defined peak, the parameter value that is most frequently reported is called the *most probable value*. While there is nearly always a histogram bin with a greater number of intercepts than

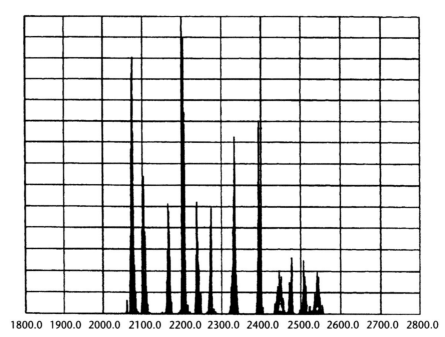

Figure 15.8 RF histogram showing channelization.

any other bin, the term "most probable value" is reserved for a well-defined peak that is well separated from other peaks (i.e., a broad distribution where the bin with the largest number of intercepts has a number of intercepts only slightly larger than several adjacent bins does not have a most probable value). Naturally, the shape of the histogram depends on the bin size selected, and the ability to locate a peak is affected by the bin size. Generally, one chooses a bin size comparable to the accuracy of the individual intercepts because smaller bins are not very meaningful (i.e., the variations shown begin to reflect measurement errors rather than the behavior of the signal of interest).

15.6 Analysis Problems

15.6.1 Signal Identification Errors

Perhaps the most difficult problem with analyzing intercept reports is dealing with those that have mistakenly identified the signal type. A few errors, even among thousands of reports, can significantly affect statistical calculations. Outlying data values must be carefully checked to see that the identification is proper. Beyond that, it is useful to make as many checks as possible to look for mistakes in identification among all the intercepts. For example, a different parameter value (as reported by only one station) should be examined for possible misidentification. Clearly, there is no guarantee that identification errors can be found. Feedback from the analyst to the interceptors can be very helpful in minimizing the problem. For example, the analyst may suspect a new mode has appeared. By asking for certain special measurements, the analyst may be able to confirm or deny that suspicion.

15.6.2 Transforming Parameters and Their Accuracies

Consider changing a parameter from one form (in which the interceptor reported it) to another form that is required for a report or user application. A simple example is changing PRF to PRI. A method that can always be used is to transform each intercept report of the parameter to the new parameter and find the histogram and limits of the new parameter directly. However, if a histogram of the old parameter exists, transforming it to the new parameter is possible. In general, if the relationship between the old parameter, x, and the new parameter, y, is single valued and monotonic, the distribution of the new parameter is given by

$$P'(y) = \frac{P(x)}{|dy/dx|} \tag{15.7}$$

where $P'(y)$ is the probability distribution of y, $P(x)$ is the probability distribution of x, and dy/dx is the derivative of y with respect to x. For example,

$$y = \frac{1}{x} \text{ and } dy/dx = \frac{-1}{x^2}$$

implies

$$P'(y) = x^2 P(x) \tag{15.8}$$

If the PRF is uniformly distributed between 1,000 and 1,100 pulses per second, $P(x)$ is 0.01/Hz for $1,000 < x < 1,100$ Hz. The distribution of the PRI is then

$$P'(y) = x^2 P(x) \text{ or } (0.01)(\text{PRF})^2 \text{ sec}^{-1} \tag{15.9}$$

These two distributions are shown in Figure 15.9. The mean PRF is 1,050 Hz, whose reciprocal is 952.3 μs. The mean PRI, however, is 953.1 μs, a difference of 0.8 μs. In other words, the statistical mean of the PRI is not the reciprocal of the mean PRF: the relationship depends on the distribution. The calculation is as follows:

$$\bar{x} = \int_{1,000}^{1,100} 0.01x \; dx = \left[\frac{0.01x^2}{2}\right]_{1,000}^{1,100} = 1,050 \tag{15.10a}$$

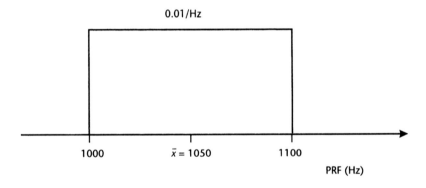

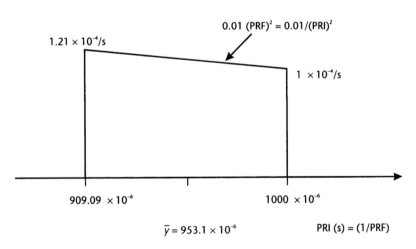

Figure 15.9 A PRF histogram transformed to a PRI histogram.

$$\frac{1}{\bar{x}} = \frac{1}{1,050} \cong 952.381 \ \mu s \tag{15.10b}$$

$$\bar{y} = \int_{1/1,100}^{1/1,000} \frac{0.01}{y^2} \, y \, dy = 0.01 \left[\log(y) \right]_{1/1,100}^{1/1,000} \tag{15.10c}$$

$$\bar{y} = 0.01 \, (-6.90776 + 7.003063) = 953.102 \ \mu s \tag{15.10d}$$

Note that $1/953.102 \ \mu s = 1,049.206$ Hz.

If the raw intercept data is expressed as PRFs and if the analyst wishes to use PRI in his report, the question of accuracy arises. If the accuracy of the PRF is $\pm \Delta F$ (Hz), then the accuracy of the PRI will be

$$\pm \Delta T = \mp \frac{\Delta F}{F^2} \tag{15.11}$$

(This is obtained by differentiating $T = 1/F$.) If the PRF reports are accurate to ± 1 Hz and if the average PRF is 1,000 Hz, then the PRI will be accurate to $\pm 1 \ \mu s$:

$$\pm \Delta T = \frac{\mp 1}{1,000} \frac{\text{Hz}}{\text{Hz}^2} = 10^{-6} \text{ sec} \tag{15.12}$$

The PRI accuracy is reported as 1 μs even though the reported PRF accuracy is 1 Hz.

Notice that a given fractional PRF error results in the same fractional PRI error:

$$\frac{\Delta T}{T} = -\frac{\Delta F}{F} \tag{15.13}$$

15.7   Histogram Analysis Summary

The following rules apply to histogram analysis:

1. Histograms require 10 to 100 points per bin for reliability (the more the better).
2. Several bin sizes must be tried to find a good representation of the PDF (keeping in mind rule 1 as the number of bins increase).
3. Conditional histograms are needed to investigate modes of operation (i.e., which PRIs go with which pulse durations?). Other uses include examining only high-accuracy reports or reports from a given geographic area or investigating parameter changes over time.
4. The accuracy values associated with the intercept data should be reported to users rather than the statistical quantities or histogram bin sizes.

5. Comments on the number of intercepts, the shape of their histogram, and any significant peaks or valleys should be provided for the user.

6. Because different users have different needs, the more information provided, the more likely that the users' needs will be met.

7. Try to verify that intercept reports have properly identified the signal. A *few* bad values can have a significant effect (see rule 8).

8. Statistical values (particularly the standard deviation) are significantly affected by outlying values. The nonparametric approach to finding limits avoids this problem.

9. Transforming reported parameter values to another form (e.g., PRI = *1/PRF*) must be done correctly.

References

[1]   Mood, A. M., *Introduction to the Theory of Statistics*, New York: McGraw-Hill, 1950, Chapter 16.

[2]   Jordan, E. C., (ed.), *Reference Data for Radio Engineers*, Indianapolis, IN: Howard W. Sams and Company, 1977, Chapter 42.

ELINT Data Files

16.1 Introduction

The users of ELINT are primarily concerned with emitter identification (as in radar warning receivers), and with countering the emitter once it is recognized. Other uses include driving simulation systems used in training. For identification, the interceptor makes use of the ELINT parameters and the parameter limits expected for each type of emitter as discussed in Chapter 15. For designing an electronic attack to counter an emitter, it is necessary to have more detailed information about how the radar operates and especially how the receiver processes signals. This is because only receivers are jammed—not transmitters. For this, detailed technical information is needed and not all of it is available from ELINT. This gives rise to dividing the information into *observed* and *assessed* categories. Observed data is largely from ELINT. Assessed data is an analyst's best estimate of how the system performs based on all information available at the time and including engineering calculations and the limitations due to target motion.

ELINT can be divided into technical information (TECHELINT) and operational information (OPELINT) as shown in Figure 16.1. This book is mainly concerned with technical ELINT. But note that emitter location, an OPELINT parameter, makes use of highly technical signal processing techniques. Of course, data files of non-ELINT data are also maintained and used. SIGINT other than ELINT includes the data listed in Figure 16.2. Non-SIGINT intelligence files are listed in Figure 16.3.

Electronic Intelligence (ELINT):
Technical and intelligence information derived from foreign noncommunications electromagnetic radiations emanating from other than atomic detonation or radioactive sources.

Technical ELINT (TECHELINT):
The category of electronic intelligence concerned with the signal characteristics, modes, functions, associations, capabilities, limitations, vulnerabilities, and technology levels of foreign noncommunications emitters and the electronics or weapons systems with which they are associated. In brief, Technical ELINT determines the capabilities and limitations of target emitters.

Operational ELINT (OPELINT):
The category of electronic intelligence concerned with the introduction, location, disposition, movement, employment, tactics and activity levels of known foreign noncommunications emitters and the weapon systems and military units/platforms with which they are associated. In brief, Operational ELINT determines the locations and readiness of target emitters.

Figure 16.1 Technical and operational ELINT.

Communications Intelligence (COMINT)
Derivation of intelligence by intercepting and analyzing communications signals

Unintentional Radiation Intelligence (RINT)
Intercepting and analyzing spurious and/or unintentional radiations

Foreign Instrumentation Intelligence (FIS)
Intercepting and analyzing telemetry and other instrumentation

Figure 16.2 Types of non-ELINT SIGINT data.

Radar Intelligence (RADINT)
Using a radar to obtain intelligence

Image Intelligence (IMINT)
Using photographic (PHOTINT) or optical sensors (IR thru UV, OPTINT) to obtain intelligence

Human Intelligence (HUMINT)
Using persons to obtain intelligence (overt or covert)

Acoustical Intelligence (ACOUSTINT)
Using passive or active acoustic means to obtain intelligence

Measurement and Signature Intelligence (MASINT)
MASINT includes RF, electro-optical, infrared, acoustic, seismic, materials, nuclear, and radar
information (i.e., data is widely diverse and complex)

Figure 16.3 Types of non-SIGINT intelligence data.

OPELINT is concerned with answering questions such as:

- What? This information is contained in the ELINT notation and this identifies the type of emitter.
- When? This information is contained in the time of intercept.
- Where? This information is obtained by emitter location.
- Who? This information is obtained by correlating the what, when, and where information with the "Order of Battle" (OB) databases. The electronic order of battle (EOB) lists emitters by type (or ELINT notation) and their locations. The ground emitters are contained in the ground EOB. It tells which military units have a particular type of emitter. For ship-borne and airborne emitters, the naval and air orders of battle contain information about the types of ships and their home ports and the types of aircraft and their units and home airfields. The *electronic fit* provides information about which electronic equipment is carried by each type of aircraft, ship, or ground unit.

Uses of OPELINT include adjusting ESM/RWR emitter identification logic, planning penetration routes, monitoring the readiness of enemy forces, and discerning enemy tactics.

Both TECHELINT and OPELINT require databases that provide a repository of the parameters that describe each signal. There are so many parameters that may be used to describe a signal, and so many different ELINT signals to describe, that it becomes a problem to maintain up-to-date and accessible files suitable for the several users of such data. To help manage this situation, it is natural to try to use computerized databases. Such databases can also help point out gaps and shortcomings in the data—in effect, blanks in the data file that should be filled or parameter values with considerable uncertainty about them. Each chapter in this book that deals with an ELINT parameter might be thought of as an outline for a list of parameter names that should be accompanied by measured values that describe the signal

For each parameter to be included in the file, there is a need to indicate the lower and upper limits that cover 95% of the emitter population for that signal. There should also be an indication of the accuracy to which the parameter is measured. For each value entered, there should be a reference to a technical report or other data containing the information so that the interested user can learn more details about the parameter value, measurement methods, and circumstances. In addition, there is a need to key the data entered to a textual commentary that is generally used to further explain or qualify the data.

To be useful, an ELINT data file must also provide a way to show how the various parameters operate together—which PRIs go with which pulse durations, or which RF channels are used for different functions. It is also necessary to indicate any relationships that may exist between this signal and other emitters in the file (e.g., a height finder that is always collocated and synchronized with a certain search radar).

Another important requirement is that the data file must contain different kinds of parameters for different kinds of signals. Airborne pulse Doppler radars require different parameters than over-the-horizon search radars; data links require different parameters than an air traffic control radar, and so on.

Finally, the data file must be useful for people interested in different levels of detail. A jammer designer has different information needs than those of a radar warning receiver designer, who has different information needs than those of the intercept receiver operator, who has different information needs than those of the mission planner, and so on.

16.2 Signal Identification

One of the critical uses of the data files is to provide a guide to the interceptor to allow him to *name the signal*. The interceptor makes basic field measurements of the fundamental ELINT parameters such as RF, PRI, pulse duration, and scan time or rate.

From these and other indications (such as PRI modulation or mode switching or an associated signal), the interceptor tries to determine which known signal this might be. This is done by comparing the observed parameters from the intercept with those listed in the interceptor's identification guide. Usually, the starting point is a list in RF order of all known signals. The observed RF may fall between the lower and upper limits of the RFs listed for a number of signals. For each of these signals, the other parameters are compared. If the interceptor finds a set of parameter limits that includes all those he observed, he concludes that the signal being intercepted can be so identified. At this stage, the interceptor attaches to his measured values of ELINT parameters the alphanumeric code from the identification guide that identifies that signal and then submits the report to the ELINT analyst responsible for analyzing all of the reported intercepts of that signal. The identification step is critical Identification must be verified by analysts in order to keep errors out of the estimates of the parameter limits (which, of course, will be used for creating the next edition of the identification guide).

Clearly, the identification guide parameter limits may be different from the true parameter limits for any particular radar. This is because the limits must include not only the variations observed across the emitter population but also the variations in the calibration errors across the population of intercept receivers. Also, the identification guide may omit a number of parameters that are not observable at the intercept sites. For instance, RF coherence or stability measures may not help identify a signal and may not be measurable except under certain conditions. As intercept equipment became more accurate and could be automatically calibrated, these differences between the measured and true population parameter values became less pronounced than when the equipment was more primitive.

Proper signal identification is perhaps the most critical step in the ELINT process.

16.3   ELINT Data for Radar Warning Receiver Design

The function of an RWR is to warn of a threat (usually to the aircraft on which it is mounted) from some radar-based weapon. In some ways this is similar to the signal identification problem faced by the intercept operator, yet there are fundamental differences in the information needed from ELINT data files. One of the main differences is that the RWR must reach a decision automatically and in a very short time. Also, because it is generally mounted in military aircraft (designed for combat or weapon delivery), it must be much smaller than the typical intercept gear used in reconnaissance aircraft. On the other hand, it need only be concerned with threats to that particular aircraft and only those that may be found in its theater of operation.

Modern RWRs are generally reprogrammable so that their threat libraries and the parameter limits contained in them can be altered as the threat changes or as the aircraft mission changes. Clearly, the parameters of most interest depend on the particular design of the RWR to be reprogrammed. If it is based on a wide-open (crystal-video) receiver, there will be little interest in RF limits except to ensure that a threat is in one of the bands covered by the RWR. On the other hand, the PRI and PRI modulation parameters will be crucial to the threat identification process. For a staggered PRI, for example, the period of the stagger (or stable sum) can be used to distinguish a threat from a nonthreat radar. There have been dozens of RWR systems deployed and tens of thousands manufactured. Therefore, they represent one of the major uses of ELINT products.

The following example helps to show how ELINT data might be used to program an RWR to identify threats. First, a crystal-video receiver provides the pulse train envelope to a processor that examines the pulse duration and PRI. The threat identification process first requires that the pulse duration be tested against a threshold to divide the signals into long and short pulse durations. For example, all pulse durations above 1 μs would be in one class and all those below 1 μs would be in the other class. For each of these pulse duration classes, a PRI table similar to Figure 16.4 would be constructed. This is constructed by using the PRI upper and lower limits from ELINT and arranging them in order of the lower limits. The x-axis is the PRI value and the y-axis is an arbitrary scale representing the emitter identification number. For this example, emitter 1 has a conventional (nominally constant) PRI with a lower limit of 40 μs and an upper limit of 130 μs. The processor also checks for stagger (emitter 3) and jitter (emitter 6) and notes the presence of these modulations. It might also measure the stagger period.

If a PRI of 100 μs is observed, the RWR identifies it as emitter 1. If it finds a PRI between 130 and 180 μs, the signal is ignored as a nonthreat. Between 180 and 220 μs, emitter 2 is identified. Between 220 and 250 μs there is confusion between emitters 2 and 3, but this can be resolved by noting whether or not stagger is present. Between 250 and 280 μs, there is confusion among emitters 2, 3, and 4. However, the presence of stagger identifies the signal as emitter 3.

To distinguish between emitters 2 and 4, some other parameter must be used. One possibility is to add a tunable narrowband superheterodyne receiver that can measure RF. This is too slow to search for signals. But because the possibilities have been narrowed to just two emitters, the superheterodyne receiver can be directed to search only the RF range used by one of the emitters. If it is found, then the RF limits of the other emitter are also checked in case both signals are present. Then the identification process is complete—unless the RF limits also overlap and the signal RF is found to be inside the overlap range. This implies the existence of an RF sort plot similar to Figure 16.4 within the identification processor. If the RF search fails to break the tie and identify the emitter, another parameter such as a scan characteristic or the simultaneous operation of an associated emitter may be used. The logic and associated sorts by parameter value are often called the emitter identification (EID) table.

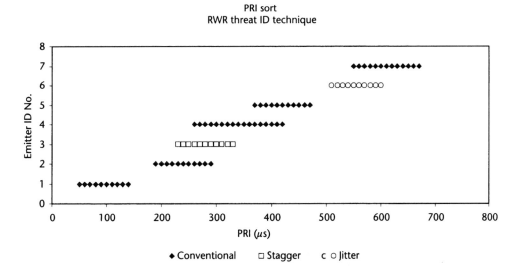

Figure 16.4   Emitter identification through PRI sort.

The construction of a good EID table is like solving a puzzle whose clues are embedded in the ELINT data files. Because the ELINT analyst never knows what little quirk of the emitter might be used to identify it, it is important to try to put as much information as possible in the database concerning everything that can be observed about the emitter. Likewise, different RWRs will make use of different ELINT parameters, depending on how each RWR operates. Note that the PRI sort is similar to inverting the ELINT data file so that instead of entering the emitter identification and finding the PRI limits, one enters the PRI limits and determines the possible emitter identifications for that PRI.

16.4 ELINT Data for Simulation and Training

Using ELINT data for simulation and training is different in that the parameter limits themselves are not used. Instead, to simulate a specific emitter requires that particular values within the parameter limits be selected. For example, one might use a random number generator to select an operating radio frequency from within the allowable range of the emitter type being simulated. In this case, simulation of a realistic radar environment containing a number of emitters of the same type requires that the distribution of the RF assignments among this class of emitters be known.

16.5 Adding Non-ELINT Data

To refine the data available for applications like threat warning and jamming applications, the observed ELINT data are augmented by any other information that may be available from sources such as photographs, manuals, and interviews with designers or operators. This might result in wider limits than are observed by ELINT and the addition of other parameters not available from ELINT. For example, jammers operate against the radar's receiver, not its transmitter. The receiver parameters are not directly observable from ELINT. Some important examples include the receiver bandwidth and the servo characteristics of range, angle, and velocity tracking loops [1]. Such non-ELINT data can be added in the same format and database organization as the ELINT data. A file that includes this non-ELINT data and also includes the assessments of the ELINT observable parameters by systems experts is a better source of the information needed for many applications than an ELINT-only file.

While it would be helpful if all of this information could be stored, retrieved, and manipulated by computers, it is usually necessary for the user to examine the data manually, at least to some extent. This happens because there are always important details that do not quite fit into the established data file entries, or else it is necessary to explain some important aspect of the system operation in sentence form. So far, the user of the file has no option but to read and understand these sentences. All of these databases are computerized and automated, but interpreting the data in the files requires human intellect.

16.6 Summary

Computerized databases are used to create many ELINT products. These include:

1. An identification guide for interceptors;
2. A comprehensive technical parameter file;
3. An augmented file that includes non-ELINT data, especially data concerning receivers.

In addition to these products, there are data files to retain the intercept reports. These require a standardized reporting format so that any intercept report can be handled, regardless of who generates it. These files would be used to create the histograms (described in Chapter 15) used to define the parameter limits.

Clearly as radar evolves, the data files used to describe their signal must also evolve. The new emphasis on multifunction radar and on electronically steered arrays means that the data files must include more capabilities to describe and catalog the signals from such radar systems.

These data files keep the ELINT data cycle working. Because this book is concerned with ELINT analysis, the starting point to describe the cycle is the analyst. The analyst reports results to various users. The users, as well as the analyst, determine the information that is lacking. These information needs are passed on to those who establish ELINT interception requirements (i.e., those who let the intercept operators know what signals to search for, what parameters to measure, and how accurately the measurements should be). These requirements direct the intercept site operators to try to obtain certain specific intercept data, which results in intercept reports arriving back to the analyst. This completes the ELINT cycle—and this book.

Reference

[1] Lothes, R. N., M. B. Szymanski, and R.G. Wiley, *Radar Vulnerability to Jamming*, Norwood, MA: Artech House, 1990.

Spectrum Widths: 3-dB and First Nulls for Trapezoidal Pulses

A.1 Introduction

The width of the spectrum of a simple pulsed radar signal is largely determined by the pulse shape. "Simple" means that there is no intrapulse FM or phase modulation for pulse compression purposes. Spectrum width is conventionally measured at the half power points, while pulse duration is measured at the half voltage points.

A.2 Rectangular Pulses

For a rectangular pulse, the power spectrum has a $\sin^2 x/x^2$ shape. In this case, the 3-dB points are 2.78 units apart, while the first nulls (which occur at $x = \pm\pi$) are 2π units apart. Therefore, the ratio of the width between the first nulls to the 3-dB spectrum width is 2.27. This means that the null-to-null spectrum width is 2.27 times the 3-dB bandwidth.

It should be noted that the peaks of the spectrum sidebands decrease in proportion to $1/f^2$ for frequencies separated by f from the carrier. The first nulls occur at a frequency of plus and minus the reciprocal of the pulse duration away from the carrier frequency. Therefore, the null-to-null spectrum width = 2/PD and the 3-dB spectrum width = 0.88/PD.

A.3 Trapezoidal Pulses: Equal Rise and Fall Times

A trapezoidal pulse with equal rise and fall times can be thought of as a narrow pulse of width τ convolved with a wider pulse of width T. This produces a total rise time of the composite pulse of τ (this is also the fall time), as shown in Figure A.1.

The spectrum of the trapezoidal pulse is simply the product of the spectra of each individual pulse, one of width τ and one of width T. Therefore, the combined spectrum is

$$|S^2(f)| = \frac{\tau^2 \sin^2(\pi f \tau)}{(\pi f \tau)^2} \cdot \frac{T^2 \sin^2(\pi f T)}{(\pi f T)^2} \tag{A.1}$$

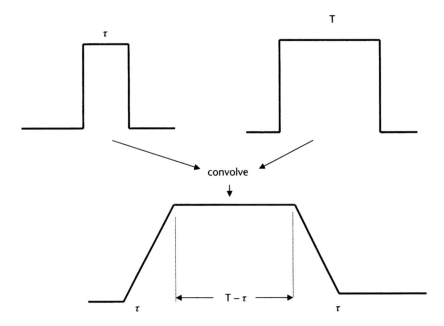

Figure A.1 Trapezoidal pulse produced by convolution of two rectangular pulses ($\tau < T$).

If $\tau \ll T$, then the first factor on the right is near 1 at the frequency for which the second factor is 0.5 (i.e., the 3-dB point). For example, if $\tau = 0.1T$, then the frequency at which the product on the right is 0.5 is still approximately $0.88/T$ and the null-to-null spectrum width is still approximately 2.27 times this. Note that the frequency at which the first null occurs is still the frequency at which the second factor on the right goes to zero (i.e., $f = 1/T$). This holds for all rise time values such that $\tau < T$. For $\tau = T$, the pulse is triangular and the spectrum becomes

$$|S^2(f)| = \frac{T^4 \sin^2 (\pi f T)}{(\pi f T)^4} \tag{A.2}$$

For this case, the 3-dB bandwidth is $0.64/T$ and the null-to-null spectrum width is still at $f = 2/T$, but this is now 3.13 times the 3-dB bandwidth.

For the case $\tau = T/2$, the 3-dB bandwidth is determined by the frequency for which

$$|S^2(f)| = \frac{(T/2)^2 \sin^2 (\pi f T/2)}{(\pi f T/2)^2} \cdot \frac{T^2 \sin^2 (\pi f T)}{(\pi f T)^2} = 0.5 \tag{A.3}$$

This occurs approximately at $f = 0.4/T$, so the 3-dB bandwidth is $0.8/T$ and the null-to-null frequency separation is 2.5 times the 3-dB bandwidth. This is summarized in Table A.1 for pulse width T.

Table A.1   Rise and Fall Times Versus the 3-dB Bandwidth and
Null-to-Null Separation

Rise/Fall Time	3-dB Bandwidth (B)	Null-to-Null Separation
0	$0.88/T$	$2.27B$
$0.1T$	$0.88/T$	$2.27B$
$0.2T$	$0.87/T$	$2.3B$
$0.3T$	$0.848/T$	$2.36B$
$0.4T$	$0.832/T$	$2.40B$
$0.5T$	$0.8/T$	$2.5B$
$0.6T$	$0.77/T$	$2.6B$
$0.7T$	$0.74/T$	$2.7B$
$0.8T$	$0.7/T$	$2.86B$
$0.9T$	$0.67/T$	$2.99B$
$1.0T$	$0.64/T$	$3.13B$

A.4   Trapezoidal Pulses: Unequal Rise and Fall Times

Often the fall time is longer than the rise time (typically three times longer). In this case, the spectrum is the spectrum of the central rectangular pulse plus the spectra of the triangular pulses representing the rise and fall times. The triangular pulses must be shifted in time to the beginning and end of the rectangular portion. Also, the time axis must be inverted for the triangle representing the rising edge. After manipulating the complex spectra to obtain the real and imaginary parts of their sum, the absolute value can be computed to locate the 3-dB points and first nulls. The complex spectrum of the sum is

$$s(f) = \frac{T \sin(\pi f T)}{(\pi f T)}$$

$$+ \frac{t_0}{4(\pi t_0 f)^2} [\cos(\pi T f) - 2\pi t_0 f \sin d(\pi T f) - \cos \pi(T + 2\pi t_0)f]$$

$$+ \frac{j t_0}{4(\pi t_0 f)^2} [-\sin(\pi T f) - 2\pi t_0 f \sin(\pi T f) + \sin \pi(T + t_0)f] \qquad \text{(A.4)}$$

$$+ \frac{t_1}{4(\pi t_1 f)^2} [\cos(\pi f T) - 2\pi t_1 f \sin(\pi T f) - \cos \pi(T + 2t_1)f]$$

$$+ \frac{j t_1 f}{4(\pi t_1 f)^2} [\sin(\pi T f) - 2\pi t_1 f \sin(\pi T f) - \sin \pi(T + 2t_1)f]$$

Here T is the duration of the rectangular portion, t_0 is the total rise time, and t_1 is the total fall time. Note that the time between the 50% points of the pulse is

$$\text{PD}(50\%) = T + 0.5(t_0 + t_1) \qquad \text{(A.5)}$$

This result can be compared with that obtained above for equal rise and fall times. For $t_1 = t_0$, the imaginary portions cancel and the real portions add, resulting in

$$s(f) = \frac{T \sin \pi TF}{\pi TF} + \frac{t_0}{2(\pi t_0 f)^2} [\cos(\pi Tf) - 2\pi t_0 f \sin \pi Tf - \cos \pi(T + 2t_0)f]$$

$$(A.6)$$

Combining terms yields

$$s(f) = \frac{\sin[\pi f(T + t_0)]}{\pi f} \frac{\sin(\pi t_0 f)}{\pi f}$$

$$(A.7)$$

Notice that this has the same form as the equation for equal rise and fall times, which was derived using convolution in Section A.3. The 50% pulse duration in this case is the width of the rectangular portion plus half of the rise and fall times. In other words, in this analysis $T + t_0$ is the 50% pulse duration, whereas in Section A.3, T is the 50% pulse duration.

For unequal rise and fall times, the first minimum of the spectrum occurs very close to the reciprocal of the 50% pulse duration on each side of the center frequency. However, this minimum is not strictly a null. In other words, the spectrum does not actually drop to zero. The situation is summarized in Table A.2 for a 50% pulse duration T.

The true spectral nulls are quite far removed from the center of the spectrum—on the order of 12 to 16 times the 3-dB bandwidth. For practical trapezoidal pulses, the separation of the first minima is a more realistic measure of the spectral width than the separation of the first nulls. As the rise and fall times become short and/or as the pulse shape becomes more symmetrical, the depth of the minima approaches that of a true null.

Table A.2 Unequal Rise and Fall Times for a 50% Pulse Duration

Rise Time	Fall Time	3-dB Bandwidth (B)	Separation of First Minima	Depth of First Minima
0.1T	0.3T	0.87/T	2.3B	−34.3 dB
0.2T	0.6T	0.82/T	2.44B	−23.2 dB
0.3T	0.9T	0.75/T	2.67B	−18.5 dB
0.01T	0.5T	0.85/T	2.38B	−24.7 dB

Some ELINT Considerations of FM Signals

John F. Green, Ph.D.

B.1 Introduction

In some ELINT applications, there is a need to collect and process signals in a manner that enables their high-fidelity demodulation. One of the most critical parameters is the minimum acceptable RF collection bandwidth. It is shown here that Carson's rule, used in communications signal engineering, can be modified appropriately to provide a relevant ELINT specification.

This modification of Carson's rule [1] is derived from an analysis of the spectrum of circular (sinusoidal) frequency-modulated signals and from an assessment of the influence of individual spectral features on demodulated signals. A quantitative evaluation is made of the degradations imposed on such signals by excessive RF spectral band limiting.

Finally, a brief derivation is presented, the results of which describe the effects of weak and moderate Gaussian noise on frequency-demodulated signals.

B.2 Effects of Sinusoidal Interference on Phase- and Frequency-Demodulated Signals

The intent of this section is to establish a quantitative basis for estimating the effects of an individual sinusoidal interference on phase- and frequency-demodulated signals.

If a sinusoidal interference is added to an arbitrary signal, the result can be expressed in the form:

$$g(t) = \cos[\phi_s(t)] + \gamma \cos(2\pi f_o t) \tag{B.1}$$

where f_o is the frequency of the sinusoidal interference and $\phi_s(t)$ is the total phase of the signal, including the carrier. In (B.1), γ is taken to be a (local) constant. The analytic signal components of the signal in (B.1) are

$$g_I(t) = \cos[\phi_s(t)] + \gamma \cos(2\pi f_o t) \tag{B.2}$$
$$g_Q(t) = \sin[\phi_s(t)] + \gamma \sin(2\pi f_o t)$$

The instantaneous phase, $\phi(t)$, of the composite signal defined by (B.2) is

$$\tan[\phi_i(t)] = \frac{g_I(t)}{g_Q(t)} \tag{B.3}$$

If (B.2) is introduced into (B.3), and γ is assumed to be small, then

$$\tan[\phi_i(t)] = \left[1 + \frac{\gamma \sin(2\pi f_0 t - \phi_s)}{\sin \phi_s \cos \phi_s}\right] \tan \phi_s \tag{B.4}$$

For a small $\delta\phi$,

$$\tan(\phi_s + \delta\phi) \approx \left[1 + \frac{\delta\phi}{\sin \phi_s \cos \phi_s}\right] \tan \phi_s \tag{B.5}$$

By comparison of (B.4) with (B.5):

$$\phi_i(t) \approx \phi_s(t) + \gamma \sin[2\pi f_0 t - \phi_s(t)] \tag{B.6}$$

Equation (B.6) indicates that the effect of the weak additive tone on the instantaneous phase is to introduce a sinusoidal ripple, with its frequency equal to the difference frequency between the two signals and amplitude equal to the ratio of their amplitudes.

If the signal-to-interference power ratio, SIR, is introduced,

$$\phi_i(t) \approx \phi_s(t) + \frac{1}{\sqrt{\text{SIR}}} \sin[2\pi f_0 t - \phi_s(t)] \tag{B.7}$$

It is necessary to emphasize that derivation of this description of additive tone effects on the instantaneous phase has assumed that the tone has low power, compared with the signal of interest. This description increasingly underestimates the effects of this additive tone as its relative power level grows. However, for SIR levels on the order of 20 dB or higher, (B.7) is a good approximation. If the SIR is 20 dB, the total range of the phase ripple is about 11.5°.

The instantaneous frequency, $f_i(t)$, of the weak additive tonal composite signal can be derived easily by direct time differentiation of (B.7).

$$f_i(t) \approx f_s(t) + \frac{1}{\sqrt{\text{SIR}}} [f_0 - f_s(t)] \cos[2\pi f_0 t - \phi_s(t)] \tag{B.8}$$

In (B.8), $f_s(t)$ is the instantaneous frequency associated with $\phi_s(t)$. This result shows that the effect of the weak additive tone is to cause a sinusoidal ripple of the instantaneous frequency (the rate of which is the frequency difference between the two signals), the same rate as affects the instantaneous composite phase.

Of greater importance, in some ELINT applications, is the amplitude of the frequency ripple. Its magnitude is the product of the frequency difference times the signal amplitude ratio. In effect, the phase ripple is *amplified* by the frequency difference. For example, if the SIR is 20 dB and the frequency difference is 20 MHz, the total range of the frequency ripple is about 4 MHz. If the SIR is 40 dB and the frequency difference is 10 MHz, the total range of the ripple is about 200 kHz.

Figure B.1(a, b) shows several examples of the demodulated phase and frequency signals, derived without the approximation from (B.3). This illustrates the effects of sinusoidal interference. Close examination shows that the demodulated signals, at an SIR level of 20 dB, have diverged only slightly from pure sinusoids.

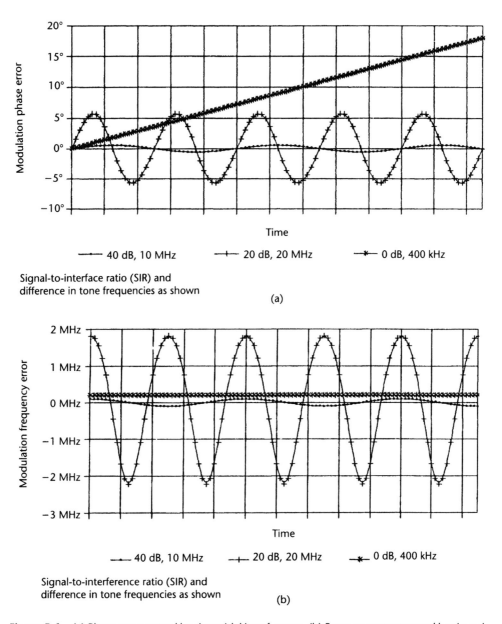

Figure B.1   (a) Phase error caused by sinusoidal interference. (b) Frequency error caused by sinusoidal interference.

Note that the phase of the 0-dB SIR output signal is linear and the frequency is constant, yielding only a beat frequency amplitude modulation.

These results indicate that the instantaneous frequency can be perceptibly altered by signal components having relative power levels as low as –40 dB within typical ELINT RF and video signal processing bandwidths. The phase signal is also degraded, but the degree of degradation is not sensitive to the frequency offset of the interference.

B.3   Signals with Sinusoidal Frequency Modulation

The intent of this section is to describe the spectra of sinusoidally frequency modulated signals. Such signals are of interest because their spectral analysis facilitates derivation of Carson's rule and its ELINT modification, which is presented in the subsequent section.

The effects of extreme spectral band limiting of FM signals, with respect to the degradation of their frequency-demodulated output, are evaluated in Section B.6.

A sinusoidal FM signal, $g(t)$, can be defined as

$$g(t) = \cos[\omega_o t + m \sin(\omega_m t)] \tag{B.9}$$

where

$$\omega_o = \text{constant carrier frequency} \tag{B.10}$$

$$\omega_m = \text{constant modulation frequency}$$

The modulation index, m, is defined as

$$m = \frac{\omega_d}{\omega_m} \tag{B.11}$$

where

$$\omega_d \equiv \text{constant peak radian frequency deviation} \tag{B.12}$$

The instantaneous radian frequency, $\omega_i(t)$, associated with the signal, $g(t)$, is

$$\omega_i(t) = \omega_o + \omega_d \cos(\omega_m t) \tag{B.13}$$

Equation (B.13) shows that the signal, $g(t)$, is indeed frequency modulated in sinusoidal fashion. By expanding the composite cosine, (B.9) can be rewritten as

$$g(t) = \cos(\omega_o t) \cos[m \sin(\omega_m t)] - \sin(\omega_o t) \sin[m \sin(\omega_m t)] \tag{B.14}$$

From [2],

$$\cos[m\sin(\omega_m t)] = J_0(m) + \sum_{k=1}^{\infty} J_{2k}(m)\cos(2k\omega_m t) \qquad (B.15)$$

and

$$\sin[m\sin(\omega_m t)] = 2\sum_{k=0}^{\infty} J_{2k+1}(m)\sin[(2k+1)\omega_m t] \qquad (B.16)$$

where $J_n(z)$ are Bessel functions of the first kind of order, n. After using (B.15) and (B.16) in (B.13), one finds

$$g(t) = +J_0(m)\cos(\omega_o t) - 2J_1(m)\sin(\omega_o t)\sin(\omega_m t) \qquad (B.17)$$
$$+ 2J_2(m)\cos(\omega_o t)\cos(2\omega_m t) - 2J_3(m)\sin(\omega_o t)\sin(3\omega_m t) + \dots$$

By multiple angle expansions, (B.17) can be rewritten as

$$g(t) = J_0(m)\cos(\omega_o t) + \sum_{k=1}^{\infty} J_k(m)\cos[(\omega_o + k\omega_m)t] + (-1)^k\cos[(\omega_o - k\omega_m)t]$$
$$(B.18)$$

Equation (B.18) demonstrates that the spectrum of a sinusoidally modulated FM signal is a discrete set of components, equally spaced at an interval of ω_m, the modulation frequency. The spectrum is centered at, and is symmetrical about, the carrier frequency, ω_o. The amplitude of a given spectral component, located at the frequency offset $(k\omega_m)$ from ω_o is the Bessel function, $J_k(m)$.

In the case of weakly modulated signals, m is small and only the lowest order Bessel functions have significant amplitudes. For such a case

$$g(t) \approx \cos(\omega_o t) + 0.5m\cos[(\omega_o + \omega_m)t] - 0.5m\cos[(\omega_o + \omega_m)t] \quad (B.19)$$

Equation (B.19) shows that the spectrum contains only the closest sideband, and most of the signal power is concentrated at the carrier frequency, for weakly modulated signals.

For more highly modulated signals, the spectrum becomes wideband, with significant power occurring at considerable displacements from the carrier frequency. This is demonstrated explicitly in the subsequent section. For certain values of the modulation index, m, $J_0(m) = 0$, causing the carrier frequency spectral region to exhibit no signal energy.

Spreading the spectrum for comparatively highly modulated FM signals can present a substantial problem to their high-fidelity ELINT capture and demodulation. Ordinarily, it is advantageous to limit the processing bandwidth of ELINT analysis systems to minimize inclusion of thermal and other noise and to minimize the probability of cochannel interference. This, in turn, improves the detectability of lower power signals.

However, the utilization of narrow bandwidths can seriously degrade the fidelity of highly modulated FM signals. This degradation can be demonstrated easily either by direct laboratory experimentation or by computer emulation. Examples of the latter are provided in Section B.6. A highly practical and efficient criterion for determining the critical minimal bandwidth necessary to avoid gross degradation of FM signals is Carson's rule, which is discussed in the next section.

B.4 Carson's Rule

Power spectra of virtually all types of FM signals generally decrease in amplitude with an increasing displacement in frequency from the carrier frequency. While details of these spectra may be diverse and complicated, as demonstrated in the subsequent section, observation shows that each spectrum tends to exhibit a more or less well-defined break point, beyond which the spectrum decreases rapidly. This break point, in the case of highly modulated CFM signals, occurs at a frequency displacement equal to the deviation frequency.

For small amounts of frequency deviation, this break point frequency is at the first sideband, a displacement equal to the modulation frequency. Both circumstances can be encompassed approximately by the specification that the minimum RF bandwidth necessary to contain the bulk of the signal energy is equal to twice the sum of the modulation and the deviation frequencies. This specification is known as Carson's rule, which has been found to be generally quite useful for communications signal applications.

Carson's rule can be derived from an examination of the spectra properties of CFM signals. It was previously shown that the spectral components of CFM signals consist of Bessel functions.

Under certain conditions, these functions can be approximated by the expression [2]

$$J_k(m) \approx \left(\frac{m}{2}\right)^k \frac{1}{k!} \tag{B.20}$$

In (B.20), m is the modulation index and k is the order of the spectral component. Stirling's approximation for $k!$ is

$$k! \approx \sqrt{2\pi k} \left(\frac{k}{e}\right)^k \tag{B.21}$$

Combining (B.20) with (B.20) gives

$$J_1(m) \approx \frac{1}{\sqrt{2\pi k}} \left(\frac{em}{2k}\right)^k \tag{B.22}$$

The term on the right-hand side of (B.22), raised to the kth power, exhibits a dramatic decrease in magnitude as k, the spectral component order (i.e., frequency displacement), is increased somewhat beyond m, the modulation index. Although the Bessel function approximation is not ideal for $m \approx k$, this roll-off in power is described well by (B.22) and is shown in Figure B.2.

This implies that the spectral components, $J_k(m)$, should decrease rapidly in amplitude if $k > m$. Because the spectral component, of order k, is displaced from the RF carrier frequency by the frequency, $k\omega_m$ (where ω_m is the FM modulation frequency), the bulk of the spectral power should be confined within a (two-sided) linear RF bandwidth, B, where

$$B \approx 2m \left(\frac{\omega_m}{2\pi} \right) \tag{B.23}$$

Equation (B.23) is deficient (in that it excludes the first order FM sidebands) at the frequency displacement ω_m, in the case of weakly modulated signals. Thus, (B.23) is commonly modified to

$$B \approx (2m + 1) \left(\frac{\omega_m}{2\pi} \right) = 2(f_d + f_m) \tag{B.24}$$

in which f_d is the frequency deviation and f_m is the modulation frequency.

Equation (B.24) is known as Carson's rule.

B.5   Modification of Carson's Rule for ELINT Applications

In Section B.2, it was shown that relatively very weak sinusoidal interference could cause a perceptible degradation in frequency-demodulated signals. The magnitude

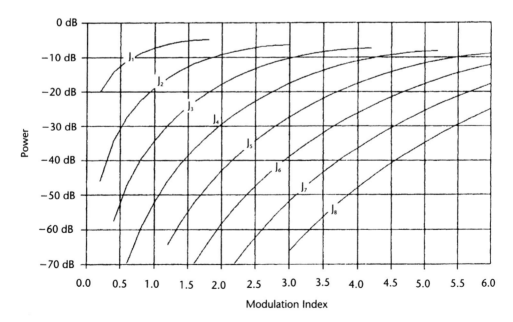

Figure B.2   Bessel function power as a function of modulation index.

of this degradation was found to be the SIR times the frequency offset of the component from the carrier frequency. This theoretical result serves as a model for modifying Carson's rule by supplying rough estimates of the contribution to the FM demodulation fidelity attributable to wideband spectral components (and to their potential inadvertent removal by underestimating acquisition bandwidths).

For example, a demodulated signal degradation with a magnitude 200 kHz was found for an additive tonal signal component having relative power of −40 dB and frequency offset of 10 MHz. If 200 kHz is the maximum level of allowable degradation, the constraint on the argument of the term raised to the kth power, in (B.22) should be stronger than that used to derive Carson's rule.

In the present case, if the component in question must be about −40 dB or smaller, the argument should be not much greater than 1/4 (compared with 1 for Carson's rule). If this modified constraint is adopted, then

$$\frac{em}{2k} \leq \frac{1}{4} \rightarrow k \geq 2em \approx 5m \tag{B.25}$$

If the weak modulation (i.e., the first sideband only) case is accommodated, as in (B.24), then the modification of Carson's rule, for this ELINT application, becomes

$$B = 5m(2f_m) + 2f_m = 10f_d + 2f_m \tag{B.26}$$

Such a modification of Carson's rule should not be taken as a requirement to be observed religiously and without contemplation, because it has been developed for a specific case, namely 200-kHz degradation. In addition, it is potentially a lot shakier than Carson's rule, given its high sensitivity to modest variations in the specifically selected cutoff ratio.

To place this speculation on a stronger foundation, Table B.1 shows, without approximation, the Bessel function order for various power thresholds as a function of the modulation index. For this range of parameters, it appears that a safe guess would be to multiply the Carson's rule–derived bandwidth by a factor of about 3, rather than 5, as indicated by (B.26). This modification of Carson's rule, for certain high-fidelity ELINT applications, has been motivated by an evaluation of the influence of small spectral components on demodulated frequency signals. Often, it is the demodulated phase, rather than the demodulated frequency, signal that is of primary ELINT interest. The question arises as to the suitability of Carson's rule, and its modification, for use with phase-demodulated signals.

Table B.1 Bessel Function Order Required for a Given Threshold Below the Peak of the Spectrum

| | | | | | Modulation Index | | | | | |
Threshold	0.2	0.4	0.6	0.8	1.0	2.0	4.0	6.0	8.0	10.0
−6 dB	1*	1*	1*	1*	1	4	5	8	—	11
−40 dB	2	3	3	4	4	6	8	12	—	16
−60 dB	3	4	4	5	5	7	10	14	—	18

* = required to maintain first sidebands.

It is easily shown that the time differentiation transfer function is $2\pi jf$, where f is spectral frequency. This transfer function can be used on the phase spectrum to generate the frequency spectrum because they are related directly by the time derivative. Because this type of transfer function increasingly emphasizes those phase spectral components, with an increasing displacement from the carrier, it is clear that the phase spectrum must, in general, be more localized than its corresponding frequency spectrum. Thus, Carson's rule is quite suitable for use with phase-demodulated signals. Indeed, it represents an overly severe criterion for phase demodulation.

Another related effect that must be considered in the selection of ELINT signal minimum collection and processing bandwidths for FM signals is that of asymmetric pass band filtering. For example, assume that the spectrum of a weakly modulated circular FM signal contains only the carrier component and the nearest sidebands. If this signal is asymmetrically filtered in such a manner as to eliminate one of the two sidebands, the resultant signal has its modulation index reduced by a factor of 2. The corresponding demodulated FM signal would be decreased, in deviation, by a factor of 2.

Generally, asymmetric filtering of the spectrum of a frequency-modulated signal tends to reduce and distort the original FM signature. The magnitude of the consequent degradation can often be estimated theoretically by variations of the approximation techniques presented here.

To accommodate all of the effects discussed concerning circular and other frequency-modulated signals, it is reasonable to select collection and processing bandwidths that are:

1. At least three times that implied by Carson's rule;
2. At least three times any linear frequency sweep;
3. Wide enough to avoid asymmetric signal filtering.

B.6 FM Demodulation Degradation by RF Band Limiting

This section develops a quantitative evaluation of the degradation caused by RF spectral band limiting to the nearest sideband of signals having sinusoidal frequency modulation. In the case of weakly modulated signals, the index of modulation m is small and only the lowest order Bessel functions have significant amplitudes.

Equation (B.19) shows that the spectrum contains only the closest sideband for weakly modulated signals and that most of the signal power is concentrated at the carrier frequency. If the higher order sidebands have very low relative power, RF band limiting of this signal, to include only the nearest sideband, results in only a little degradation of the resultant demodulated frequency signal. However, the derivations in the previous sections have demonstrated that even weakly frequency modulated signals deposit enough energy in the higher order sidebands to perceptibly affect the fidelity of the resultant demodulated signals.

To gain an initial theoretical estimate of the degradation caused by bandpass filtering to retain only the RF carrier and first order sidebands, assume such filtering of a sinusoidal FM signal having an arbitrary modulation index, m. From (B.18),

$$\frac{1}{J_0(m)} g(t) = \cos(\omega_o t) + \frac{\lambda}{2} \cos[(\omega_o + \omega_m)t] - \frac{\lambda}{2} \cos[(\omega_o - \omega_m)t] \quad \text{(B.27)}$$

where

$$\lambda(m) = 2 \frac{J_1(m)}{J_0(m)}$$

Equation (B.27) reduces to the form

$$\frac{1}{J_0(m)} g(t) = \cos(\omega_o t) - \lambda \sin(\omega_o t) \sin(\omega_m t) \quad \text{(B.28)}$$

An examination of (B.28) reveals that the quadrature signal has the form

$$\frac{1}{J_0(m)} g(t) = \sin(\omega_o t) - \lambda \cos(\omega_o t) \sin(\omega_m t) \quad \text{(B.29)}$$

After using (B.28) and (B.29) to extract the demodulated frequency signal,

$$\omega_i(t) = \omega_o + \lambda \omega_m \frac{\cos(\omega_m t)}{1 + \lambda^2 \sin^2(\omega_m t)} \quad \text{(B.30)}$$

where $\omega_i(t)$ is the instantaneous frequency. This is to be compared with the actual (unfiltered) instantaneous frequency, described by (B.13).

For a small modulation index, m, $\lambda(m) = m$, so that the narrowband-filtered signal frequency and the actual instantaneous frequency are nearly identical. For larger values of the modulation index, a considerable degradation in the frequency deviation can result from the narrowband filtering. As was noted in Section B.3, the zero-th order Bessel function is zero for certain (physically achievable) values of the modulation index. In these cases, $\lambda(m)$ becomes infinite, demonstrating that narrowband-filtered CFM signals can generate uncontained demodulated frequencies.

The mean square expectation of the sinusoidal instantaneous frequency deviation about the carrier frequency, as computed from the actual signal, (B.13), is

$$\langle [\omega_i(t) - \omega_o]^2 \rangle = \frac{1}{2} m^2 \omega_m^2 \quad \text{(B.31)}$$

In the case of the narrowband-filtered signal, the mean square expectation is found, from (B.30), to be

$$\langle [\omega_i(t) - \omega_o]^2 \rangle = \frac{1}{2} \frac{\lambda^2}{\sqrt{1 + \lambda^2}} \omega_m^2 \quad \text{(B.32)}$$

A comparison of (B.30) with (B.13) for the demodulated frequency, and (B.32) with (B.31) for the mean square demodulated frequency, indicates that narrowband filtering of CFM signals can indeed cause serious degradation.

Figures B.3 and B.4 show examples of CFM signal magnitude spectra and demodulated frequency signals for a variety of modulation indexes. Provided in each plot of demodulated frequency are results obtained by bandpass filtering to the first sideband only and the results with no bandpass filtering. The induced distortions can be seen to be extreme where the modulation index is in the neighborhood of zeroes of $J_0(m)$, the carrier frequency spectral contribution.

Table B.2 lists the ratios of the maximum deviation frequency and rms frequency deviation for the bandpass limited to unfiltered CFM signals for a range of modulation indexes.

Figure B.5(a, b) shows graphs of the ratios of the maximum deviation frequency and RMS frequency amplitude for bandpass limited to unfiltered sinusoidal FM

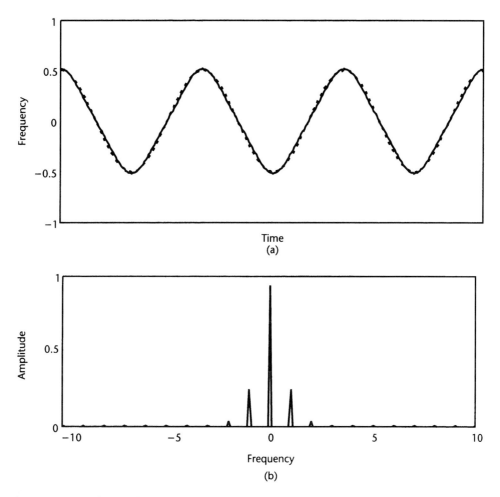

Figure B.3   (a) Effects of RF band limiting on demodulated sinusoidal FM signal ($m = 0.5$). (b) Spectrum of sinusoidal FM signal ($m = 0.5$). (c) Effects of RF band limiting on demodulated sinusoidal FM signal ($m = 1.0$). (d) Spectrum of sinusoidal FM signal ($m = 1.0$).

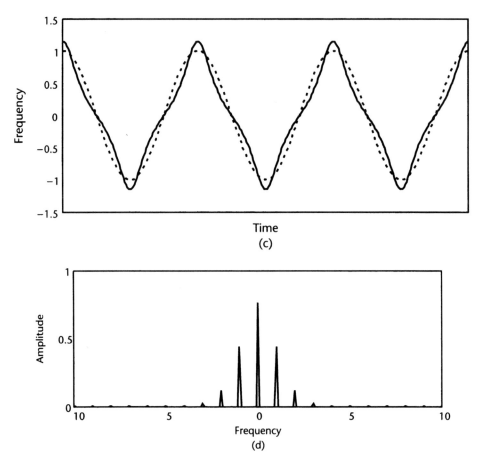

Figure B.3   (continued.)

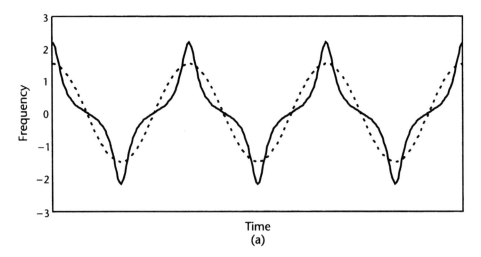

Figure B.4   (a) Effects of RF band limiting on demodulated sinusoidal FM signal ($m = 1.5$). (b) Spectrum of CFM signal ($m = 1.5$). (c) Effects of RF band limiting on demodulated sinusoidal FM signal ($m = 2.0$). (d) Spectrum of sinusoidal FM signal ($m = 2.0$).

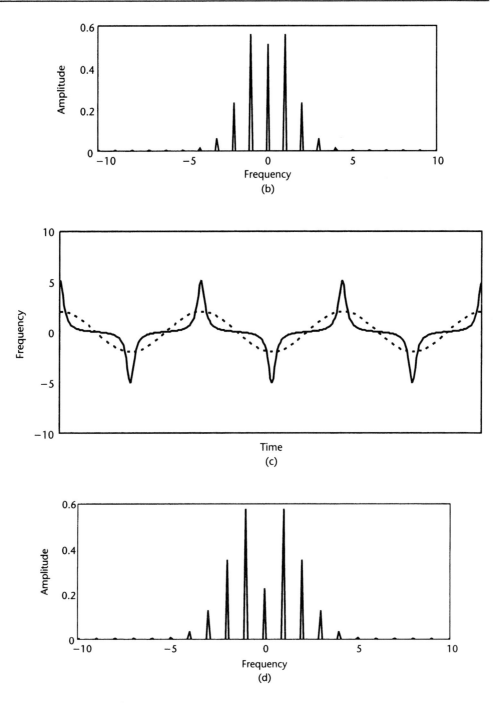

Figure B.4   (continued.)

signals for a range of modulation indexes. These theoretical results can be used to specify the required minimum ELINT signal collection and analysis bandwidths necessary to maintain a desired degree of demodulated signal fidelity. In circumstances where detailed spectral information is not available before signal collection, it is recommended that the collection be made using the widest bandwidth feasible.

Table B.2   Ratios of the Maximum Frequency Deviation and RMS
Frequency Deviation for Signals that Are Band Limited to the
First Sideband, Compared to Unfiltered Sinusoidal FM Signals

Modulation Index	Frequency Peak	RMS Frequency Deviation
0.0	1.0	1.0
0.2	1.01	1.0
0.4	1.02	0.98
0.6	1.05	0.96
0.8	1.09	0.95
1.0	1.15	0.93
1.2	1.24	0.92
1.4	1.37	0.93
1.6	1.56	0.95
1.8	1.90	1.01
2.0	2.58	1.12
2.2	4.58	1.44
2.4	172.85	8.49
2.6	3.74	1.20
2.8	1.58	0.74
3.0	0.87	0.52
3.5	0.21	0.19
4.0	0.08	0.08
4.5	0.32	0.24
5.0	0.74	0.38
5.5	18.14	1.82
6.0	0.61	0.31
6.5	0.18	0.15
7.0	0.00	0.00
7.5	0.14	0.11
8.0	0.34	0.20
8.6	1.53	0.42
9.0	0.60	0.26
9.5	0.18	0.13
10.0	0.04	0.03

B.7   Effects of Noise on FM Demodulated Signals

Assume that a signal, $A \cos[\phi_s(t)]$, is degraded by additive, band limited, white Gaussian noise, $n(t)$. The composite of signal plus noise can then be denoted as $g(t)$, where

$$g(t) = A \cos[\phi_s(t)] + n(t) \tag{B.33}$$

The total noise (circular) bandwidth is denoted as ω_N. The noise power spectral density is denoted as μ so that the total noise power, N, is

$$N = \mu \omega_N \tag{B.34}$$

The noise can be resolved into two mutually orthogonal components, $n_I(t)$ and $n_Q(t)$, which are instantaneously in phase and in quadrature, respectively, with the signal. Equation (B.34) can then be rewritten as

$$g(t) = [A + n_I(t)] \cos[\phi_s(t)] + n_Q(t) \sin[\phi_s(t)] \tag{B.35}$$

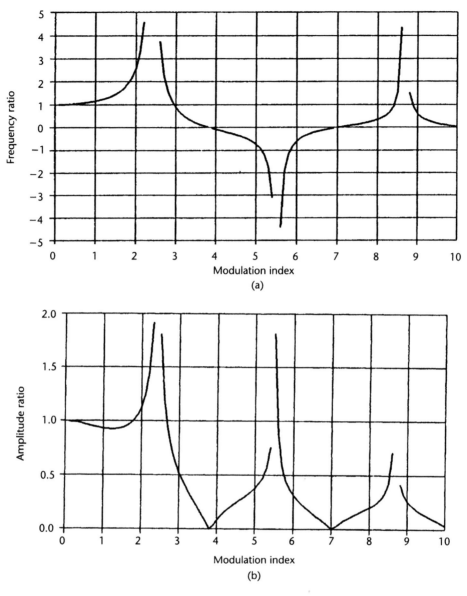

Figure B.5 (a) Ratio of the maximum frequency deviation for signals that are band limited to the first sideband, compared to unfiltered sinusoidal FM signals. (b) Ratio of the rms frequency deviation for signals that are band limited to the first sideband, compared to unfiltered sinusoidal FM signals.

Both n_I and n_Q are also Gaussian distributed, having the spectral power density, μ. Equation (B.35) can be expressed in the form

$$g(t) = a(t) \left\{ \left(\frac{A + n_I(t)}{a(t)} \right) \cos[\phi_s(t)] + \frac{n_Q(t)}{a(t)} \sin[\phi_s(t)] \right\} \qquad (B.36)$$

where

$$a(t) = \sqrt{[A + n_I(t)]^2 + n_Q^2(t)} \tag{B.37}$$

A phase angle, $\phi_n(t)$, can be defined, where

$$\cos[\phi_n(t)] \equiv \frac{A + n_I(t)}{a(t)} \tag{B.38}$$

$$\sin[\phi_n(t)] \equiv \frac{n_Q(t)}{a(t)} \tag{B.39}$$

Equations (B.38) and (B.39) are used in (B.36); one finds

$$g(t) = a(t)\{\cos[\phi_s(t) + \phi_N(t)]\} \tag{B.40}$$

If the noise is weak, compared with the signal amplitude, A, then (B.37), (B.38), and (B.39) can be combined to yield

$$\phi_n(t) \approx \sin[n_Q(t)] = \frac{n_Q(t)}{A} \tag{B.41}$$

The instantaneous frequency, ω_i, is defined as

$$\omega_i(t) \equiv \frac{d\phi}{dt} \tag{B.42}$$

If (B.41) is introduced into (B.42), then

$$\omega_i(t) = \omega_s(t) - \frac{1}{A}\frac{dn_Q(t)}{dt} \tag{B.43}$$

in which $\omega_s(t)$ is the instantaneous frequency of the signal, $A\cos[\phi_s(t)]$.

In the frequency domain, the transfer function for the differentiation, $d(\ )/dt$ = $j\omega$. Consequently, the noise term in (B.43) possesses a spectral noise power density, $p(\omega)$, where

$$p(\omega) = \frac{\mu\omega^2}{A^2} \text{ for } |\omega_N| \le \frac{\omega_N}{2} \tag{B.44}$$

Equation (B.44) reveals that the noise is preferentially concentrated (i.e., increases quadratically) in the higher frequency regions of the spectrum of the demodulated (video) signal. Consequently, if the demodulated FM (video) signal is lowpass filtered (without significant information loss), much of the noise can be eliminated. However, any such lowpass filtering process must be designed and used with some caution, especially if the (video) signal of interest is wideband. Suppression of signal information by the lowpass filtering process can substantially degrade the content, appearance, and characteristics of the FM video signal.

An estimate can be obtained for the rms amplitude of the PM noise described by (B.44). If the frequency noise amplitude is denoted as ω_N, then

$$\langle \omega_n^2 \rangle = \frac{1}{A^2} \int_{-\omega_N/2}^{\omega_N/2} \mu \omega^2 \, d\omega = \frac{2\mu}{3A^2} \left(\frac{\omega_N}{2} \right)^3 \tag{B.45}$$

Because the signal power $S = A^2/2$ and the noise power [from (B.34)] $N = \mu \omega_N$, (B.45) can be rewritten as

$$\langle \omega_n^2 \rangle = \frac{\omega_N^2}{24 \, \text{SNR}} \tag{B.46}$$

in which SNR $= S/N = A^2/(2N)$. Thus,

$$\sqrt{\langle \omega_n^2 \rangle} = \frac{\omega_N}{\sqrt{24 \, \text{SNR}}} \approx 0.2 \frac{\omega_N}{\sqrt{\text{SNR}}} \tag{B.47}$$

Converting (B.47) to frequency gives

$$f_n \, (\text{rms}) \approx 0.2 \frac{B_N}{\sqrt{\text{SNR}}} \tag{B.48}$$

where

$$f_n = \frac{\omega_n}{2\pi} \tag{B.49}$$

$$B_N = \frac{\omega_N}{2\pi}$$

Equation (B.48) shows that the effect of white noise on the frequency-demodulated (strong) signal is directly proportional to the input RF bandwidth and inversely proportional to the square root of the total input RF SNR. Note that the output video bandwidth of the demodulated signal is exactly equal to one-half of the input noise bandwidth because no video filtering has been imposed. The effects of lowpass filtering of the demodulated video signal are considered next.

If the FM-demodulated video signal is lowpass filtered such that the resultant maximum video (circular) frequency, $\omega_N = 2\pi f_v$ is less than $\omega_N/2$, then the rms frequency noise is found to be

$$f_n \, (\text{rms}) = f_v \sqrt{\frac{1}{3 \, \text{SNR}} \frac{f_v}{B_N}} \tag{B.50}$$

where

$$f_v = \frac{\omega_v}{2\pi} \qquad (B.51)$$

Equation (B.50) demonstrates that there is a nonlinear gain in the output demodulated signal video SNR by lowpass filtering, as long as no signal contributions are eliminated.

If the noise is moderate, rather than weak, the first order approximations of (B.38), (B.39), and (B.40) must be extended to the higher order. If this is done, the phase noise is found to be

$$\phi_n(t) \approx -\frac{n_Q}{A} + \frac{n_Q n_I}{A^2} \qquad (B.52)$$

After introducing (B.52), it is found that

$$f_n \text{ (rms)} = f_v \sqrt{\frac{1}{3\,\text{SNR}} \frac{f_v}{B_N} \left[1 + \frac{3}{4\,\text{SNR}} \left(1 - \frac{3f_v}{2B_N} + \frac{3f_v^2}{5B_N^2} \right) \right]} \qquad (B.53)$$

Equation (B.53) is reasonably valid for essentially all ELINT purposes for all input SNR levels at or above 10 dB.

References

[1]   Carson, J. R., "Notes on the Theory of Modulation," *Proc. of IRE*, Vol. 10, February 1922, pp. 57–64.

[2]   Abramowitz, M., and I. Stegun, (eds.), *Handbook of Mathematical Functions*, New York: Dover Publications, 1972.

A Frequency Hop Radar Example

A hypothetical interception situation can serve to illustrate a probability of intercept calculation without using the window function approach. Assume that there is a radar whose radio frequency (RF) is changed randomly to one of N frequencies on a pulse-to-pulse basis, and that we wish to intercept this signal and learn as much as possible using a relatively slowly sweeping narrowband receiver. The receiver sweeps a band, B, corresponding to the band over which the radar's frequency hops. The receiver instantaneous bandwidth β is equal to the spacing of the N radar frequencies (i.e., $N \sim B/\beta$). Suppose further that the receiver's sweep time is such that it stays open to receive a specified frequency for a time during which R pulses are transmitted by the radar.

The frequency scanning receiver can receive signals at essentially one frequency at a given time. If such a receiver is operating in the frequency range of a random frequency hopping radar, certain information concerning this radar will be available to the receiver. The calculations below predict the type and amounts of data that would be expected.

The only pulses of interest to the scanning receiver are those occurring at the frequency to which the receiver is tuned. Suppose the radar transmits on N discrete frequencies (or the receiver resolves N frequencies). Assuming the receiver is tuned to one of the N frequencies, then the probability that a particular pulse occurs at the frequency to which the receiver is tuned is

$$p = 1/N \tag{C.1a}$$

Of course, the probability that the pulse occurs at some other frequency is

$$q = 1 - p = (N - 1)/N \tag{C.1b}$$

Utilizing these probabilities, some interesting calculations can be made. In the examples that follow, it is assumed that the receiver scans through a particular frequency in a time necessary for the transmitter to emit R pulses.

C.1 Probability That One or More Pulses Occur at the Desired Frequency

One of the simplest and yet most interesting requirements necessary for an interception is that at least one of the R pulses transmitted must occur at the frequency (bin) to which the receiver is tuned. The easiest way to make a computation of this probability is to calculate the probability that the observer receives zero pulses and then subtract this probability from 1. Since the frequency of each pulse is independent of the others, the probability that all R pulses go into some other bin is

$$P_o = q^R \tag{C.1c}$$

Then the probability of getting *at least* one pulse in the desired bin is

$$P = 1 - P_0 = 1 - q_r \tag{C.2}$$

A plot of this function is shown in Figure C.1.

Equation (C.2) may be interpreted in an interesting manner. If the probability of success on a trial is p, then a certain number of trials, R, will be needed to have P probability of obtaining at least one success. Thus, if P and p are specified, the number of trials needed may be calculated using (C.2).

C.2 Probability That Exactly One Pulse Occurs at the Desired Frequency

If the event of interest was that "*exactly* one pulse occurred in the desired bin out of the R pulses transmitted," the probability of this event would be the probability of obtaining one pulse in the desired bin times the probability of getting all the other $R - 1$ pulses in some other bin times the number of ways (orders) in which this can happen. (Since the first pulse could be of the desired frequency, and the

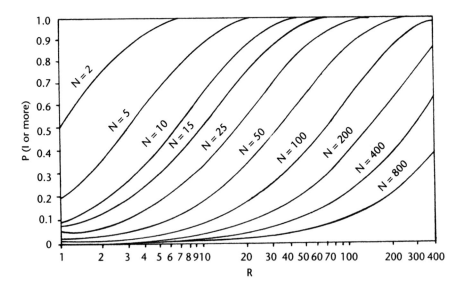

Figure C.1 Probability of receiving one or more pulses out of R at a particular frequency (N is the number of discrete radar frequencies).

rest of some other frequency, or the second could be of the desired frequency, and so on, there are R ways to get exactly one pulse of the desired frequency.) Thus, the probability of getting *exactly* one pulse of R in the desired bin is

$$P_1 = Rp(1-p)^{R-1} \qquad (C.3)$$

C.3 Probability That Exactly *k* Pulses Occur at the Desired Frequency

This reasoning can be extended to determine the probability of k pulses out of R occurring in the desired frequency bin. The result is the familiar formula for the binomial distribution:

$$P_1 = \binom{R}{k} p^k (1-p)^{R-k} \qquad (C.4)$$

where

$$\binom{R}{k} = \frac{R!}{k!(R-k)}$$

Note that the expression for P_k is recognized as the kth term of the expansion $(p+q)^R$, hence the name "binominal distribution."

C.4 Probability That Several Pulses Occur at the Desired Frequency Less Than *G* Pulses Apart

For some scanning receivers, it might be useful to require that a minimum of two pulses occur at the desired frequency before deciding that a valid interception had occurred so that PRI measurements could be made.

In this case, a certain maximum time will be allowed between the two pulses. This maximum allowable PRI will usually be some fraction of the total time the receiver is open to a particular frequency. The number of pulses transmitted during the maximum allowable PRI will be taken as G. Clearly, G is less than or equal to R. In this case, the event of interest can be defined as the case when two or more pulses out of R occur at the desired frequency and the pulses are separated by fewer than G pulse intervals.

To illustrate the requirements for an event in this case, suppose that the pulses are occurring at a steady rate and are numbered 1, 2, 3, . . . , R. In a total of R pulses, an event occurs when two or more pulses arrive in the frequency bin to which the receiver is tuned and when their numbers differ by less than G.

The probability of an event can be found by subtracting from unity the probabilities of all the results of the R trials which *do not* cause an event.

The event cannot occur if the trial of R pulses ends with zero pulses in the desired bin, and it cannot occur if the trial ends with only one pulse in the desired bin. Thus, two terms that must be subtracted are P_0 and P_1 from the binomial distribution, (C.4).

However, if two pulses occur at the desired frequency, there will be some cases that will cause an event and some that will not. It is necessary to count the ways in which two pulses may occur at the desired frequency and *not* cause an event. The pulses transmitted are numbered 1, 2, ..., G, ..., R. Let A mean that a pulse has occurred at the desired frequency and let B mean that a second pulse has occurred at the desired frequency. How many arrangements are there such that B occurs G or more pulses later than A?

If A occurs on the first pulse, then B may occur on the $(G + 1)$st pulse, or $(G + 2)$nd, and so on up to R. Hence, there are $R - G$ positions in which B may occur and still be G or more pulses from A, provided A occurs on pulse one. If A occurs on pulse two, then B may occur on pulse number $(G + 2)$, $(G + 3)$, and so on up to R. Hence, there are $R - G - 1$ positions in which B may occur if A occurs at pulse two. This process can be continued until A occurs on the $(R - G)$th pulse. Then there is only one position for B which will not cause an event: namely, B must occur on the Rth pulse.

Evidently, the total number of ways in which two pulses may occur in the same frequency bin and *not* cause an event is the sum $(R - G) + (R - G - 1) + (R - G - 2) + \ldots + 1$. The sum of the integers from 1 to k is given by

$$k(k + 1)/2 \tag{C.5}$$

Thus, the total number of ways to receive two pulses out of R at the desired frequency and have them be more than G pulses apart is

$$(R - G)(R - G + 1)/2$$

Similar methods can be used to calculate the probability that three pulses occur at the desired frequency and do not cause an event. Let A mean that the first of the three pulses has occurred at the desired frequency. Similarly, let B and C mean that the second and third of the three pulses have occurred at the desired frequency.

The event will not occur if A occurs on the first pulse, B occurs on the $(G + 1)$st pulse, and C occurs on the $(2G + 1)$st pulse. In fact, if A occurs on the first pulse, then B and C may occur in any manner among the last $R - G$ pulses such that they are at least G apart. However, this number has already been calculated above (for R pulses). Hence, the number of ways three pulses cannot cause an event if A occurs on pulse one can be obtained by substituting $R - 2G$ for k in (C.5):

$$(R - 2G)(R - 2G + 1)/2$$

In like manner, if A occurs on pulse two, then B and C may be arranged among the last $R - G - 1$ pulses in any way so as not to cause an event. This number would be

$$(R - 2G - 1)(R - 2G)/2$$

If A occurs on the jth pulse, then the number of arrangements is

$$\frac{(R - 2G - j + 1)(R - 2G - j + 2)}{2}$$

Then this counting process can be continued until A occurs on the $(R - 2G)$th pulse. Then B may occur on the $(R - G)$th pulse and C may occur on the Rth pulse, not causing an event. Thus, the total number of ways not to get an event when three pulses occur at the desired frequency is given by

$$\sum_{S=2}^{R-2G+1} \binom{S}{2} \tag{C.6}$$

The relation for the sum of the binomial coefficients is

$$\sum_{S=M}^{N} \binom{S}{M} = \binom{N+1}{M+1} \tag{C.7}$$

Thus, the total number of ways three pulses out of R may occur at the desired frequency and *not* cause an event is

$$\binom{R - 2G + 2}{3} = \frac{(R - 2G + 2)(R - 2G + 1)(R - 2G)}{6} \tag{C.8}$$

Then, to account for those situations in which three pulses occur but do *not* cause an event, the probability is

$$p^3 q^{R-3}(R - 2G + 2)(R - 2G + 2)(R - 2G)/6 \tag{C.9}$$

This process can be generalized to give the probability of getting no event if k pulses out of R occur at the desired frequency as follows.

Let $P(k)$ be the probability that k pulses out of R occur at the desired frequency and do *not* cause an event. Then,

$$P(k) = p^k q^{R-k} \binom{R - (k - 1)G + k - 1}{k} \tag{C.10}$$

It should be observed that $P(k)$ is equal to the binomial term P_k for $k = 0$ or 1. Thus, the complete expression for $P(E)$ is

$$P(E) = 1 \sum_{k-0}^{L} p^k q^{R-k} \binom{R - (k - 1)(G - 1)}{k} \tag{C.11}$$

where

$$L = \left[\frac{R + G - 1}{G} \right] \tag{C.12}$$

The brackets around the above expression designate L as the integer value of the expression. In other words, the value of $(R + G - 1)/G$ will be of the form $A.B$, where A is some integer and B is some decimal fraction; then $L = A$. For example, if $R = 9$, $G = 3$, then $(R + G - 1)/G = 3.667$. The corresponding value of L is, therefore, 3.

As a check, in the limiting case where $G = R$, $L = 2 - 1/R$, or $L = 1$, the formula reduces to $P(E) = I - P_0 - P_1$ as is expected.

A simple method of summing (C.11) into a closed form has not been found. In the particular cases of most interest, however, R is not very much larger than G ($R = 2G$, $3G$, $4G$, $5G$, for example). For this reason, calculations using the above formula are not overly cumbersome.

For $R = 3G$, the formula reduces to

$$P(E) = 1 - q^R - Rpq^{R-1} - p^2 q^{R-2} (R - G)(R - G + 1)/2 \tag{C.13}$$
$$- p^3 q^{R-3} (R - 2G + 2)(R - 2G + 1)(R - 2G)/6$$

This formula for $P(E)$ with $R = 3G$ has been numerically evaluated and the results are shown in Figure C.2.

A computer simulation has been made to check the validity of (C.13). It was comprised of 450 trials of 39 pulses per trial. There were 50 frequency bins among which the pulses were divided ($R = 39$, $G = 13$, $N = 50$). As a result, 47 events were counted. This leads to an empirical probability of 47/450, or 0.1044. The formula given by (C.13) gives 0.1150, a favorable comparison. The most probable (expected) number of events in 450 trials, according to the calculated probability, would be 51, or four more than the results of the simulation. However, since the standard deviation of the number of events in this case is 7, and the probability that between 44 and 58 intercepts will be obtained from the simulation is about 0.85, the result of 47 is quite reasonable.

C.5 Probability Distribution of the Interval Between Two Pulses

In view of the importance of the PRI in signal identification, the distribution of the interval between the two pulses making up such an intercept is of interest. This distribution is derived below, and the results for the case where the maximum pulse interval is one-third of the receiver dwell time are shown in Figures C.3 and C.4. The expected value of the PRI observed is also given. (For $G = 12$, the maximum allowed PRI is 12 of the radar's PRIs.)

The probability that two consecutive radar pulses make up the intercept is of interest, since in that case the observed PRI is the actual PRI. This function is shown in Figure C.5.

For a receiver requiring two pulses from a frequency hop radar before a valid intercept is reported, it is useful to calculate the distribution of the spacing between

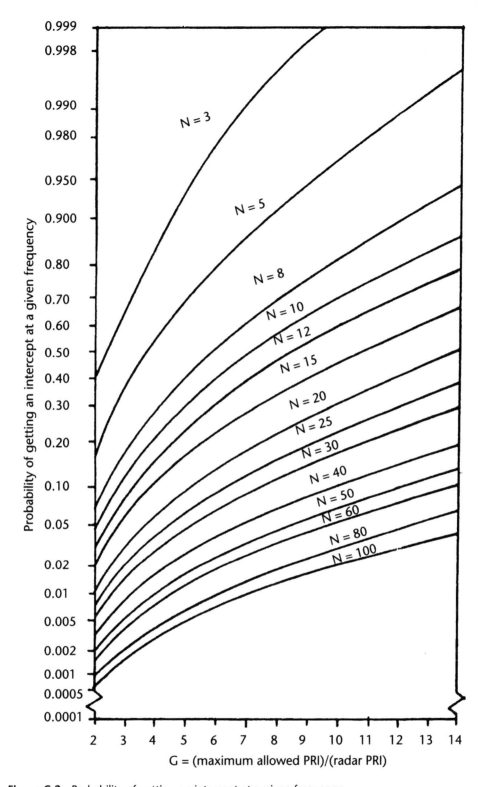

Figure C.2   Probability of getting an intercept at a given frequency.

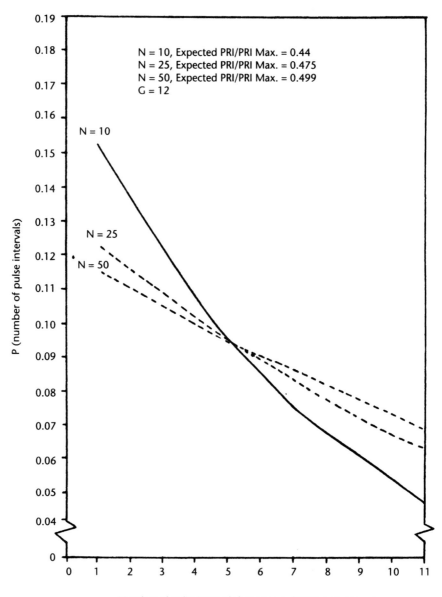

Figure C.3 Distribution of PRI values ($G = 12$).

the two pulses which combine to make up an intercept at a given frequency. The present problem is then to find out how many pulse intervals occur between the pulses making up such an event. Assume that the radar transmits R pulses during the time the receiver dwells at one frequency. If the R pulses are numbered 1, 2, 3, ... , R and occur in order, let the first pulse of the two necessary to make up an event be numbered k_2 and the second k_2, Then the number of pulse intervals between the two pulses is $I = k_2 - k_1$. Clearly, I must be an integer from 1 to $(G - I)$; G being the maximum allowable PRI divided by the PRI of the radar.

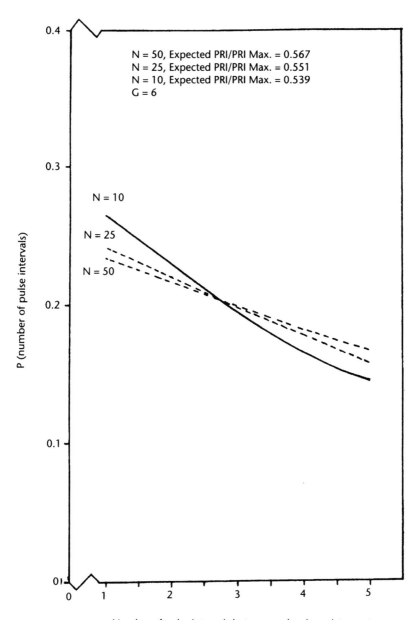

Figure C.4   Distribution of PRI values ($G = 6$).

The methods used lead to a formulation that gives the probability distribution for the number of pulse intervals between the pulses that make up an event for any values of R and G. However, inasmuch as the specific case of $R = 3G$ is typical, numerical calculations have been confined to values of R that are three times the values of G.

The distribution of I over the values 1, 2, 3, ..., $(G - 1)$, can be found by first counting all the ways in which two successive pulses can cause an event

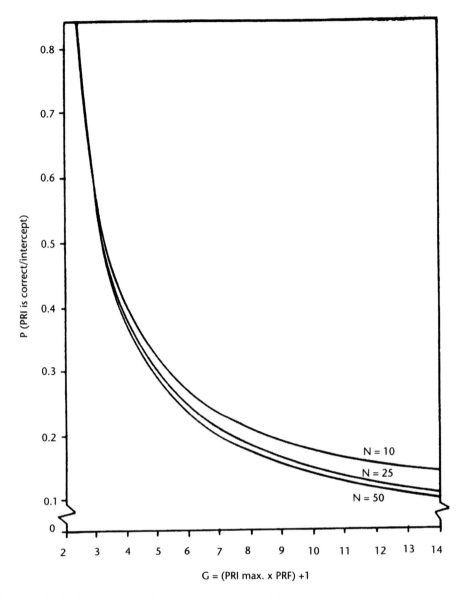

Figure C.5 Probability that an intercept has the correct PRI.

$(I = 1)$, then by counting all the ways in which the second pulse of the event is one pulse removed from the first pulse of the event $(I = 2)$, and continuing to $(G - 1)$. Suppose, for example, $G = 3$ and $R = 9$. What is the probability of an event with $I = 1$? This could happen if the first two pulses both occurred at the correct frequency. The rest of the R pulses would not matter.

Thus, an event can occur with $I = 1$ if the first two pulses occur at the correct frequency with probability p^2. It could also happen if the second and third pulses occurred at the correct frequency and the first did *not* occur at the correct frequency. The probability of this happening is pq^2. Similarly, the third and fourth pulses could occur at the desired frequency, and the first and second at some other frequency. The probability of this happening is q^2p^2. Notice that the case where the first pulse occurs at the desired frequency, the second at some other frequency, and the third and fourth at the correct frequency (pqp^2) cannot be counted because it would cause an event with $I = 2$ after the third pulse and, once an event occurs in the R trials, none after it is considered.

The next case to be considered is when the fourth and fifth pulses occur at the desired frequency. Then the second and third pulses must occur at some other frequency (since $G = 3$) in order to not cause an event. However, the first pulse may or may not occur at the desired frequency since it is G pulses away from the fourth. Thus, for this case, the probability expression contains two terms: $q^3p^2 + q^2p^3$.

Table C.1 shows this situation ("*" indicates a pulse at the desired frequency; "—" indicates a pulse at a different frequency; and "?" indicates a pulse that may or may not be at the desired frequency).

The probabilities numbered 1, 2, 3, and 4 have been discussed. Number 5 comes about in the following way. There are now two "question" pulses, numbers 1 and 2. One of them may occur at the desired frequency, or both may occur at some other frequency. However, the case in which both occur at the desired frequency is not permitted. (This would repeat the situation in case 1.) Thus, the probability expression comprises the following terms: $qqq^2p^2 + pqq^2p^2 + pqp^2p^2 = q^4p^2 + 2q^3p^3$. This counting process may be continued until the eight probabilities are obtained. The total probability of obtaining an event with $I = 1$ is, then, the sum of all these probabilities. Notice that the first G probabilities have one term, the next G probabilities have two terms, and the next $(G - I)$ have three terms. Also, there are $(R - I)$ probabilities in all.

For the case $G = 3$, there will also be events with $I = 2$. These cases are illustrated in Table C.2.

The total probability that $I = 2$ is the sum of all the above terms. Notice that the first G terms have only one term, the next G terms have two, and the remaining $(G - I)$ terms have three.

Consideration of the mechanics of the counting process, with the help of some further examples, has led to the general form for the probabilities. Let k be an index that numbers the rows, and let j be an index that numbers the columns. There are $(R - I)$ rows and L columns; L having been found to be

$$L = \left[\frac{R + G - I - 1}{G} \right] \tag{C.14}$$

Table C.1 Ways to Get an Intercept with $I = 1$

Pulses	1	2	3	4	5	6	7	8	9	Probability
	*	*								1. p^2
	—	*	*							2. qp^2
	—	—	*	*						3. q^2p^2
	?	—	—	*	*					4. $q^3p^2 + q^2p^3$
	?	?	—	—	*	*				5. $q^4p^2 + 2q^3p^3$
	?	?	?	—	—	*	*			6. $q^5p^2 + 3q^4p^3$
	?	?	?	?	—	—	*	*		7. $q^6p^2 + 4q^5p^3 + q^4p^4$
	?	?	?	?	?	—	—	*	*	8. $q^7p^2 + 5q^6p^3 + 3q^5p^4$

Table C.2 Ways to Get an Intercept with $I = 2$

Pulses	1	2	3	4	5	6	7	8	9	Probability
	*	—	*							1. qp^2
	—	*	—	*						2. q^2p^2
	—	—	*	—	*					3. q^3p^2
	?	—	—	*	—	*				4. $q^4p^2 + q^3p^3$
	?	?	—	—	*	—	*			5. $q^{5}p^2 + 2q^4p^3$
	?	?	?	—	—	*	—	*		6. $q^5p^2 + 3q^5p^3$
	?	?	?	?	—	—	*	—	*	7. $q^7p^2 + 4q^6p^3 + q^5p^4$

The brackets here mean to take L as the integer value of the expression. The determination of L is discussed later.

The probabilities in this matrix are given by the general term

$$P(I; j, k) = \binom{k - (G - 1)(j - 1) - 1}{j - 1} p^{j+1} q^{k-j+1-1} \qquad (C.15)$$

in which k ranges between I and $(R - I)$, and j ranges between I and L.

In order to derive an expression for the probability of the occurrence of an event with a specified I, it is necessary to sum all the possibilities in this array:

$$P(I) = \sum_{j=1}^{L} \sum_{k-1}^{R-1} P(I, j, k) \qquad (C.16)$$

Before performing this sum, it is helpful to discuss the derivation of the expression for L. In the $P(I, j, k)$ terms, the binomial coefficient that appears has a value only for values below a certain j, after which it becomes zero. This value of j occurs when the "upper" number becomes equal to the "lower" number, or when

$$K - (G - 1)(j - 1) - 1 = j - 1 \qquad (C.17)$$

Thus, the largest value of j to be assigned in the calculation is given by

$$L = \left\lceil \frac{k + G - 1}{G} \right\rceil \qquad (C.18)$$

However, in the above expression for $P(I)$, the summation eliminating the value k has already been performed; hence, the largest j value is given when k assumes its largest value, namely $(R - I)$. Thus,

$$L = \left\lceil \frac{R - k + G - 1}{G} \right\rceil \qquad (C.19)$$

Notice that j must become at least as large as L, but since the binomial term becomes a zero for j values larger than L, any limit for j that is larger than L will give the same answer.

The initial step is to perform the inner sum:

$$\sum_{k-1}^{R-1} \binom{k - (G - 1)(j - 1) - 1}{j - 1} p^{j=1} q^{k-j+I-1} \tag{C.20}$$

For purposes of this sum, the values of j, G, and I are constants. The terms from $k = 1$ to $k = (j - 1)G + I$ are zero, since for these values of k the upper number in the binomial coefficient is smaller than the lower number. Thus, the sum can just as well be performed from $k = (j - 1)G + 1$ to $k = R - I$:

$$\sum_{k=(j-1)G+1}^{R-1} \binom{k - (G - 1)(j - 1) - 1}{j - 1} p^{j+1} q^{k-j+I-1} \tag{C.21}$$

The usefulness of this last step can be seen by making a change of variable. Let

$$K - (G - 1)(j - 1) - 1 = h \tag{C.22}$$

Then the sum is given by

$$\sum_{h=j-1}^{R-I-(G-1)(j-1)-1} \binom{h}{j - 1} p^{j+1} q^{h+(G-1)(j-1)-j+1} \tag{C.23}$$

If the constant factors are moved outside the summation sign, the simplified sum to be performed is

$$p^{j+1} q^{1+(G-1)(j-1)-j} \sum_{h=j-1}^{R-I-(G-1)(j-1)-1} \binom{h}{j - 1} q^h \tag{C.24}$$

In order to simplify the notation, it is useful to make the following definitions:

$$S = \sum_{h=\alpha}^{\beta} \binom{h}{a} q^h \tag{C.25}$$

$$\alpha = j - 1 \tag{C.26}$$

$$\beta = R - I - (G - 1)(j - 1) - 1 \tag{C.27}$$

The summation may be performed by using a generating function. One defines a function $A(t)$ as

$$A(t) = 1 + qt + q^2 t^2 + q^3 t^3 + \ldots + q^\beta t^\beta \tag{C.28}$$

By applying the formula for the sum of a geometric series, $A(t)$ can be expressed in closed form:

$$A(t) = \frac{1 - (qt)^{\beta-1}}{1 - qt} + \sum_{b=0}^{\beta} q^b t^b \qquad (C.29)$$

The α derivative of $A(t)$ is

$$\frac{d^\alpha A(t)}{dt^\alpha} = \sum_{b=\alpha}^{\beta} (b)_\alpha q^t t^{-\alpha} \qquad (C.30)$$

in which $(b)_\alpha = b(b-1)(b-2) \ldots (b - \alpha + 1)$. This sum, with $t = 1$, is identical to S above with the exception of the $\alpha!$ factor, which should be in the denominator. Thus,

$$S = \frac{1}{\alpha!} \frac{d^\alpha A(t)}{dt^\alpha} \bigg|_{t=1} \qquad (C.31)$$

In terms of the original problem, the inner sum in the expression for $P(I)$ can now be replaced. Equation (C.16) becomes

$$P(I) = \sum_{j=1}^{L} \frac{p^{(j+1)(G-1)(j-1)-j}}{(j-1)!} \frac{d^{j-1} A(t)}{dt^{j-1}} \bigg|_{t=1} \qquad (C.32)$$

in which

$$A(t) = [1 - (qt)^{R-I-(G-1)(j-1)}]/(1 - qt) \qquad (C.33)$$

Although this formula is somewhat cumbersome, the value of L is practically never more than 5 or 6. The specific case for $R = 3G$ was selected as an example, hence the formula for $P(I)$ was obtained in closed form for this relationship between R and G.

If $R = 3G$, then L is found from

$$L = \left[\frac{3G + G - I - 1}{G} \right] \qquad (C.34)$$

$$L = 4 \left[\frac{I + 1}{G} \right] \qquad (C.35)$$

Since I must be between 1 and $(G - 1)$, L must always be an integer equal to or greater than 3 and less than or equal to $(4 - 2/G)$. This means that L is always 3 for any value of I if $R = 3G$. In this case, $P(I)$ is given by

$$P(I) = \sum_{j=1}^{J} p(I : j) \qquad (C.36)$$

or

$$P(I) = pq^{I-j}\left(1 - q^{3G-1} + q^{G-1}[1 - q^{G-1}(2G - I + 1) - (2G - I)q]\right.$$

$$+ q^{2G-2}\left\{1 + \frac{q^{G-1}}{2}[2(G - I)(G - I + 2)q - (G - I + 1)(G - I + 2)\right.$$

$$\left.\left.- q^2(G - I)(G - I + 1)]\right\}\right) \tag{C.37}$$

Figures 3.3 and 3.4 show several characteristic "I-distributions" for some typical values of G and N (for $R = 3G$). The distributions have been normalized so that the values of $P(I)$ obtained from the curves are given by

$$\frac{P(I)}{\displaystyle\sum_{j=1}^{G-1} P(I)}$$

The summation of the $P(I)$ terms over all possible values of I must, of course, equal the probability of getting an event, as given by $P(E)$.

C.6   Determining an Optimum Receiver Sweep Rate

The similar shapes of the curves in Figures C.3 and C.4 lead one to consider combining them and expressing the probability of intercept in terms of the ratio G/N. Recall that G is the maximum number of radar PRIs that the receiver allows for a valid intercept, and N is the number frequencies in the radar's frequency hop sequence. The result is shown in Figure C.6.

Recall that we have also required that the receiver dwell time be three times the maximum permitted PRI, and therefore $R = 3G$. Therefore, the faster the receiver sweep speed, the smaller are R and G. The best strategy may be to choose the receiver sweep speed to maximize the expected number of intercepts. Figure C.7 shows the expected number of intercepts divided by G versus G/N, which is based on the empirical curve in Figure C.6. The optimum value for G/N is near 1.0, for which the expected number of intercepts is $0.7G$. Thus, the correct strategy is to set the sweep speed so that the dwell time at any given frequency is the product of the number of frequency slots used by the radar (N) times 3 times the radar's PRI. Similar results could presumably be obtained for ratios of R/G other than 3.

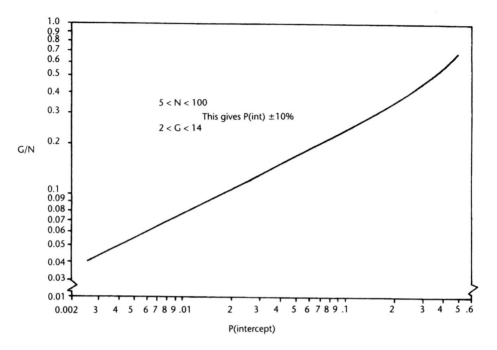

Figure C.6 G/N versus probability of intercept.

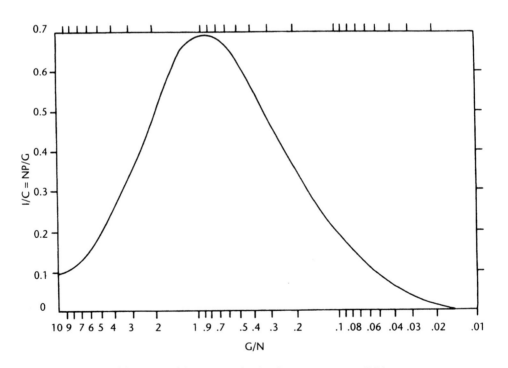

Figure C.7 Expected fractions of frequencies having intercepts versus G/N.

History and Fundamentals of the IFM

The use of crystal video receivers for the interception of radar signals was a natural development because of the high power of radar transmitters combined with the low cost and small size of crystal video receivers. Their main shortcoming was that they did not indicate the carrier frequency. After all, a small antenna, detector, video-audio amplifier, and recorder could preserve the PRI and scan information and be packaged in a pocket-sized unit. The missing parameters were pulse duration and the carrier frequency. Techniques to encode the pulse duration in a way that could be stored on an audio recorder were of interest, as were similar methods for encoding the carrier frequency.

During the late 1950s, work on *instantaneous frequency indication* (IFM) techniques was in progress in at least three locations: Syracuse Research Corporation (SRC),[1] Stanford Electronic Laboratory (SEL), and in the United Kingdom at Mullard Research Laboratories. Mullard began such research in 1954 [1]. During 1957 and 1958, work at SRC emphasized a switched stub technique [E.M. Williams, private communication with E. M. Williams, 1960]. In this technique, after each pulse, an RF switch was opened or closed to either insert a shorted transmission line stub into the RF circuit or not. This affected the voltage standing wave ratio (VSWR) and hence the amplitude. Variations could then be related to the carrier frequency by means of a calibration chart. Clearly, this technique was affected by pulse-to-pulse amplitude changes from other causes (e.g., scanning), and pulse-to-pulse frequency agility could not be handled. Similar work at SEL was also in progress [2].

If audio recording of the RF was not necessary, then an oscilloscope could be used to display the pulsed signal data. Otherwise, pulse width modulation or pulse amplitude modulation was used to encode the RF information within an audio bandwidth signal. The next step was to use two channels, one with a frequency sensitive element (e.g., shorted stub) and one without, as shown in Figure D.1. SRC delivered such a device in 1962: a 5.25″ × 8.25″ box covering the 50- to 250-MHz band [3]. Dynamic range was 15 dB and the frequency accuracy was 13%. The RF output was encoded as pulse duration, as shown by the calibration curves in Figure D.2.

Events were moving in what turned out to be a more fruitful direction in the United Kingdom. In 1957, S. J. Robinson of Mullard Research Laboratories conceived the basic microwave phase discriminator circuit, which became the basis of the modern IFM [3]. By 1959, Mullard delivered a device called Pendant [4].

---

1.   Then called Syracuse University Research Corporation.

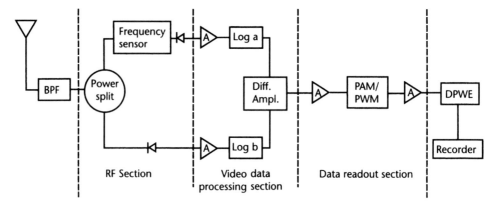

Figure D.1   An early two-channel IFM approach.

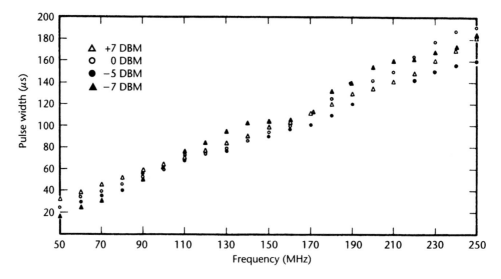

Figure D.2   Early IFM calibration curve.

This used the polar type of CRT display, which had been demonstrated in 1958. Each pulse was displayed as a vector drawn from the center of the CRT. The length of the vector was proportional to the received signal strength and the angle was proportional to the RF. A mechanical cursor could be rotated to lie along any observed radial strobes, and the RF frequency could be read from a scale around the outer edge of the CRT.[2] In contrast to the miniature IFM described above, Pendant was very large (a full 19-inch rack of equipment) and was based on vacuum tube circuitry (by contrast, much of the energy of U.S. electrical engineers at the time was spent transistorizing circuitry of all kinds). Nonetheless, to those who saw it operate, Pendant was clearly a very useful device which performed admirably.

Dr. Thomas F. Curry of SRC recognized the potential of this device and, no doubt influenced by SRC's requirements at the time for compact instantaneous frequency measurement gear, assigned Mr. Grover M. Boose the task of converting

2.   Pendant arrived at SRC for evaluation purposes at about the same time as the author.

the S-band waveguide microwave circuitry contained in the Mullard receiver into stripline form. By 1963, an S-band receiving unit had been completed using a stripline discriminator module. The calibration curve for this module is shown in Figure D.3 [5]. (This was the first stripline IFM. The nonlinearity of the curve was

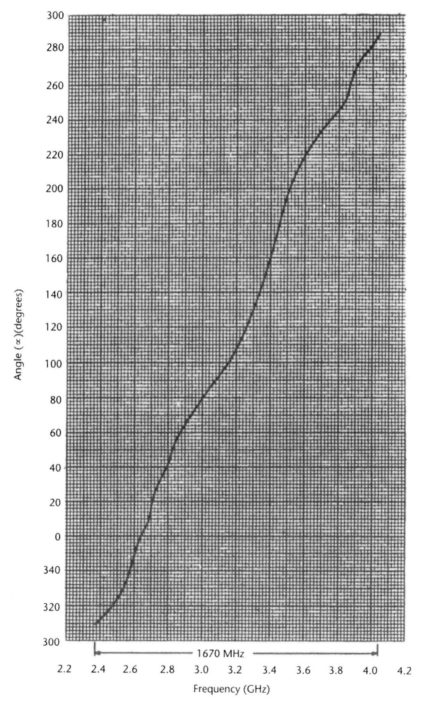

Figure D.3   Calibration curve for the first stripline IFM.

partially due to the use of a frequency sensitive phase-shift element. Later this element was replaced by a broadband 90° hybrid.) The next IFM design by the workers at SRC included a tunnel-diode limiting preamplifier on the front end for wide dynamic range and an analog-to-digital converter connected to the video outputs. This unit, completed in late 1964, was a U.S. ancestor of modern digital IFMs.[3]

The work done in 1964 considered nearly every IFM problem: simultaneous signal effects [6], CW bias problems, the need for limiting, and the utility of a digital output of the RF on a single pulse. The analog IFMs to follow were concerned with increasing the dynamic range and sensitivity, providing frequency coverage to 12 GHz (or to 18 GHz), warning the operator of the presence of simultaneous signals, CW signals, and other system aspects of the IFM.[4] The two IFM systems produced at SRC in 1965 and 1967 are shown in Figures D.4 and D.5. Frequency coverage from 0.5 to 18 GHz with high sensitivity was provided. TWT preamplifiers with leveling were used because low-noise solid-state amplifiers were very costly or unavailable at the higher frequencies. The 12.4- to 18-GHz band discriminators were made from waveguide components. The large "black box" on top of the CRT in Figure D.4 is a mechanical device that converted the angular position of the cursor to an RF value. The cursor was manually tuned to coincide with the signal

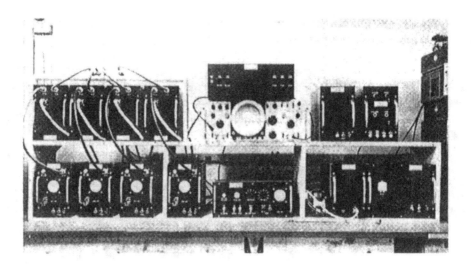

Figure D.4   An analog IFM receiving system produced in 1965. (Courtesy of SRC.)

3.   The issue of ambiguity elimination was a hot topic of conversation at SRC during the summer of 1964. Dr. Peter Knoke argued forcefully for the inclusion of ambiguity elimination circuitry to avoid the problem of ambiguous frequency indication due to IFM nonlinearity when the outputs of a number of IFMs with different frequency sensitivities are combined to provide a digital word. He was overruled at the time, partly due to the cost of logic circuits in 1964 and partly due to the belief that sufficient discriminator linearity could be achieved to avoid the ambiguity problem. The digital discriminator without ambiguity elimination was successfully built, but endless hours of delay line cutting and calibration were required. Today's digital IFMs all incorporate ambiguity elimination logic, which is indispensable to their successful operation.

4.   It was believed by many at SRC that the golden age of IFMs was at hand in 1965, the author included; but it was several years before the AN/FSQ-65(I) was even tested in the field.

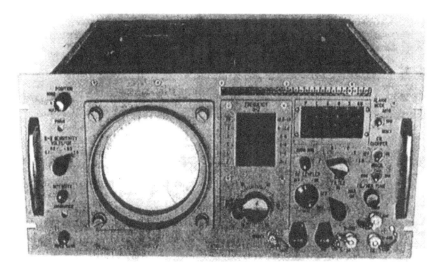

Figure D.5 AN/FSQ-65(V) analog IFM receiving system produced in 1967. (Courtesy of SRC.)

of interest using a knob. This also moved a 35-mm film calibrated with the RF frequency behind the rectangular "windows" cut in the panel of the black box. The appropriate window was illuminated, depending on which band was selected. The same calibration technique was used in the AN/FSQ-65(V) shown in Figure D.5 with considerable improvement in the implementation. The AN/FSQ-65(V) included a matrix of alarm indicators. Each band had an activity indicator, as well as indicators for CW and for simultaneous signals in the same band (intra) and in different bands (inter). The unit included a chopper to make CW signals visible on the display and an expansion scale (longer delay line), which could be used in each band as desired to show frequency hopping or FMOP more clearly. Narrowband tunable filters could also be inserted in each band as needed, along with calibrated marker signals.

The 1967 stripline design shown in Figure D.6 is believed to be the first such stripline unit to operate above 12 GHz [7].[5]

D.1 The Broadband Microwave Frequency Discriminator

The basic building block in the instantaneous frequency discriminating (IFD) receiver is shown in Figure D.7. Consider the operation of the circuit when the input is a pulsed signal of the form

$$e_2 = \frac{E_2(t)}{\sqrt{2}} \cos(\omega t - \beta L_1) \tag{D.1}$$

---

5. In 1964, several SRC engineers started another company (then called Curry, McLaughlin, and Len, and later called Microwave Systems, Inc.). One of the engineers was Robert L. McLaughlin who developed the 12- to 18-GHz stripiine IFM shown in 1967. Another engineer, James E. Secord, built in the late 1960s a waveguide IFM unit that operated to 40 GHz.

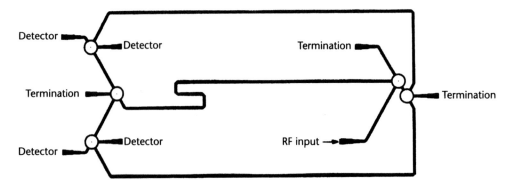

Figure D.6   12- to 18-GHz discriminator stripline layout.

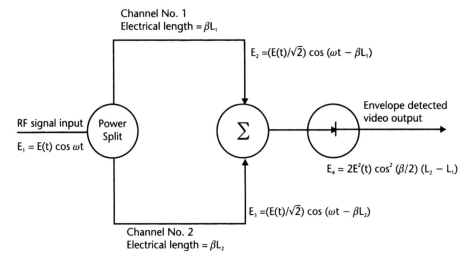

Figure D.7   Illustrating the basic principle of the operation of the microwave discriminator IFD.

in which $E_0(t)$ is the pulse modulating function. This signal is divided by a power splitter into two channels, commonly comprising two lengths of transmission line of different electrical lengths, βL_1 and βL_2 (β represents the phase constant of the lines). The signal at the outputs of the two channels are

$$e_2 = \frac{E_0(t)}{\sqrt{2}} \cos(\omega t - \beta L_1) \tag{D.2}$$

and

$$e_3 = \frac{E_0(t)}{\sqrt{2}} \cos(\omega t - \beta L_2) \tag{D.3}$$

These two signals are added at the input to the detector. Using the identity

$$\cos a + \cos b = 2 \cos \frac{1}{2}(a + b) \cdot \cos \frac{1}{2}(a - b) \qquad (D.4)$$

the input to the detector may be expressed as

$$e_4 = 2E_0(t) \cos \frac{\beta\Delta}{2} \cos \omega t - \frac{\beta(L_1 + L_2)}{2} \qquad (D.5)$$

in which ΔL is $L_2 - L_1$.

The detector output, assuming a square-law detection characteristic and neglecting the detector figure or merit, is

$$e_S = 2\left[E_0(t) \cos \left(\frac{\beta\Delta L}{2}\right)\right] \qquad (D.6)$$

This output is always positive and a function of frequency because of the frequency dependence of β. A more convenient bipolar frequency-dependent output may be obtained by combining the circuit of Figure D.7 with a second circuit in which an additional delay of 180° has been introduced between the two channels. The resultant circuit is shown in Figure D.8. The 180° delay ($\lambda/2$ length at the design center frequency) in the L_2 arm of the second channel combination results in a sine-squared output at the second channel detector output unlike the cosine-squared output of the first channel. A video output e_6 is then formed from the difference of the two channel outputs:

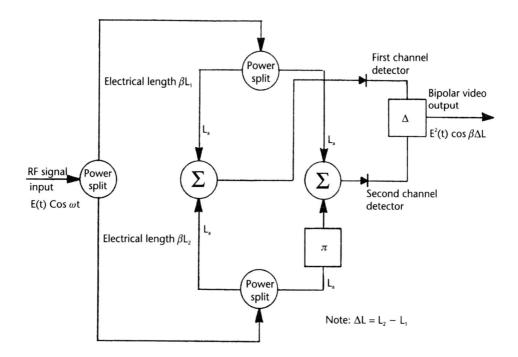

Figure D.8   Combination of two units to obtain a bipolar video output as a function of frequency.

$$e_6 = 2\left[E_0(t) \cos\left(\frac{\beta\Delta L}{2}\right)\right]^2 - 2\left[E_0(t) \sin\left(\frac{\beta\Delta L}{2}\right)\right]^2 \qquad \text{(D.7a)}$$

or

$$e_6 = E_0^2(t) \cos(\beta\Delta L) \qquad \text{(D.7b)}$$

A second bipolar output e_7 can be formed by a circuit identical to that of Figure D.8, except that the first split is obtained by the use of a backward wave coupler, so that a phase shift of 90° appears at the coupled arm output to line L_2. For this case, with the same signal input, the video output, e_7, is

$$e_7 = 2E_0^2(t) \sin \beta L \qquad \text{(D.8)}$$

The two outputs, e_6 and e_7, may be used to form a frequency-indicating display in a number of ways. The most obvious approach is to feed e_6 and e_7 to the horizontal and vertical plates of a CRT. The resultant display is a polar strobe with an angle θ dependent upon frequency and amplitude dependent upon signal amplitude. The angle θ of the strobe is actually

$$\theta = \beta\Delta L \qquad \text{(D.9)}$$

The variation of θ cannot exceed 360° without ambiguity. The frequency range, $\omega_2 - \omega_1$, then, is limited by the requirement that

$$\frac{\omega_2}{V_0}\Delta L - \frac{\omega_1}{V_0}\Delta L = 2\pi \qquad \text{(D.10a)}$$

or

$$\omega_2 - \omega_1 = \frac{2\pi V_0}{\Delta L} \qquad \text{(D.10b)}$$

in which V_0 is the velocity in the transmission lines. By appropriate choice of the delay differential, $\Delta L/V_0$, either a wide or narrow range of frequencies can be covered without ambiguity.

The entire RF assembly, as implied earlier, is sometimes termed a *microwave discriminator*; the term IFD is less ambiguous and will be used hereafter.

A wide variety of designs for IFD units have been successfully implemented. Perhaps the most favorable results are obtained with variations of an all-quadrature hybrid configuration, such as shown in Figure D.9. The couplers (quadrature hybrids) are connected so that the power at each detector has the proper phase and amplitude relationship. The four detectors receive signals as follows:

$$E_1 = \frac{E_0}{2\sqrt{2}} \angle -180° - \beta L + \frac{E_0}{2\sqrt{2}} \angle -270° \qquad \text{(D.11a)}$$

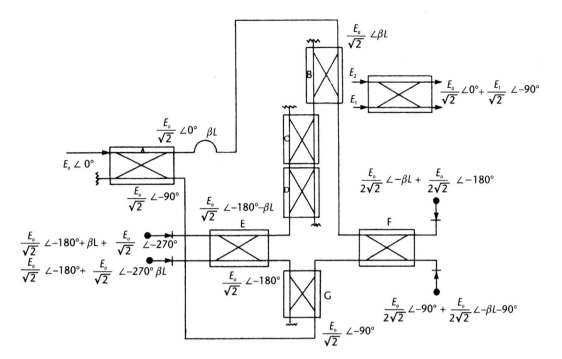

Figure D.9   Seven-coupler IFD.

$$E_2 = \frac{E_0}{2\sqrt{2}} \angle -180° + \frac{E_0}{2\sqrt{2}} \angle -270° - \beta L \qquad \text{(D.11b)}$$

$$E_3 = \frac{E_0}{2\sqrt{2}} \angle -\beta L + \frac{E_0}{2\sqrt{2}} \angle -180° \qquad \text{(D.11c)}$$

$$E_4 = \frac{E_0}{2\sqrt{2}} \angle -90° + \frac{E_0}{2\sqrt{2}} \angle -\beta L - 90° \qquad \text{(D.11d)}$$

in which E_0 is the input signal and βL the differential delay. The square-law detection process results in

$$V_1 = \frac{E_0^2}{4} (1 - \sin \beta L) \qquad \text{(D.12a)}$$

$$V_3 = \frac{E_0^2}{4} (1 - \cos \beta L) \qquad \text{(D.12b)}$$

$$V_2 = \frac{E_0^2}{4} (1 + \sin \beta L) \qquad \text{(D.12c)}$$

$$V_4 = \frac{E_0^2}{4} (1 + \cos \beta L) \qquad \text{(D.12d)}$$

The signals for the display are obtained by combining these detection output voltages to form

$$V_y = V_2 - V_1 = \frac{E_0^2}{4} \sin \beta L \tag{D.13a}$$

$$V_x = V_4 - V_3 = \frac{E_0^2}{4} \sin \beta L \tag{D.13b}$$

As shown in Figure D.9, couplers C and D are connected in tandem to produce a 90° shift in one leg of the IFD. This is done in order to set the power output of coupler 0 equal to the power passing from B to G. Couplers C and D can be replaced by a single coupler with shorted or open terminations, as shown in Figure D.10. Because it is difficult to obtain a perfect open or short circuit termination, a phase shifter (Schiffman) is added to trim the phase. The couplers B, C, and 0 can be replaced with a single 180° coupler or hybrid, as shown in Figure D.11. In this case, the diamond hybrid shown is limited to about an octave in bandwidth. The proper RF processing can also be achieved with four 180° couplers and one 90° phase shifter in place of couplers B through G. In the original U.K. units and earliest stripline versions, a quarter-wavelength line length difference at the design center frequency was used to achieve the 90° shift. Because of the resulting dependence of phase shift upon frequency, this design is limited to less than one octave in bandwidth, and the θ versus frequency characteristic is nonlinear.

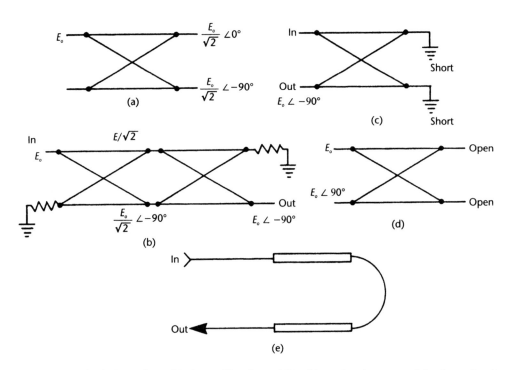

Figure D.10 Techniques for achieving a 90° phase shift with no loss in power: (a) schematic of 90° hybrid; (b) two hybrid phase shifter; (c) shorted hybrid as 90° phase shifter; (d) open hybrid as 90° phase shifter; and (e) Schiffman phase shifter.

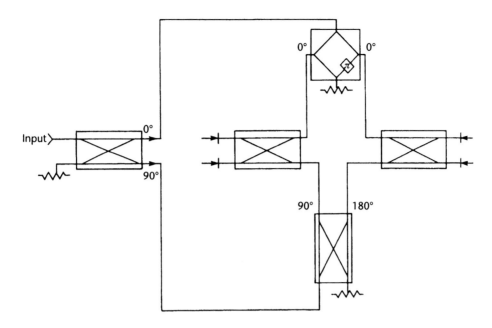

Figure D.11   IFD using a diamond hybrid.

References

[1]   Goddard, N. E., "Instantaneous Frequency-Measuring Receivers," *IEEE Trans. on Microwave Theory and Techniques*, April 1972, pp. 292–293.

[2]   May, B. B., "A Wide Band Frequency Discriminator Using Open and Shorted Stubs," Technical Report 791-2, Stanford Electronic Lab, November 1960 (SEL Project 791 K).

[3]   Boose, G. M., *E-l/O Instantaneous Frequency Indicating Receiver*, Final Report, Task Order 7, Subtask 1, Contract 582RR2, October 31, 1962.

[4]   Robinson, S. J., "Comment on Broadband Microwave Discriminator," *IEEE Transactions Professional Group on Microwave Theory and Techniques*, March 1964.

[5]   Boose, G. M., "Wide Open Frequency Indicating Receiver," paper presented to a joint meeting of the Syracuse Chapters of the IEEE Professional Groups Microwave Theory and Techniques and Antennas and Propagation, April 1964.

[6]   Boose, G. M., "Wide Open Frequency Indicating Receiver," Syracuse University Research Corporation, Electronics Research Laboratory, Technical Memo No. 24, August 1964.

[7]   Wiley, R. G., and E. M. Williams, "Spectrum Display Is Instantaneous," *Microwave Journal*, August 1970, pp. 44–47.

Emitter Location Partial Derivatives

This appendix includes expressions for partial derivatives of TOA and FOA measurements due to position, heading, and velocity in the notation used in the MathCad software tool. The subscripts i and k denote indices for a cross and downrange x-y emitter position, respectively.

Partial Derivative of TOA with Respect to Aircraft Position

$$\frac{\partial \text{TOA}_N}{\partial X_{M_{i,k}}} = \frac{-1}{c} \frac{(x_i - X_M)}{\sqrt{(x_i - X_M)^2 + (y_k - Y_M)^2}}, \quad M = N$$
$$0 \qquad\qquad\qquad\qquad\qquad\qquad\qquad \text{otherwise} \tag{E.1}$$

$$\frac{\partial \text{TOA}_N}{\partial Y_{M_{i,k}}} = \frac{-1}{c} \frac{(y_k - Y_M)}{\sqrt{(x_i - X_M)^2 + (y_k - Y_M)^2}}, \quad M = N$$
$$0 \qquad\qquad\qquad\qquad\qquad\qquad\qquad \text{otherwise} \tag{E.2}$$

Partial Derivative of FOA with Respect to Aircraft Position

$$\frac{\partial \text{FOA}_N}{\partial X_{M_{i,k}}} = \frac{-1}{\lambda} \left\{ \frac{-(y_k - Y_M)^2 \cdot (V_M \cos \theta_M) + (x_i - X_M) \cdot (y_k - Y_M) \cdot (V_M \sin \theta_M)}{[(x_i - X_M)^2 + (y_k - Y_M)^2]^{3/2}} \right\}, \quad M = N$$
$$0 \qquad\qquad\qquad\qquad\qquad\qquad\qquad\qquad\qquad\qquad \text{otherwise} \tag{E.3}$$

$$\frac{\partial \text{FOA}_N}{\partial Y_{M_{i,k}}} = \frac{1}{\lambda} \left\{ \frac{(x_i - X_M) \cdot (y_k - Y_M) \cdot (V_M \cos \theta_M) + (x_i - X_M)^2 \cdot (V_M \sin \theta_M)}{[(x_i - X_M)^2 + (y_k - Y_M)^2]^{3/2}} \right\}, \quad M = N$$
$$0 \qquad\qquad\qquad\qquad\qquad\qquad\qquad\qquad\qquad\qquad \text{otherwise} \tag{E.4}$$

Partial Derivative of FOA with Respect to Aircraft Heading

$$\frac{\partial \text{FOA}_N}{\partial \theta_{M_{i,k}}} = \frac{1}{\lambda} \left\{ \frac{-(x_i - X_M) \cdot (V_M \sin \theta_M) + (y_k - Y_M) \cdot (V_M \cos \theta_M)}{\sqrt{(x_i - X_M)^2 + (y_k - Y_M)^2}} \right\}, \quad M = N$$
$$0 \qquad\qquad\qquad\qquad\qquad\qquad\qquad\qquad\qquad \text{otherwise} \tag{E.5}$$

Partial Derivative of FOA with Respect to Aircraft Velocity

$$\frac{\partial \text{FOA}_N}{\partial V_{M_{i,k}}} = \frac{1}{\lambda} \left\{ \frac{(x_i - X_M) \cdot \sin \theta_M + (y_k - Y_M) \cdot \cos \theta_M}{\sqrt{(x_i - X_M)^2 + (y_k - Y_M)^2}} \right\}, \quad M = N$$

$$0 \qquad\qquad\qquad\qquad\qquad\qquad\qquad\qquad\qquad \text{otherwise}$$

$$(\text{E.6})$$

The following are the expressions for partial derivatives of emitter x-y location coordinates due to position and AOA measurements. The AOA measurement is computed as

$$\text{AOA}_{M_{i,k}} = \tan^{-1}\left(\frac{x_i - X_M}{y_k - Y_M}\right) \tag{E.7}$$

Partial Derivative of Emitter x-y Location with Respect to Aircraft Position

$$\frac{\partial x}{\partial X1_{i,k}} = \frac{1}{1 - \dfrac{\tan(\text{AOA}1_{i,k})}{\tan(\text{AOA}2_{i,k})}} \tag{E.8}$$

$$\frac{\partial y}{\partial X1_{i,k}} = \frac{1}{\tan(\text{AOA}2_{i,k}) - \tan(\text{AOA}1_{i,k})} \tag{E.9}$$

$$\frac{\partial x}{\partial Y1_{i,k}} = \frac{-\tan(\text{AOA}1_{i,k})}{1 - \dfrac{\tan(\text{AOA}1_{i,k})}{\tan(\text{AOA}2_{i,k})}} \tag{E.10}$$

$$\frac{\partial y}{\partial Y1_{i,k}} = \frac{-\tan(\text{AOA}1_{i,k})}{\tan(\text{AOA}2_{i,k}) - \tan(\text{AOA}1_{i,k})} \tag{E.11}$$

$$\frac{\partial x}{\partial X2_{i,k}} = \frac{-\dfrac{\tan(\text{AOA}1_{i,k})}{\tan(\text{AOA}2_{i,k})}}{1 - \dfrac{\tan(\text{AOA}1_{i,k})}{\tan(\text{AOA}2_{i,k})}} \tag{E.12}$$

$$\frac{\partial y}{\partial X2_{i,k}} = \frac{\dfrac{-\tan(\text{AOA}1_{i,k})}{\tan(\text{AOA}2_{i,k})}}{\tan(\text{AOA}2_{i,k}) - \tan(\text{AOA}1_{i,k})} - \frac{1}{\tan(\text{AOA}2_{i,k})} \tag{E.13}$$

$$\frac{\partial x}{\partial Y2_{i,k}} = \frac{\tan(\text{AOA}1_{i,k})}{1 - \dfrac{\tan(\text{AOA}1_{i,k})}{\tan(\text{AOA}2_{i,k})}} \tag{E.14}$$

$$\frac{\partial y}{\partial Y2_{i,k}} = \frac{\tan(\text{AOA}1_{i,k})}{\tan(\text{AOA}2_{i,k}) - \tan(\text{AOA}1_{i,k})} + 1 \tag{E.15}$$

Partial Derivative of Emitter x-y Location with Respect to AOA

$$\frac{\partial x}{\partial AOA1_{i,k}} = \left\{ \frac{\dfrac{\tan(AOA2_{i,k})}{\cos(AOA1_{i,k})^2}}{[\tan(AOA2_{i,k}) - \tan(AOA1_{i,k})]^2} \right\} \cdot X1$$

$$+ \left\{ \frac{\dfrac{\tan(AOA2_{i,k})}{\cos(AOA1_{i,k})^2}}{[\tan(AOA2_{i,k}) - \tan(AOA1_{i,k})]^2} \right\} \cdot X2$$

$$+ \left\{ \frac{\dfrac{-\tan(AOA2_{i,k}) \cdot [\tan(AOA2_{i,k}) - \tan(AOA1_{i,k})] - \tan(AOA1_{i,k}) \cdot \tan(AOA2_{i,k})}{\cos(AOA1_{i,k})^2}}{[\tan(AOA2_{i,k}) - \tan(AOA1_{i,k})]^2} \right\} \cdot Y1$$

$$+ \left\{ \frac{\dfrac{-\tan(AOA2_{i,k}) \cdot [\tan(AOA2_{i,k}) - \tan(AOA1_{i,k})] + \tan(AOA1_{i,k}) \cdot \tan(AOA2_{i,k})}{\cos(AOA1_{i,k})^2}}{[\tan(AOA2_{i,k}) - \tan(AOA1_{i,k})]^2} \right\} \cdot Y2$$

$$(E.16)$$

$$\frac{\partial x}{\partial AOA2_{i,k}} = \left\{ \frac{\dfrac{-\tan(AOA1_{i,k})}{\cos(AOA2_{i,k})^2}}{[\tan(AOA2_{i,k}) - \tan(AOA1_{i,k})]^2} \right\} \cdot X1$$

$$+ \left\{ \frac{\dfrac{\tan(AOA1_{i,k})}{\cos(AOA2_{i,k})^2}}{[\tan(AOA2_{i,k}) - \tan(AOA1_{i,k})]^2} \right\} \cdot X2$$

$$- \left\{ \frac{\dfrac{-\tan(AOA1_{i,k}) \cdot [\tan(AOA2_{i,k}) - \tan(AOA1_{i,k})] + \tan(AOA1_{i,k}) \cdot \tan(AOA2_{i,k})}{\cos(AOA2_{i,k})^2}}{[\tan(AOA2_{i,k}) - \tan(AOA1_{i,k})]^2} \right\} \cdot Y1$$

$$+ \left\{ \frac{\dfrac{-\tan(AOA1_{i,k}) \cdot [\tan(AOA2_{i,k}) - \tan(AOA1_{i,k})] - \tan(AOA1_{i,k}) \cdot \tan(AOA2_{i,k})}{\cos(AOA2_{i,k})^2}}{[\tan(AOA2_{i,k}) - \tan(AOA1_{i,k})]^2} \right\} \cdot Y2$$

$$(E.17)$$

$$\frac{\partial y}{\partial AOA1_{i,k}} = \left\{ \frac{\dfrac{1}{\cos(AOA1_{i,k})^2}}{[\tan(AOA2_{i,k}) - \tan(AOA1_{i,k})]^2} \right\} \cdot X1$$

$$+ \left\{ \frac{\dfrac{-\tan(AOA2_{i,k}) \cdot [\tan(AOA2_{i,k}) - \tan(AOA1_{i,k})]}{\cos(AOA1_{i,k})^2 \cdot \tan(AOA2_{i,k})^2} - \dfrac{\tan(AOA1_{i,k})}{\cos(AOA1_{i,k})^2 \cdot \tan(AOA2_{i,k})}}{[\tan(AOA2_{i,k}) - \tan(AOA1_{i,k})]^2} \right\} \cdot X2$$

$$+ \left\{ \frac{\dfrac{-\tan(AOA2_{i,k})}{\cos(AOA1_{i,k})^2}}{[\tan(AOA2_{i,k}) - \tan(AOA1_{i,k})]^2} \right\} \cdot Y1$$

$$+ \left\{ \frac{\dfrac{[\tan(AOA2_{i,k}) - \tan(AOA1_{i,k})] + \tan(AOA1_{i,k})}{\cos(AOA1_{i,k})^2}}{[\tan(AOA2_{i,k}) - \tan(AOA1_{i,k})]^2} \right\} \cdot Y2$$

$$(E.18)$$

$$\frac{\partial y}{\partial AOA2_{i,k}} = \left\{ \frac{\dfrac{-1}{\cos(AOA2_{i,k})^2}}{[\tan(AOA2_{i,k}) - \tan(AOA1_{i,k})]^2} \right\} \cdot X1$$

$$+ \left\{ \frac{\dfrac{\tan(AOA1_{i,k}) \cdot [\tan(AOA2_{i,k}) - \tan(AOA1_{i,k})]}{\cos(AOA2_{i,k})^2 \cdot \tan(AOA2_{i,k})^2} + \dfrac{\tan(AOA1_{i,k})}{\cos(AOA2_{i,k})^2 \cdot \tan(AOA2_{i,k})}}{[\tan(AOA2_{i,k}) - \tan(AOA1_{i,k})]^2} \right\} \cdot X2$$

$$+ \left\{ \frac{1}{\cos(AOA2_{i,k})^2 \cdot \tan(AOA2_{i,k})^2} \right\} \cdot X2$$

$$+ \left\{ \frac{\dfrac{\tan(AOA1_{i,k})}{\cos(AOA2_{i,k})^2}}{[\tan(AOA2_{i,k}) - \tan(AOA1_{i,k})]^2} \right\} \cdot Y1$$

$$+ \left\{ \frac{\dfrac{-\tan(AOA1_{i,k})}{\cos(AOA2_{i,k})^2}}{[\tan(AOA2_{i,k}) - \tan(AOA1_{i,k})]^2} \right\} \cdot Y2$$

$$(E.19)$$

About the Author

Richard G. Wiley received a B.S. in electrical engineering in 1959 and an M.S. in electrical engineering in 1960, both from Carnegie Mellon University. After employment at Syracuse University Research Corporation (SURC), he served in the U.S. Army from 1961 to 1963. Dr. Wiley continued employment at SURC from 1963 to 1967. From 1967 to 1974, he was employed at Microwave Systems, Inc. He continued his education during this period at Syracuse University and received a Ph.D. in electrical engineering in 1975. He was employed by Syracuse Research Corporation from 1975 to 1985 as a staff consulting engineer and then became a cofounder and chief scientist of Research Associates of Syracuse, Inc., a position he still holds. He was elected a Fellow of the IEEE and is a member of the Association of Old Crows (AOC). In 2001 Dr. Wiley was named chairman of the AOC's Professional Development Committee and is in charge of the AOC's Continuing Education Program.

Index

Millimeter-Wave Radar Targets and Clutter, Gennadiy P. Kulemin

Modern Radar System Analysis, David K. Barton

Multitarget-Multisensor Tracking: Applications and Advances Volume III, Yaakov Bar-Shalom and William Dale Blair, editors

Principles of High-Resolution Radar, August W. Rihaczek

Principles of Radar and Sonar Signal Processing, François Le Chevalier

Radar Cross Section, Second Edition, Eugene F. Knott et al.

Radar Evaluation Handbook, David K. Barton et al.

Radar Meteorology, Henri Sauvageot

Radar Reflectivity of Land and Sea, Third Edition, Maurice W. Long

Radar Resolution and Complex-Image Analysis, August W. Rihaczek and Stephen J. Hershkowitz

Radar Signal Processing and Adaptive Systems, Ramon Nitzberg

Radar System Analysis and Modeling, David K. Barton

Radar System Performance Modeling, Second Edition, G. Richard Curry

Radar Technology Encyclopedia, David K. Barton and Sergey A. Leonov, editors

Range-Doppler Radar Imaging and Motion Compensation, Jae Sok Son et al.

Signal Detection and Estimation, Second Edition, Mourad Barkat

Space-Time Adaptive Processing for Radar, J. R. Guerci

Theory and Practice of Radar Target Identification, August W. Rihaczek and Stephen J. Hershkowitz

Time-Frequency Transforms for Radar Imaging and Signal Analysis, Victor C. Chen and Hao Ling

For further information on these and other Artech House titles, including previously considered out-of-print books now available through our In-Print-Forever® (IPF®) program, contact:

Artech House	Artech House
685 Canton Street	46 Gillingham Street
Norwood, MA 02062	London SW1V 1AH UK
Phone: 781-769-9750	Phone: +44 (0)20 7596-8750
Fax: 781-769-6334	Fax: +44 (0)20 7630-0166
e-mail: artech@artechhouse.com	e-mail: artech-uk@artechhouse.com

Find us on the World Wide Web at: www.artechhouse.com

LaVergne, TN USA
09 February 2011

215901LV00005B/1-4/P